Proceedings of the

FIFTH NEW ENGLAND BIOENGINEERING CONFERENCE

ALSO PUBLISHED BY PERGAMON PRESS . . .

THE JOURNAL OF BIOENGINEERING

In four sections

Chemical Engineering

Electrical Engineering

Materials Science & Engineering

Mechanical Engineering

each with its own Editor and International Editorial Board.

Modeled after Pergamon's highly successful rapid publication journal, *Thrombosis Research,* the *Journal of Bioengineering* serves as a rapid publication forum for high quality papers in all major areas of bioengineering.

The Journal is published bimonthly by a photoduplication process from camera-ready copy. Each paper is reviewed by at least two peer reviewers. Accepted papers are generally published within four months of receipt of acceptable camera-ready copy.

Proceedings of the FIFTH NEW ENGLAND BIOENGINEERING CONFERENCE

April 14-15, 1977
University of New Hampshire
Durham, New Hampshire

EDITOR
Michael R. Cannon

PERGAMON PRESS
NEW YORK / TORONTO / OXFORD / SYDNEY / FRANKFURT / PARIS

Pergamon Press Offices:

U.K.	Pergamon Press Ltd., Headington Hill Hall, Oxford OX3 OBW, England
U.S.A.	Pergamon Press Inc., Maxwell House, Fairview Park, Elmsford, New York 10523, U.S.A.
CANADA	Pergamon of Canada, Ltd., 207 Queen's Quay West, Toronto 1, Canada
AUSTRALIA	Pergamon Press (Aust.) Pty. Ltd., 19a Boundary Street, Rushcutters Bay, N.S.W. 2011, Australia
FRANCE	Pergamon Press SARL, 24 rue des Ecoles, 75240 Paris, Cedex 05, France
WEST GERMANY	Pergamon Press GmbH, 6242 Kronberg/Taunus, Frankfurt-am-Main, West Germany

Library of Congress Cataloging in Publication Data

New England Bioengineering Conference, 5th, University
of New Hampshire, 1977.
Proceedings of the Fifth New England Bioengineering
Conference, April 14-15, 1977, University of New
Hampshire, Durham, New Hampshire.

Includes index.
1. Biomedical engineering--Congresses. 2. Bio-
engineering--Congresses. I. Cannon, Michael R.
R856.A1N47 1977 610'.28 77-4196
ISBN 0-08-021937-3 pbk.

Printed in the United States of America

Sponsored by:	College of Engineering and Physical Sciences and the Division of Continuing Education, University of New Hampshire

Conference Chairman:	Michael R. Cannon, Ph.D. Department of Electrical Engineering University of New Hampshire Durham, New Hampshire 03824

Program Committee:	Prof. J. Bronzino	Ms. C. Murphy
	Ms. M. Erskine	Prof. R. Peura
	Prof. D. Jaron	Prof. S. Saha
	Prof. L. Katz	Ms. J. Scheibel
	Ms. P. Lindsey	Prof. J. Strohbehn
	Prof. R. McLaughlin	Prof. I. Young

CONTENTS
and
TECHNICAL PROGRAM

Opening of the Program — Thursday, April 14, 1977

PLENARY SESSION

Greetings
Michael R. Cannon, Conference Chairman
University of New Hampshire

David W. Ellis, Vice Provost for Academic Affairs
University of New Hampshire

Guest Lecturer
Sponsored by
The Institute of Electrical and Electronics Engineers
Dr. Charles D. Ray, M.D.
Vice President
Medtronics, Incorporated
Minneapolis, Minnesota
"Neuro-Augmentation -- Shades of the Bionic Man"

Session 1A — BIOMATERIALS

Chairman: Prof. S. Saha, Yale University, New Haven, CT

Session 5B MONITORING INSTRUMENTATION

Chairman: Prof. R. Peura, Worcester Polytechnic Institute

Session 6B INSTRUMENTATION

Chairman: Prof. D. Jaron, University of Rhode Island

PREFACE

The Fifth Annual New England Bioengineering Conference was held on April 14 and 15, 1977, at the New England Center in Durham, New Hampshire. The New England Center supports continuing education and conference programs for the region's six state universities and is a fitting location for this event. This conference is the regional meeting for engineers, physicians, and scientists interested in bioengineering. The purpose of this gathering is to provide a rapid means of communication of bioengineering research, development, and education in the New England area. Presented papers have been selected from those contributed by interested individuals and are published in this conference proceedings for worldwide distribution.

The conference consists of approximately seventy presentations. Of particular interest is the plenary session lecture entitled "Neuro-Augmentation - Shades of the Bionic Man" presented by Dr. Charles D. Ray, M.D., Vice President, Neuro/Rehabilitation and Medical Research, Medtronics, Incorporated (sponsored by the Institute of Electrical and Electronics Engineers). A two-session presentation entitled "Computer Courses for Biomedical Engineering Education" was organized by the American Society for Engineering Education. Other sessions covered topics such as biomaterials, biomechanics, signal processing, computers in medicine, modeling and simulation, and instrumentation.

I greatly appreciate the contributions of the authors who have made this conference a success. Especially appreciated are the efforts of the reviewers, session chairmen, and the Conference Group of the Division of Continuing Education, as well as the support provided by the College of Engineering and Physical Sciences. Presenting a conference would be an impossible task if it were not for the combined efforts of all those involved. Thank you.

April, 1977
University of New Hampshire
Durham, New Hampshire 03824

Michael R. Cannon
Program Chairman

BIOLOGICAL RESPONSE TO PLASMA FORMED POLYMERIC FILMS

Michael F. Nichols, J. Roger Easley, Kenneth G. Mayhan and Allen W. Hahn

The John M. Dalton Research Center, University of Missouri, Columbia, Missouri (Nichols, Easley and Hahn) and the Graduate Center for Materials Research, University of Missouri, Rolla, Missouri (Mayhan)

Many types of polymers used in biological applications require potentially toxic catalysts or high temperatures to accomplish polymerization. Nevertheless, numerous studies have shown many of these materials to have excellent biological compatibility. A new class of polymers formed in a high energy radio frequency (RF) electromagnetic field have been studied in our laboratories over the past four years (1). While the detailed physical and chemical mechanisms are still unresolved (2) these polymers can be deposited as conformal coatings on various surfaces and thus hold potential for use in a variety of biological applications. The purpose of this paper is to report our findings on the effect of the plasma formed polymeric films on biological tissue.

Previous studies on these films have shown them to have different physical and chemical properties when compared with bulk polymers prepared from similar monomers polymerized in a conventional fashion (1,3). When tested for biocompatibility by growing cells on their surface in tissue culture (4,5) and in preliminary implantation studies in rabbits (5), the films showed only a slight tendency toward inhibiting growth in cell culture, better adhesion to substrates, and minimal reactivity when implanted. With these initial encouraging results, we embarked on a long range study of biocompatibility in a large group of rats. By using a well known material (polymethylsiloxane-Silastic) as both control and substrate we have also been able to quantitate the biological response to both the control Silastic and the plasma formed films. Of particular interest has been the quantification of the fibrous capsule response and grading of the inflammatory response as a function of time; the later a variable far too often ignored in compatibility studies.

MATERIALS AND METHODS

Polymer Film Synthesis

All of the polymeric films studied were polymerized in a specially designed "glow discharge" reactor vessel as shown in figure 1. The material to be coated was placed in the chamber which was then evacuated to a pressure of 4P (30 μmHg.). A monomer gas (in our case, ethylene, E, styrene, S, or chlorotrifluroethylene, CTFE) was introduced into the chamber after purging with argon. A radio frequency electromagnetic field (∿ 4MHz) was introduced around the chamber and a plasma initiated by spark-gap induction. The sustained plasma was adjusted and stabilized by varying the monomer flow rate, chamber pressure and power input to the RF field. Our previous studies (1) showed that relatively low power input levels (∿ 6 watts) allowed film type polymers to form. Film thickness can be varied by allowing longer times for polymerization.

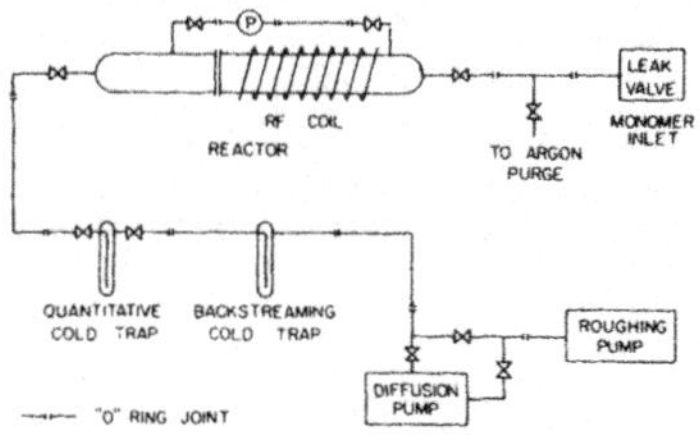

FIGURE 1. *Schematic of reactor system.*

Implant Materials

Polymer films of E, S, and CTFE were deposited upon substrates of polymethylsiloxane tubing, 1.65 mm OD (Dow Corning batch HH2897 Silastic Medical Grade tubing), cut into 6-7 mm. lengths and cleaned ultrasonically. The reactor was evacuated to 0.13 P (0.1 μmHg) and prepurified argon gas "bled" in until pressure stabilized at 1.3 P (10 μmHg). The RF power input to the coil was adjusted to 6 watts and deposition carried out for 15 hours to allow a conformal polymer deposition of approximately 5 μm. This procedure was repeated until 60 substrates had been coated.

The individual coated substrates and uncoated controls, cleaned similarly, were loaded onto stainless steel rods and sterilized by exposure to ethylene oxide gas at 30°C for 1 hour. The sterilized material was allowed to air wash under vacuum for 96 hours prior to implantation.

Implantation in Rats

Seventy-five male Wistar rats (Charles River Biol. Lab.) were used for polymer implant evaluation. The group had an average body mass of 200 grams at the onset of the project. Each rat was anesthetized with Halothane and 2 test lengths and 2 control lengths implanted in the lumbar paravertebral muscle mass. The technique used was modified from ASTM and USP methods. The rats were allowed to recover and then housed in separate cages.

After surgery, twenty rats receiving the same implant were randomly divided into five groups for necropsy. Necropsy intervals were at 2,4,8,12 and 24 weeks respectively. Of the remaining rats, 8 were necropsied at intervals of 75 weeks (2 with PS), 77 weeks (3 with PE) and 84 weeks (3 with PCTFE) and these tissues subjected to scanning electron microscopy (SEM).

Pathologic Studies

At necropsy, the implant and surrounding tissue was removed and fixed in 10% neutral buffered formalin. After fixation, the tissue was sectioned and stained with hematoxylin/eosin in preparation for histologic evaluation. Other tissues were examined grossly and histologically for the presence of abnormalities.

Histologic examination evaluated 2 features of the tissue surrounding the implant; the average number of acute inflammatory cells per high power field (hpf) and the thickness of the connective tissue capsule in μm (see legend figure 2).

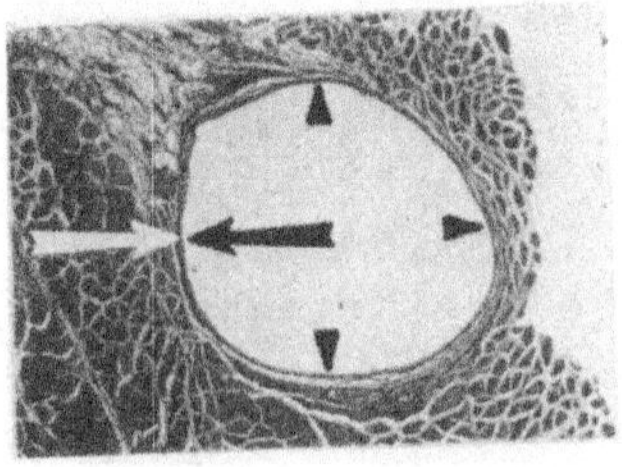

FIGURE 2. Grade 0, capsule thickness 10 μm. The inflammatory reaction was graded on a scale of 0-3 as follows: (1) Grade 0: minimal, < 1 cell/high power field (400 X); (2) Grade 1: mild, >1<3 cells/hpf: (3) Grade 2: moderate, >3<10 cells/hpf; and (4) Grade 3: severe, >10 cells/hpf. The grade was based on the average of 10 representative high power fields. The arrows indicate sites of capsule thickness measurement.

Tissue and implants removed after 75 weeks were fixed in a 4% gluteraldehyde/5% DMSO solution, mounted and coated with gold subsequent to examination by scanning electron microscopy.

RESULTS

A. Response of Rat Skeletal Muscle to the Polymeric Films.

Necropsy examination of the rats revealed no evidence of systemic toxicity as evaluated by gross and microscopic examination of major organs.

Grossly, the implant site showed only a glistening connective tissue capsule surrounding the implant. The capsule characteristics were grossly indistinguishable between control and test polymers.

Histologic evaluation of each implant site however showed some striking features, particularly when viewed in the time domain. The acute inflammatory cell response comparison of P-CTFE and its paired control showed no statistically significant difference ($p \leq .10$) at any of the sampling intervals. Polystyrene (PS) showed a statistically significant ($p \leq .10$) increase in the number of acute inflammatory cells at 2 and 4 weeks after implantation but there was no difference after 4 weeks. On the other hand, polyethylene (PE) had significantly ($p \leq .10$) fewer acute inflammatory cells than the control at 2 weeks. By 4 weeks the number of cells was still less, but the difference was not statistically significant.

The fibrous connective tissue reaction elicited by each of the test polymers was of a greater magnitude than the control for all polymers during the early phases, but by 24 weeks both control and test groups had capsule thicknesses of the same order of magnitude.

FIG 3 CAPSULE RESPONSE TO PLASMA FORMED POLYMERS

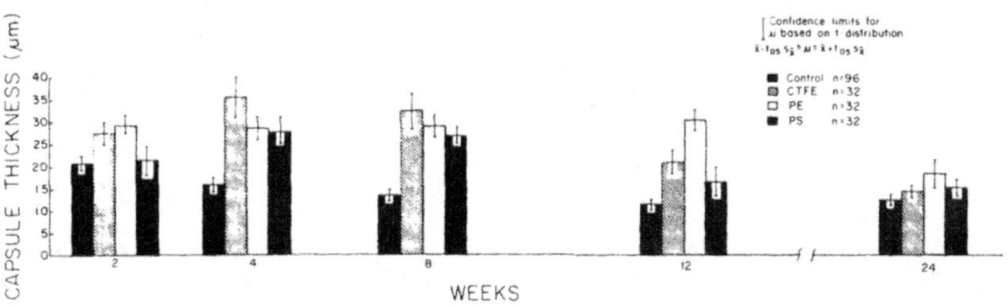

Figure 2 is a photomicrograph (X 40) showing the general morphology of the implant site and the points of the capsule at which thickness was measured. The results of our measurements on this capsule thickness are summarized in figure 3.

SEM done on tissue surrounding the test and control polymers appeared morphologically similar. Figure 4a is a SEM, at 20,000 magnification, of the connective tissue capsule surrounding the implant site and in direct contact with the test polymer. A "flake" of the film (in this instance PS), is on the left and the collagen fibers of the capsule can be observed on the right. Figure 4b is the capsule around the control at the same magnification. The collagenous tissue appears somewhat more "spongy" at this magnification.

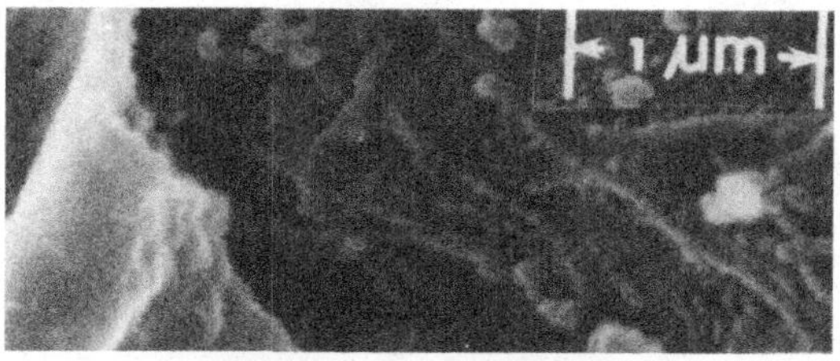

Fig. 4a. Capsule Surrounding PS (X20,000)

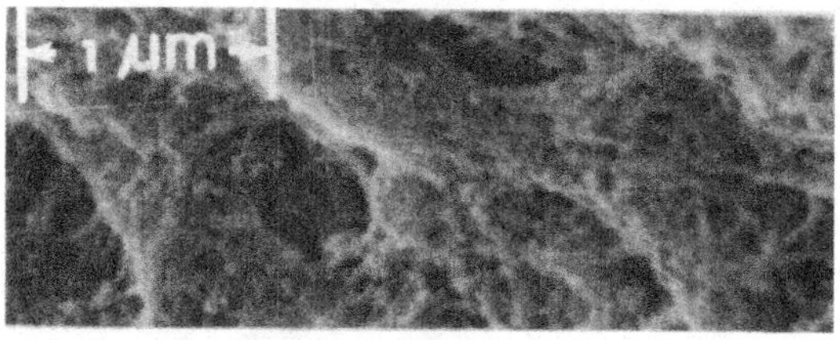

Fig. 4b. Capsule Surrounding Silastic (X20,000)

B. Effects of Implantation on the Polymeric Surface.

While the polymer's presence in skeletal muscle tissue was well tolerated with no attendent problems, the long term effect of being implanted took its toll on the polymer. Figure 5 is a SEM of the polymeric surface of PS that had not undergone implantation. Note that the surface is uniform and has "pores" approximately 0.1 μm in diameter distributed randomly over the surface (X20,000). Figure 6 is a SEM of the surface of PS after 75 weeks implantation (X2000). Note that the film has undergone cracking into 5-10 μm plates. These small plates were observed throughout the tissue capsules of all three types of polymer implants.

DISCUSSION

This new class of plasma polymerized films have been shown to have chemical and physical properties widely different from their monomers polymerized by standard means. Biologically, at least in skeletal muscles of rats, they elicit a tissue response not too different from Silastic, a material relatively inert in biological tissue. Of particular interest in the biological reactivity was the fact that in the "steady-state" (i.e., after 12-24 weeks) the fibrous tissue response is almost indistinguishable from that of Silastic. From these data we can conclude that these types of films do not elicit an untoward foreign body reaction at a skeletal muscle implant site. Our preliminary results with tissue culture are in general accord with the data of Homsey et al. (6), and these in vivo results are also generally in accord with published data on implantation responses (7). From a biological point of view, we feel we have been able to satisfactorily demonstrate the lack of general toxicity of these films in rats (previously in rabbits) and the very favorable and gentle response of tissue to their presence.

Fig. 5 PS Before Implantation (X20,000)

Fig. 6 PS After 75 Weeks Implantation (X2000)

We have of course been concerned about the inability of the plasma polymerized films to maintain their mechanical integrity during the course of the 2 year implantation. Recent work by Rice and O'Kane (3) has shown that films synthesized within the inductively coupled end of the reactor are very brittle and contain considerably more free radicals than conventional polymers. They were able to alleviate this problem by allowing polymerization to take place downsteam from the RF coil. Previous studies by our group has also shown higher concentration of free radicals in these films. A decrease in power input to the reactor vessel has modified this somewhat.

We were particularly impressed by the change in fibrous capsule response to the films as a function of time. It appears that caution needs to be taken where evaluating datum of this type, especially in what appears to be a dynamic situation. We are presently formalizing this reaction with a 2-3 parameter model which so far seems to characterize material and species in a simple, straightforward manner. Such being the case, polymeric implants could then be assigned "figures of merit" on the basis of these parameters. We do feel that when long term implants are desired, there is little difference between these substances and this should allow material choice to be based on material properties alone.

REFERENCES

1. Mayhan, K. G., Hahn, A. W., Havens, M. R., Peace, B. W.: "Plasma formed polymers for biomedical application: Part 1. Synthesis and fundamental studies." N.B.S. Spec. Publ. 415:1-12, 1975.

2. Bell, A. T.: "Fundamentals of plasma chemistry" in Techniques and Applications of Plasma Chemistry," J. H. Hollarhan and A. J. Bell, editors, (pp 1-56), John Wiley & Sons Inc., New York (1974).

3. Rice, D. W. and O'Kane, D. F: "Glow discharge polymerization of tetrafluorethylene, 1,1 difluoroethylene, and chlorotrifluoroethylene." J. Electrochem. Soc:123:1308-1312, 1976.

4. Yasuda, H., Bumgarner, M. O., and Marsh, H. C: "Ultrathin coating by plasma polymerization applied to corneal contact lens." J. Biomed. Mater. Res. 9:629-643, 1975.

5. Hahn, A. W., Mayhan, K. G., Easley, J. R., and Sanders, C. W: "Plasma formed polymers for biomedical application: Part II. Biocompatibility and applications". N.B.S. Spec. Publ. 415:13-17, 1975.

6. Homsy, C. A., Ansevin, K. D., O'Bannon, W. Thompson, S. A. Hodge, R. and Estrella, M. E.: "Rapid *in vitro* screening of polymers for biocompatibility" in Biomedical Polymers A. Rembaum and M. Shen, ed., Marcel Dekker, Inc. New York (1971). pp 121-140.

7. Laing, P. G., Ferguson, A. B. and Hodge, E. S.: "Tissue reaction in rabbit muscle exposed to metallic implants." J. Biomed. Mater. Res. 1: 135-149, 1967.

We would like to express our thanks and appreciation to Mr. Robert Pinnel, USDA Biol. Control of Insects Lab., Columbia, Mo., for his excellent help with the SEM.

This research supported in part by Grant No. DMR73-07641 from the National Science Foundation.

METAL ALLERGY AND ITS ROLE IN THE BIOCOMPATIBILITY OF ORTHOPAEDIC IMPLANTS

Brown, S.A.; Shafer, J.W.; Mayor, M.B.; Merritt, K.
Department of Surgery, Dartmouth Hitchcock Medical Center
Hanover, New Hampshire 03755

Allergic reactions to metals are usually associated with repeated skin exposure to metals or metal salts, resulting in a contact dermatitis (7). Several recent reports have implicated allergic reactions in complications associated with metal implants composed of cobalt, nickel or chromium, especially in situations where mechanical wear or abrasion rates and hence tissue levels of the alloy constituents are high (5,6). The contact dermatitis of cell mediated or type IV hypersensitivity is caused by the reaction of T-lymphocytes with the metal salt-skin protein complex to which they are sensitive. In the case of implant reactions the salts are probably bound to other tissue proteins.

We have developed an in vitro test for metal sensitivity and the presence of an allergic reaction to an implant which does not have the possible risk of sensitization associated with skin patch testing (3,4). The method challenges the patient's cells with metal-protein complexes and measures the inhibition of white cell migration as an indication of MIF production by sensitized T-lymphocytes. In a previous report (4) we have discussed the potential of the test for documenting specific sensitivities to cobalt, nickel and chromium, as well as the detection of in vivo production of MIF as part of an allergic reaction to an implant. In this report we will discuss the correlation between patient histories, allergic reactions, and the metallographic appearance of the retrieved implants. All of the implants were made of cast cobalt alloy (27-30% Cr, 5-7% Mo, 0-2.5% Ni, balance Cobalt).

METHODS. The MIF Test (4). Ten ml of the patient's blood is drawn, allowed to settle, and the white cell layer drawn off. Ten µl samples of resuspended cells are incubated with the patient's serum and subjected to five test conditions with PHA (a T-cell stimulant), CrCl, NiCl, CoCl, or saline as a control. Aliquots of each are then placed in duplicate in holes punched in agarose plates and incubated for 18-24 hours. After cell fixation, plate drying and cell staining, the plates are examined microscopically for cell migration out of the wells.

Implant Analysis: Implants were examined for evidence of mechanical deformation or abrasion. Sections were cut with a carbide wheel, then mounted in bakelite, sanded and polished for metallographic examination. Polished samples were etched for 15 seconds with a saturated solution of $FeCl_3$ in concentrated HCl to which a little HNO_3 is added and examined for microstructure and the presence of nonmetallic inclusion using a Zeiss Photo III microscope with epi-illumination. The characteristics of the implant surfaces which were exposed to the in vivo environment were studied with an AMI 1200 scanning electron microscope after ultrasonic cleaning.

CASE HISTORIES Three case histories are presented to demonstrate the correlation we have been observing between the patient's response to an implant and the metallographic appearance of the implant after removal.

Case #1. Patient #1 received an Austin-Moore endoprosthesis in March 1960. In December 1973 the patient began to complain of slight pain in the involved hip and bronchial asthma. By December 1974 the pain was severe and he was on medication for asthma. In October 1976 an MIF test was performed which demonstrated no cell migration from any of the wells, implicating allergy as a possible cause of pain. The endoprosthesis was removed and replaced with a Charnley total hip arthroplasty in November 1976 using bone cement and #8 Dacron suture for trochanteric reattachment. After removal the patient experienced relief of pain. An MIF test performed 4 weeks after removal demonstrated slight cell migration; by 5 weeks the migration was normal with MIF production to PHA, Cr, Ni, and cobalt, and he expressed relief from asthmatic symptoms and required no further medication.

Case #2. Patient #2 received a Thompson endoprosthesis in February 1972. Beginning in January 1975 the patient experienced increasing pain in the involved hip; subsequent radiographs demonstrated migration of the prosthesis. In July of 1976 an MIF test demonstrated no cell migration, implicating allergy as a possible cause of the symptoms. In August the joint was aspirated. No organisms were cultured; the WBC was 620 (64% polys, 24% lymphocytes, 12% monocytes). By November 1976 the pain was incapacitating and the patient developed asthma. The prosthesis was removed and replaced with a Charnley total hip arthroplasty in December 1976, with bone cement and #8 Dacron for trochanteric reattachment. The patient experienced immediate relief of pain and asthma. The MIF test performed 6 weeks later showed good migration in the control with MIF to PHA, Co and Cr.

Case #3. Patient #3 received a UCI total knee arthroplasty of the right knee in another hospital in August 1973. In January 1974 the patient complained of increasing pain in the knee; no organism was cultured on repeated cultures. The device was removed and the knee fused in February 1975. In December 1975 a correctional osteotomy of the opposite knee was performed using two staples. By April 1976 the patient complained of severe pain in the vicinity of the staples. An MIF test performed in November 1976 demonstrated no cell migration. The staples were removed in November 1976, at which time the osteotomy was clinically solid. The pain was relieved within two weeks after removal and an MIF test suggested slight cell migration from the control well, and questionable migration in the presence of Chromium. An MIF test performed 2 months after staple removal showed excellent migration in control and in the presence of Chromium, no migration with PHA and Nickel and inhibition with Cobalt.

IMPLANT ANALYSIS Case #1. At low magnification the implant showed extensive burnishing under the collar, on the medial aspect of the stem proximally and on the lateral portion of the tip, indicative of mechanical abrasion between bone and metal. A longitudinal section was cut from the proximal portion of the stem and the metallographs are shown in Figures 1a, 1b and 1c. The polished section (1a) shows a fine interdendritic carbide pattern typical of wear resistant cast cobalt alloys with no evidence of voids or inclusions. In the lightly etched state (1b and 1c), one sees minimal evidence of grain boundary attack, a few intergranular carbides and what would appear to be a very clean and electrolytically stable metal.

Case #2. At low magnification the implant showed evidence of abrasion, although not to the extent seen in Case #1. A metallographic section was cut and prepared in an identical manner as in the first case, and comparable photographs are shown in Figures 2a-c. In the polished section (2a) a significant number of black holes indicative of voids in the metal can be seen. Many of the voids are along the boundaries between the crystal grains which can be seen in the polished section. This would indicate that impurities had segregated out at the boundaries as the implant casting solidified. After a light etch the grain boundaries seen in 2b and 2c have a black appearance especially in contrast with Figures 1b and 1c. The extent to which the grain boundaries were attacked by the etchant suggests that similar metal degradation probably occurred on the implant surface from the corrosive *in vivo* environment.

SEM Analysis. Figure 3 shows comparable SEM pictures of the surfaces of implant 1 (3a) and implant 2 (3b). The irregular surface of 3a is typical for an investment casted implant. The cracks and pits on Figure 3b would suggest that the grain boundaries had been attacked in vivo and that the implant was deteriorating.

Case #3. The staples removed from Case 3 did not show any significant features at low magnification. Samples were cut from the staple and examined metallographically. In the polished condition a very pronounced grain boundary pattern could be seen. The etched section from the corner region shown in Figures 4a and 4b demonstrate an extremely high intergranular inclusion content. SEM examination of the implant surfaces showed pitting similar to that in Figure 3b, suggesting that the metal was deteriorating in vivo.

DISCUSSION One of the objectives of the metallographic analysis was to identify implant characteristics indicative of metallic deterioration (abrasion or corrosion) resulting in elevated tissue levels of cobalt, nickel and/-or chromium which could lead to patient sensitization. The implant in Case 1 was in situ for 14 years prior to the onset of significant pain; that in Case 2 was in situ for only three years. The metallography demonstrated that the implant in Case 1 was more electrolytically stable than that in Case 2. The surfaces of implant 2 showed significantly more deterioration in less than 5 years than did #1 after 16 years. One can conjecture that the relatively rapid onset of symptoms in Case 2 was related to the quality of the metal used for implant manufacture, while the onset of symptoms in Case 1 may have been related to the production of metal particles from abrasion. Both cases were interesting in that they are among the small number of patients to date with chromium sensitivities, both developed symptoms of asthma with the onset of pain, and both experienced relief from asthma after prosthesis removal.

The patient in Case 3 complained of pain even more quickly (4 months) than patient 2, as one might predict from the metallography. However, the history raises an additional question. A cast stainless steel knee prosthesis had been implanted 2 years previously, and been associated with symptoms of pain and loosening. Although we were not doing the MIF test at this time, one can conjecture that the symptoms were related to the nickel sensitivity, since the blade of the device had not been completely covered with bone cement and hence was in contact with trabecular bone. It is possible that previous sensitization accelerated the patient's reaction to the nickel in the staples, or that both sensitivities existed prior to the first procedure.

These studies have raised a number of questions.The first is the sequence of events and the cause and effect relationships between patient sensitization, the onset of pain, implant loosening and abrasion and the quality of metal used for implants. We hope to answer these in a proposed prospective study of patients being admitted to our hospital for prosthesis implantation and metallic implant removal, and a parallel study in rabbits. The other question is that of alternatives for use in sensitive patients. As mentioned, the use of a non-metallic interface between implant and bone as with bone cement and plastic acetabular components seems to minimize complications (5). As previously reported (3), the use of #8 Dacron rather than wire for trochanteric reattachment has proved effective in our experience. In another study in our laboratory we have found the use of polymeric pins and plates to be very effective for internal fixation of fractures (2), and have had one clinical experience where polymeric pins were used for fixation of a finger fusion in a sensitive patient (1).

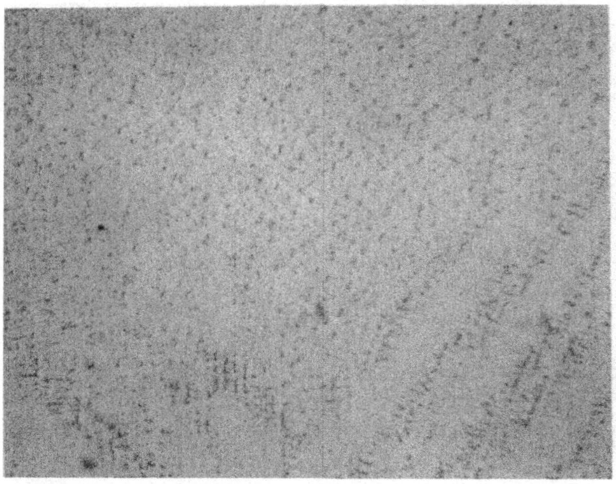

FIGURE 1a, Case 1, Polished, 32x

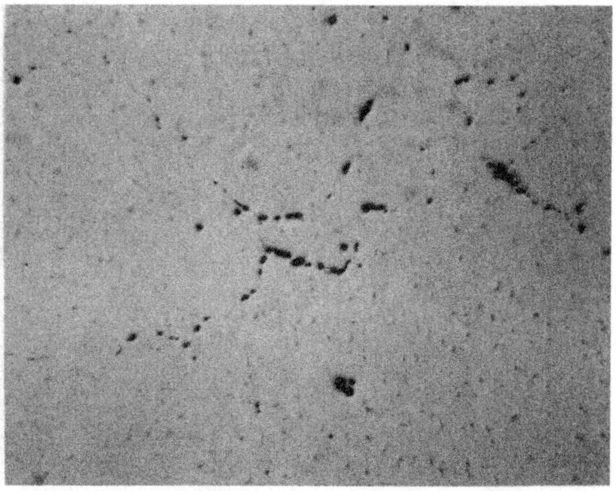

FIGURE 2a, Case 2, Polished, 32x

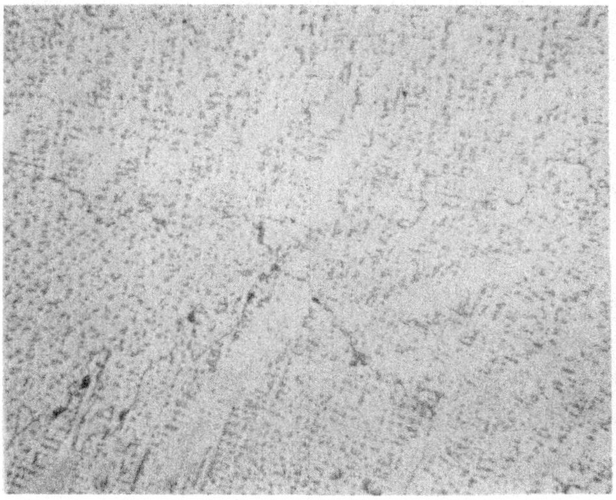

FIGURE 1b, Case 1, Etched, 32x

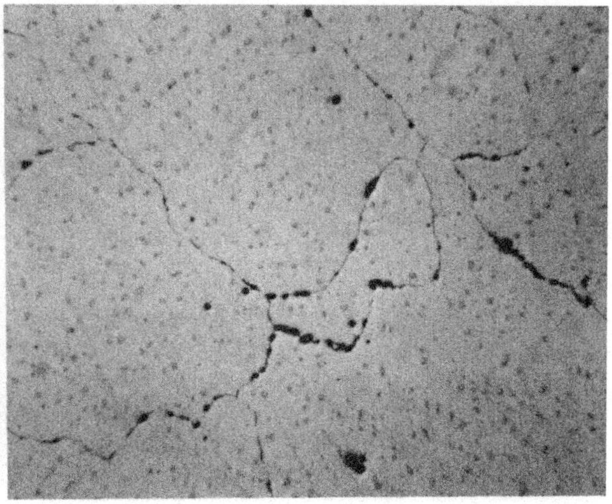

FIGURE 2b, Case 2, Etched, 32x

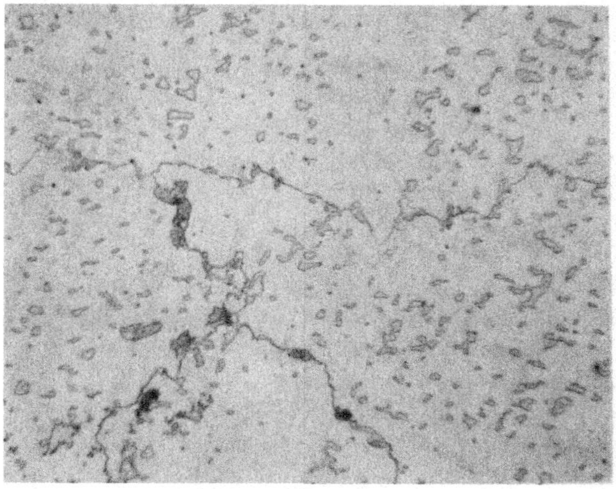

FIGURE 1c, Case 1, Etched, 100x

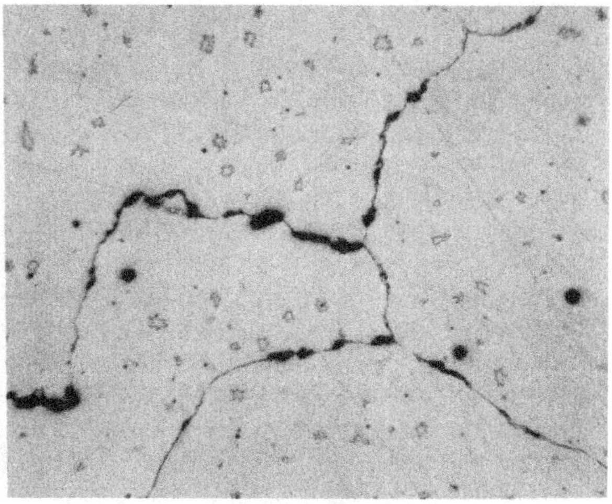

FIGURE 2c, Case 2, Etched, 100x

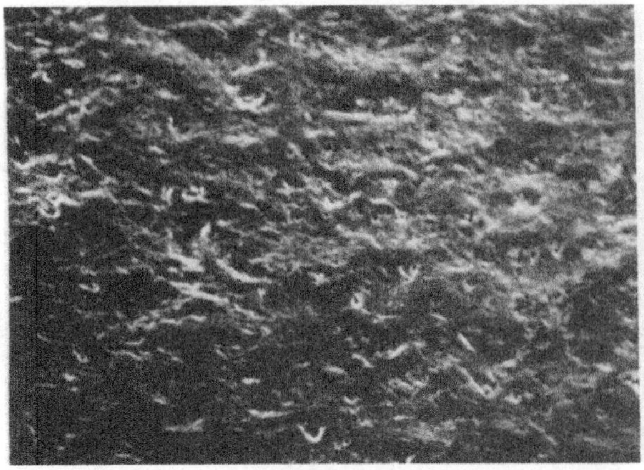

FIGURE 3a, Case 1, SEM 300x

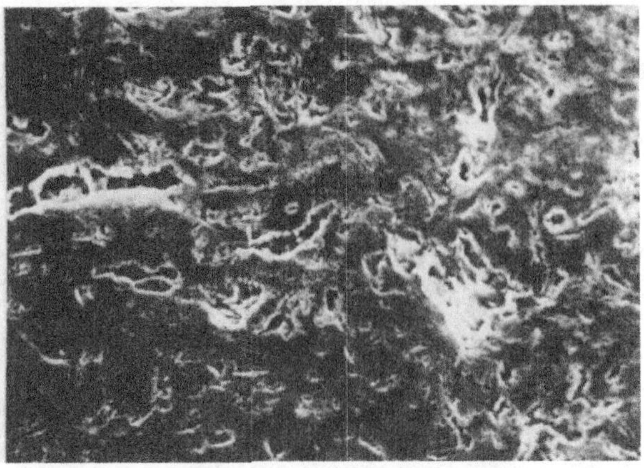

FIGURE 3b, Case 2, SEM 300x

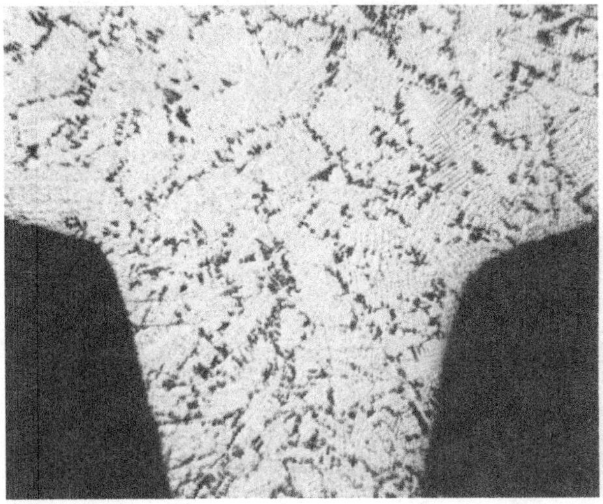

FIGURE 4a, Case 3, 15 sec. etch, 32x

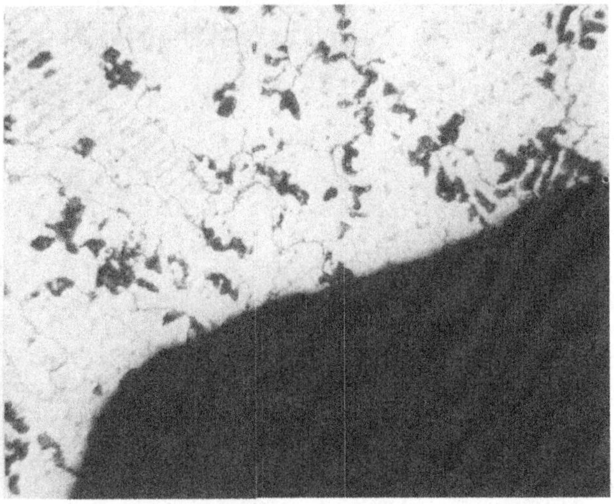

FIGURE 4b, Case 3, 15 sec. etch,100x

REFERENCES

1. Brown, F., Merritt, K, Brown, SA, "Nickel sensitivity and joint fusion with plastic 'K-wires'." Presented to the NE Society of Plastic and Reconstructive Surgery, Hanover, N.H., April 1976.

2. Brown, SA, Polymeric fixation and ultrasonic assessment of fracture healing. Bull. NY Acad. Med. In press.

3. Brown, SA, Mayor, MB and Merritt, K. Dacron as an alternative to wire for trochanteric osteotomy. Presented at 2nd annual meeting of Biomaterials Society. Phila. April 9-13, 1976.

4. Brown, SA, Shafer, JW, Mayor, MB and Merritt, K. "Leukocyte Migration Inhibition Test for Metal Sensitivity," in Problems of Biocompatibility, Brunel Univ. Sept. 1976. Proceedings, In press.

5. Evans, EM, Freeman, MAR, Miller, AJ and Vernon-Roberts, B. Metal sensitivity as a cause of bone necrosis and loosening of the prosthesis in total joint replacement. JBJS 56B:626-42 (1974).

6. Elves, MW, Wilson, JN, Scales, JY and Kemp, HBS. Incidence of metal sensitivity in patients with total hip replacements. Br.Med.J. 4:376-8 (1975).

7. Fisher, AA. Contact Dermatitis. Phila. Lea and Febiger (1976).

ACKNOWLEDGEMENTS: This study was supported in part by USPHS Grant AM17749, NSF Grant DMR 7513596 and gifts from the Gillette Corp.

EXPERIMENTAL AND ANALYTICAL STUDIES OF CLASSICAL AND STRESS GRADIENT PIEZOELECTRICITY IN BONE

Daniel J. Holt, R. Bruce Martin, and Sunder Advani
West Virginia University, Morgantown, W. Va. 26506

Since its discovery in 1957 by Fukada and Yasuda, the piezoelectric effect in bone has promoted much research and many papers. Our prime objective is the development of physiologic theories in order to explain the possible role of strain related potentials in bone remodelling control. We have undertaken the formulation of a complete matrix of piezoelectric coefficients. Our early work on cantilevered beams of dry bone (3) demonstrated a definite histological dependence of the piezoelectric effect. This prompted the current work on dry bone cubes with consideration (and correcting) for bone's structural organization as quantified by collateral porosity, histology, and x-ray diffraction studies as shown in Figure 1. We feel that we should achieve an understanding of the piezoelectric behavior of dry bone before the problems of wet bone is approached.

EXPERIMENTAL METHODS

The bone cubes used were machined from bovine femur and their orientation is shown in Figure 2. A set of cubes consists of an aligned cube and cubes rotated 45° about each of the 3 axes. The nominal cube dimensions were 8 x 8 x 8 mm. The bone specimens were washed in acetone for 24 hours and dried at 38^oC under a vacuum of 29 in Hg for 48 hrs. Platinium foil electrodes were affixed to all six sides with Micro Measurements M-Bone 200 adhesive. The tests were performed in a chamber containing Dri-rite at room temperature. The cubes were subjected to compression at physiologic strain rates ($\dot{\epsilon}$ = .02 sec^{-1}) up to a maximum stress of 16 x 10^6 N/m^2. Records of charge and force vs time, and voltage or charge vs force are utilized to obtain piezoelectric coefficients which are corrected for decay due to the bone's conductance. A Keithley Model 616 electrometer and a Tektronix Oscilloscope Model 549 with the polaroid camera attachment was used to record the data as shown in Figure 3.

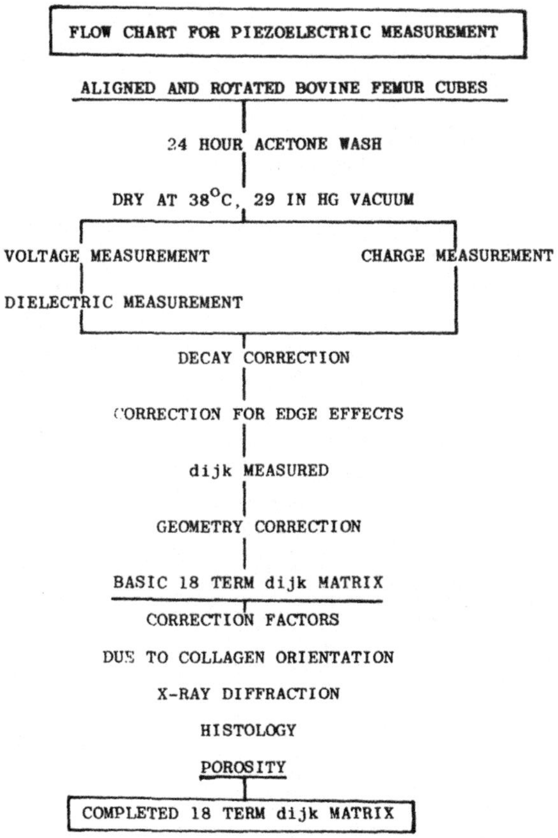

Figure 1

Figure 2

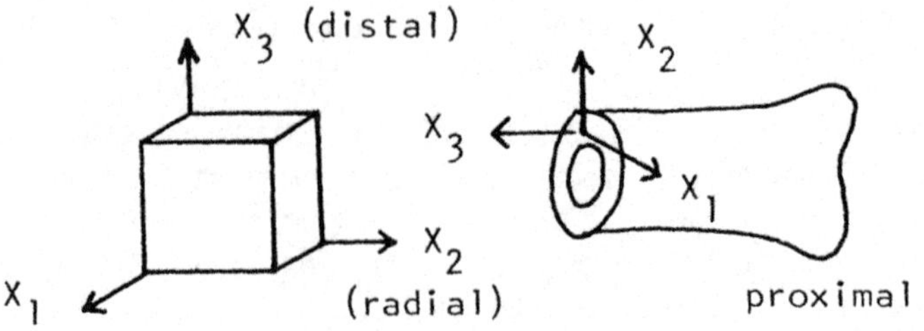

Figure 3

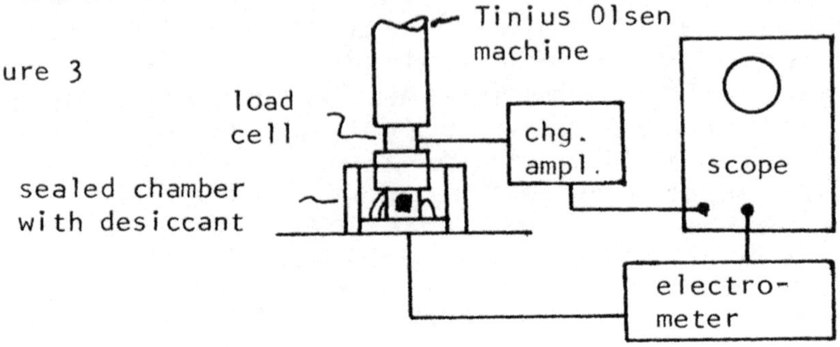

NOTE: Specimen is loaded via teflon pads to reduce shear stresses due to Poisson effect

RESULTS

The results have been quite reproducible in terms of measurements on a single specimen, but more difficult to interpret when different specimens are compared. A comparison of the initial portion of the voltage vs force records has uncovered four basic repeatable waveforms common to all specimens. An example of each is given in Figure 4 along with the relative percentage of occurance of each waveform.

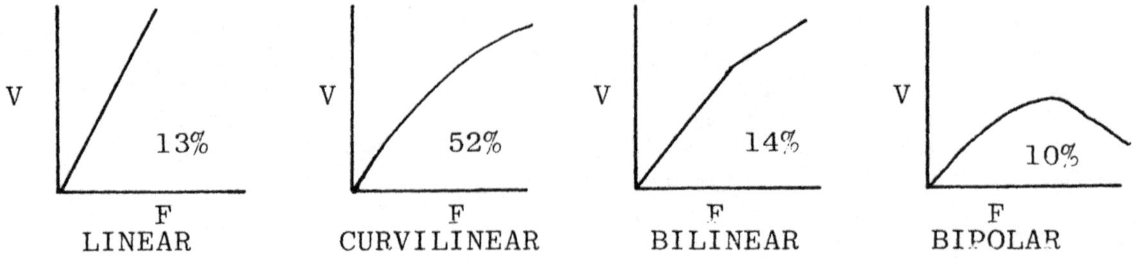

Figure 4

The most interesting of these are the bilinear, which changes slope with increasing stress, and the bipolar, in which the voltage-stress relationship actually changes sign with increasing stress. Figure 5A shows a matrix of such waveforms for one of the aligned cubes.

An interesting phenomenon has been encountered in the testing of the specimens which would not be observable in dynamic tests. Reversing the electrode connections to the electrometer usually does not simply reverse the polarity of the charge measurement. This may be seen in Figure 5B which presents the same measurements as Figure 5A except that the electrodes were reversed. We have spent a great deal of time trying to locate the source of this behavior, and are now satisfied that it does not represent an experimental malfunction. We believe it is a real property of the bone. Testing a quartz crystal in our apparatus results in a voltage which behaves as one would expect, changing polarity when the electrodes are reversed.

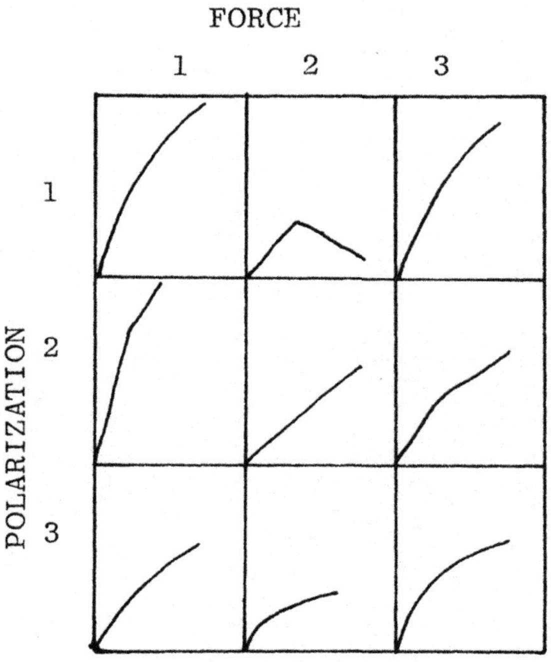

Figure 5A

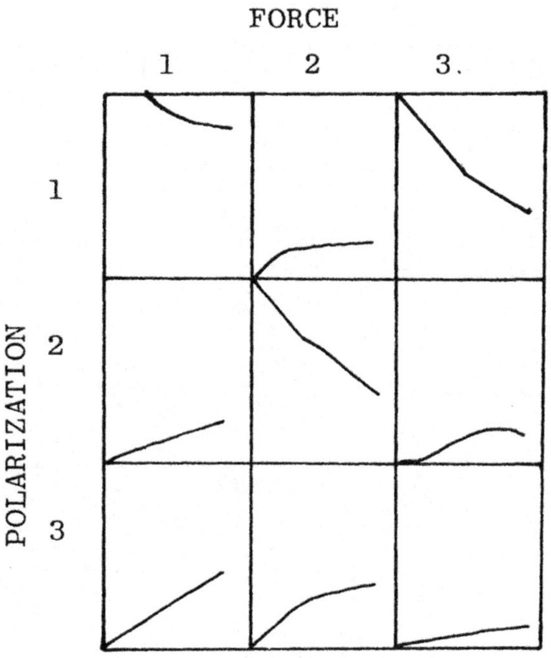

Figure 5B

The aligned cubes yield the front nine peizoelectric coefficients and the three rotated specimens supply the remaining nine coefficients to obtain a complete matrix. To date the available results in terms of the voltage measurement on two sets of cubes are given below, where $E_i = g_{ijk}\,\sigma_{jk}$. E_i = electric field; σ_{jk} = stress

$(10^{-12}$ cgsesu)

$$g_{ijk} = \begin{vmatrix} 6.11 & .66 & 6.64 & 5.83 & 2.30 & 17.40 \\ 9.98 & 5.55 & 1.55 & -20.01 & -7.90 & 18.58 \\ 3.60 & .66 & 2.66 & 12.50 & -2.90 & 9.10 \end{vmatrix}$$

$$g_{ijk} = \begin{vmatrix} 1.55 & .55 & .44 & .78 & 1.60 & -3.09 \\ .66 & .88 & 2.31 & -1.84 & 2.92 & 5.49 \\ .88 & .77 & 4.84 & 1.22 & 1.44 & -.18 \end{vmatrix}$$

DISCUSSION

The variation in the two sets of cubes, which were from the same bovine femur, demonstrates the need to transform the data into the reference frame of the mean collagen fiber, or to otherwise eliminate structural differences between cubes.

We are currently attempting to do this using the x-ray diffraction methods suggested by Gundjian and Chen (2), as well as histologic observations.

The electrode reversal anomaly is, we believe, caused by the special nature of bone's piezoelectricity. We have established that the classical, linear theory of peizoelectricity (that is, a formulation such as $E_i = g_{ijk}\,\sigma_{jk}$) correctly predicts the behavior of a quartz crystal in this regard, but not bone. Neither does the stress gradient theory of Breger (1) and Williams (5) predict this effect in bone. There are, however, effects due to rotating the cube in the apparatus which are analagous to the problems these authors have described in cantilever beams. Since switching the electrode leads is equivalent to rotating a portion of the apparatus about the cube, we believe the electrode reversal anomaly is related in principle to bone's other piezoelectric peculiarities, but we have not yet developed a theory which can explain the differences between Fig. 5A and Fig 5B. A description of this work will be available shortly (4).

The bilinear and bipolar voltage-stress relationships are not understood at this point. We believe they may represent sticking and slipping phenomena at the cube-loading machine interface, or perhaps some sort of buckling behavior within the substructure of the bone. The repeatability of these effects causes us to lean toward the latter of these possibilities. We suspect that an appreciation of the interplay between classical and stress gradient piezoelectric contributions will prove important to understanding these observations, too.

In general we have found the piezoelectric behavior of dry bone to be very complex. There are many observations which we still do not understand, but our experience indicates that even the most confounding of these problems may be solved when theoretical analysis and experimental work are pursued in concert with one another.

References

1. Breger, L., Piezoelectricity in Tendon and Bone, Ph.D. Thesis, University of Illinois, Urbana, 1973.
2. Gundjian, A.A., and Chen, H. L., Standardization and Interpretation of the Electromechanical Properties of Bone, IEEE Transactions on Biomedical Engineering, Vol BME 21, No. 3 May 1974, pp 177-182.
3. Holt, D. J., Martin, R. B., and Advani, S., 1976, Experimental Investigation of Bone as a Piezoelectric Cantilever Beam. Proceedings of 4th New England Bioengineering Conference, 33036
4. Martin, R. B., Holt, D.J., Advani, S., An Electrode Reversal Anomaly in Measurements of Bones Piezoelectric Effect, in preparation.
5. Williams, W.S., CRC Critical Rev. Bioengr., 95, Feb, 1974.

Acknowledgements

This research was supported by NIH grant No. AM16612.

IMPORTANCE OF DYNAMIC PARAMETER τ IN EVALUATING BILAYER MEMBRANES EXPOSED TO LOCAL ANESTHETICS

By
Rajendra P. Nanavati
Syracuse University
Syracuse, N.Y. 13210

INTRODUCTION

From electrical studies (i.e. voltage clamp experiments) done on gaint axon of squid [1] and other biological cells, it is now well accepted that voltage-dependent time-variant conductance is the key property underlying excitability in biological membranes. Rudin and Mueller [2] have shown that lipid-hydrocarbon bilayer membranes, when exposed to certain compounds, namely EIM (excitability inducing material) and two antibiotics, alamethicin and monazomicin show this key property. Furthermore with alamethicin and EIM they have shown "action potentials" which are remarkably similar to the action potentials in biological cells. Thus the bilayer membranes in the presence of one of these compounds is a good model system to study excitability.

Nanavati and Maura [3] have shown that bilayer membranes exposed to alamethicin show current response to a step of voltage excitation given below

$$i(t) = I_o(1-e^{-t/\tau}) \qquad (1)$$

where i = instantaneous current through the membrane, t = time, I_O = steady-state current, τ = time-constant, a dynamic parameter.

Recently there has been a good deal of interest in the study of bilayer membranes exposed to anesthetics. Most of these studies involve the local anesthetics; moreover only a static parameter namely d-c resistance is measured before and after the addition of the local anesthetic [4]. Within the last year Nanavati [5][6] has shown that the static parameter I_O and the dynamic parameter τ are both important in evaluating the effect of anesthetics on bilayer membranes. In the present paper it will be shown that the dynamic parameter τ is crucial to this study.

MEASUREMENT TECHNIQUE

Our bilayer membranes were formed from a solution of 2 percent egg lecithin and 1 percent cholesterol in normal decane. These membranes were formed over a circular hole 1 mm^2 in area in a teflon partition which separates two halves of a lucite chamber each having 3.5 c.c. of 0.1 M NaCl. Into the IN compartment was introduced 10 μl of aqueous saturated solution of alamethicin. Voltage clamp experiments were then done on bilayer membranes. Voltage across the membrane was measured with commertially available calamel electrodes with the help of a digital voltmeter. Current through the membrane from the IN to the OUT chamber was measured with Ag-AgCl electrodes made in our laboratory. The voltage and current were observed on a dual beam oscilloscope. Different step voltages were applied across the membrane and the current response in each case was measured. In each case both the steady state reponse I_o and the dynamic parameter were noted. This constituted our control data. Then 20μl of the local anesthetic tetracain-hydrochloride was introduced into the OUT compartment. Solution of both chambers was stirred for three minutes. After the current response reached steady state condition, again data was taken on I_o and τ. These data are tabulated in Table A and plotted in Fig. 1. Another experiment of precisely the same nature as above was carried out where the anesthetic used was 100μl of 3% carbocaine in the OUT compartment. The data is tabulated in Table B and plotted in Fig. 2.

Table A

Data with and without TETRACAINE-HYDROCHLORIDE

Voltage step (mV)	steady-state current I_o (nA)	Time-Constant τ (sec.)	
81.9	50	0.52	without tetracaine-hydrochloride (control)
82.9	75	0.46	
84	95	0.40	
87.4	45	1.73	with 20μl of 0.5% tetracaine-hydrochrolide
87.9	70	1.15	
89	103	1.04	

DISCUSSION

It can be readily seen from Tables A and B that with anesthetics tetracaine-hydrochloride and carbocaine that the steady-state current I_o diminishes for the same voltage or that the same current requires bigger step voltage to be applied. This indicates that the membrane resistance increases in the presence of the anesthetics consistent with physiological data. (Ohki's experimental results showed that the membrane resistance decreases in the presence of anesthetic which is contrary to physiological data on excitable membranes). Thus from the standpoint of d-c resistance both tetracaine-hydrochloride and carbocaine behave essentially similarly. However these two anesthetics are different in kind as can be seen from a comparison of Fig. 1 with Fig. 2. Here it can be seen that the curve of I_o as a function of the dynamic parameter time-constant τ shifts to the left of control data for carbocaine and to the right for tetracaine-hydrochloride. The time-constant is longer in the presence of one anesthetic whereas it is shorter in the presence of the other anesthetic. We see therefore that the dynamic parameter τ allows us to distinguish between the two anesthetics considered here. This distinction would be impossible from the static parameter. The parameter

Table B

Data with and without Carbocaine

Voltage step (mV)	Steady-State current I_o (nA)	Time-constant τ (sec.)	
70	45	6.35	without carbocaine (control)
72	72	3.75	
73	98	2.74	
75	125	2.31	
79	45	1.30	with 100μl of 3% carbocaine
80	72	1.23	
80+	98	1.23	
82	122	0.94	

τ may play a similar role when comparing other anesthetics with each other. It is thus demonstrated with experimental data that the dynamic parameter is a most important parameter to measure. This may turn out to be a very significant parameter for anesthesiologist to know and possibly to use.

References

1. Hodgkin, A. L. and Huxley, A. F. (1952), J. Physiol; 117, 500-522.
2. Mueller, P. et al (1962) Circulation, 26, 757-766.
3. Nanavati, R. P. and Mauro, A. (1974) Proc. 2nd Annual New England Bioengineering Conf., 81-84.
4. Ohki, S. (1970), Biochimica et Biophysica Acta, 219.
5. Nanavati, R. P. (1976) Proc. 4th Annual New England Bioengineering Conf.
6. Nanavati, R. P. (1976), Proc. 29th ACEMB.

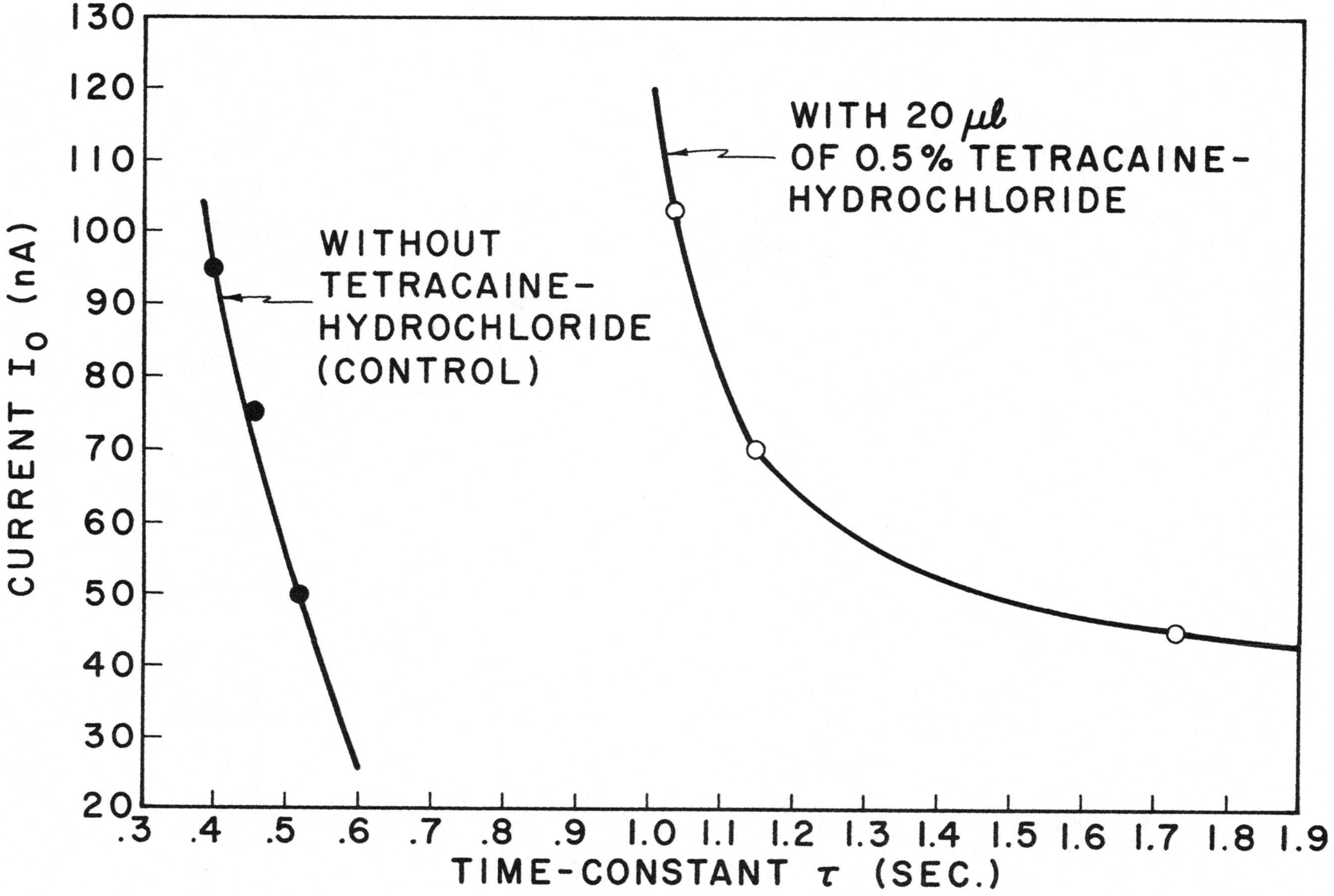

FIG. 1 STEADY-STATE CURRENT AS A FUNCTION OF τ (TETRACAINE)

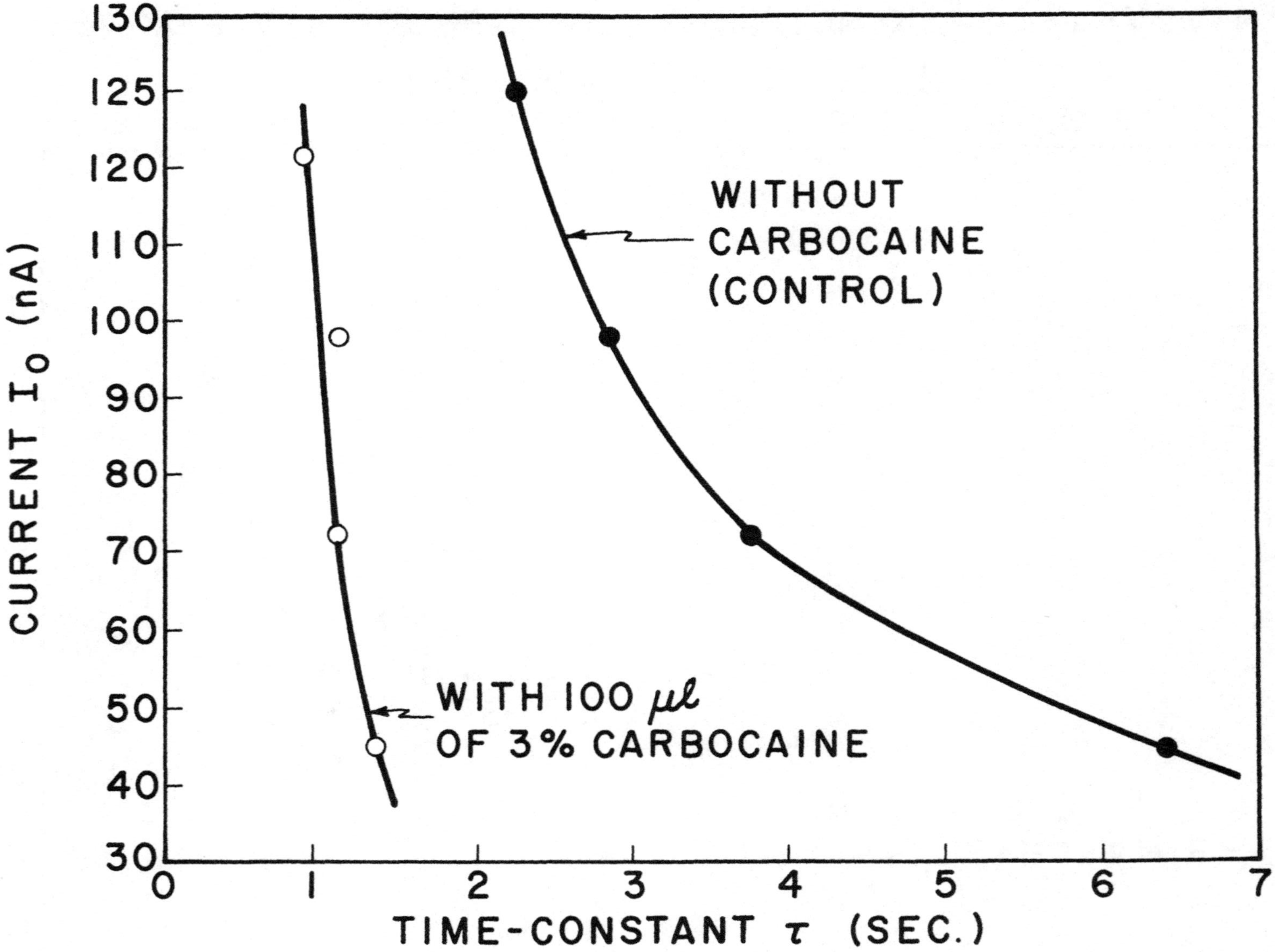

FIG. 2 STEADY-STATE CURRENT AS A FUNCTION OF τ (CARBOCAINE)

THE HANDLING CHARACTERISTICS OF SURGICAL SUTURES

Allen H. Hoffman[1], John B. Herrmann[2], David V. Lesneski[1] & James L. O'Connor[1]

1 Dept. of Mechanical Engineering
Worcester Polytechnic Institute
Worcester, MA 01609

2 Dept. of Surgery
Univ. of Massachusetts Medical Center
Worcester, MA 01605

ABSTRACT

The flexural rigidity of four sizes of fourteen different types of sutures were measured using Peirce's ring loop method. The suture materials were ranked according to the lowest value of flexural rigidity. This ranking strongly correlated (r=0.93) with the ranking of handling characteristics obtained thru a survey of practicing surgeons. Thus flexural rigidity can be used as a significant criterion for quantifying the handling characteristics of surgical sutures.

INTRODUCTION

A surgical suture must possess several desirable physical qualities. These properties are evaluated qualitatively by the surgeon and include strength, knot security, stiffness, tissue drag, lack of chatter, and knot tying properties. The total qualitative evaluation of these properties is often described as a suture's handling characteristics. The handling characteristics of a suture are dependent upon the material, construction, and physical properties of the suture. A major problem exists in attempting to quantify desirable handling characteristics in terms of measurable physical properties.

In an attempt to quantify handling characteristics, Herrmann (1) conducted a survey among 32 surgeons at Georgetown University in 1970. In this survey surgeons ranked sutures according to their handling characteristics. Results of the survey also rank correlated the qualities of flexibility (r=0.77) and lack of resiliency (r=0.78) with the handling characteristics of these sutures. Other properties such as strength, knot security, and coefficient of friction were not strongly correlated with the handling characteristics.

Based upon these results, work was initiated to develop and test a relationship between the handling characteristics of sutures and a measurable physical property related to the qualitative term "flexibility". A suture is used primarily in a bending mode. It is bent and flexed around the surgeon's fingers, surgical instruments, body tissue, and about itself when tied. Thus the term "flexibility" must be related to a suture's lack of resistance in the bending mode. Since the mechanical property flexural rigidity is a physical measure of a material's resistance to bending, it was chosen to quantitatively represent the "flexibility" of a suture. Flexural rigidity (G) can also be defined in terms of Young's modulus (E) and the area moment of inertia (I) by the equation

$$G = EI \qquad (1)$$

Flexural rigidity is superior to Young's modulus as a measure of a suture's resistance to bending since it is independent of the suture construction (braid or monofilament). The use of Young's modulus would require the calculation of the area moment of inertia which is not well defined for a braid construction.

THEORY

The ring loop method (2,3) was developed to determine the flexural rigidity of very flexible materials, such as textiles. The ends of a segment of material are tied together to form a circular loop (Figure 1). Opposing forces are exerted on the loop, deflecting it, and changing its curvature. The loop structure creates sufficient resistance to bending (changes in curvature) so that loads and deflections are measurable, even with very flexible materials. Development of the equations governing the ring loop method begins by analyzing an element in the deflected loop using the beam equation

$$\frac{1}{\rho} = \frac{M}{G} \qquad (2)$$

where ρ is the radius of curvature and M is the applied bending moment. Use of this equation assumes that the curvature is small ($2\rho/D > 10$ where D is the diameter of the suture). Sutures easily satisfy this criterion. The weight of the loop is assumed to be small in comparison to the applied load.

After a rather lengthy derivation in which a constant of integration is evaluated by using experimental data from feeler steel (2,4), the flexural rigidity can be represented by

$$G = \frac{0.0047 \cos\theta \; FL^2}{\tan\theta} \quad \text{gm-mm}^2 \qquad (3)$$

$$\text{where } \theta = \frac{493d}{L} \quad \text{degrees} \qquad (4)$$

In these equations F is the force to deflect the loop, L is the circumferential length of the undeformed loop, and d is the loop deflection. The results are most accurate for θ values between 16° and 60° (3).

EXPERIMENTAL METHODS

The validity of the ring loop method was checked using an elastic material (music wire). The flexural rigidity was determined using the ring loop method and then Young's modulus was determined from Equation (1). These values were compared to published values with highly satisfactory agreement.

Three ring loops were tested for each of four sizes (0 thru 4-0) of the fourteen different sutures listed in Table I. Loops were tied around a 7.88 mm glass rod. The ends of the knot were snipped off and the loops were placed in a temperature and humidity controlled environment for at least two hours prior to testing. Stress relaxation tests had shown that this was a sufficient period of time for stresses induced by tying to fully relax.

Testing was performed using an Instron universal testing machine. Glass hooks were fabricated to hold the loop. The knot was positioned in a region of minimum change in curvature (3). The deflection was set to yield a θ value of 56.9°. After some preliminary testing, a strain rate of 100 cm/minute was chosen. The applied loads were measured using a load cell and recorded on a chart recorder. The load signal could be amplified to achieve a full scale deflection of 2 grams on the chart recorder.

RESULTS AND DISCUSSION

All the suture materials except steel exhibited viscoelastic behavior in the form of relaxation of the load as a function of time (stress relaxation). Figure 2 shows a typical load curve. The finite time necessary to deflect the loop is clearly evident. The effects of the viscoelastic response and the finite strain rate were investigated using a method similar to that of Cokelet and Smith (5). The load relaxation curves are essentially exponential and therefore plot as straight lines on semi-logarithmic paper. This straight line can be extrapolated back to zero time (Figure 2). Since only a small amount of load relaxation takes place while the loop is being deformed, it can be concluded that the finite strain rate has little effect on the results. Time constants for these materials ranged from one-half second to five seconds. Higher order time constants do not effect the handling characteristics. Flexural rigidity is directly proportional to load (Equation 3) and is therefore an exponential function of time.

In order to rank the flexural rigidity of the different sutures a handling characteristic time must be chosen. Values between zero and four seconds yield essentially the same ranking. A handling time of one second was chosen and flexural rigidity values were computed using the average for three tests. Typically the standard deviation was 15% of the average value. Sutures of a given size were ranked in terms of lowest values of flexural rigidity. Rankings were then averaged over the four sizes to obtain the final ranking shown in Table I. Flexural rigidity values for size 2-0 sutures are also listed and do not exactly follow the average ranking. The relative ranking for a given size is effected by the considerable diameter variation (up to 25%) among different types of sutures of that size.

In the 1970 survey, sutures were classified only by material and not by coating. For comparison purposes it was necessary to combine silicone silk, wax silk, and silk into a single category. Similarly, Mersilene, Tycron and Silky Polydex II were combined and classified as dacron. The braided nylons including Surgilon and Nurolon were not included in the survey.

Sutures were ranked according to the lowest value of flexural rigidity (greatest flexibility). This ranking strongly correlated (r=0.93) with the ranking of handling characteristics obtained in the survey (Table II). While the flexural rigidity values do not produce significant differences between all of the sutures listed in Table I, the differences are significant for all of the materials listed in Table II.

FIGURE 1. RING LOOP

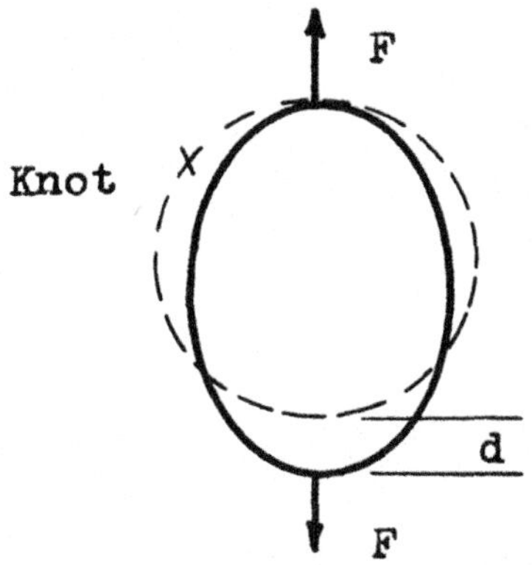

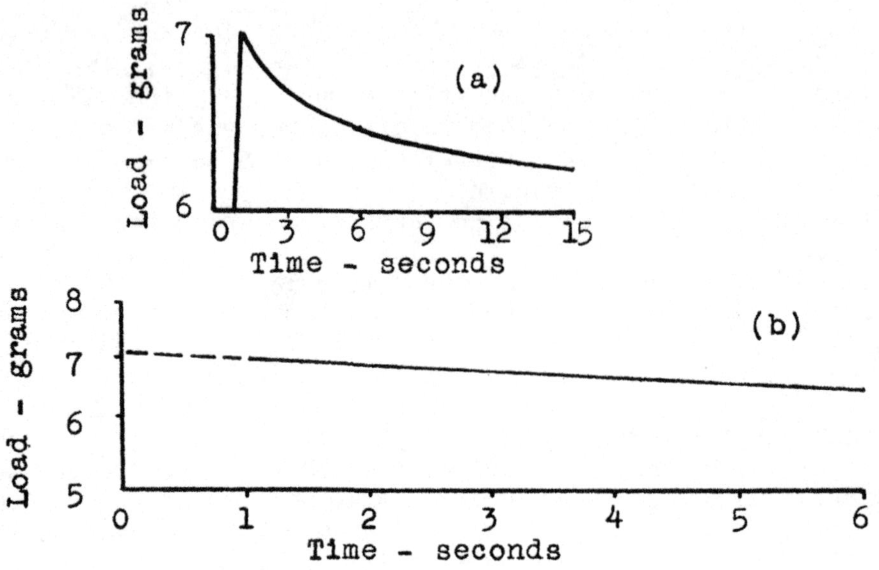

FIGURE 2. LOAD RELAXATION FOR SIZE 2-0 COTTON SUTURE
(a) Rectangular Coordinates (b) Semi-logarithmic Coordinates

TABLE I SUTURE RANK USING FLEXURAL RIGIDITY VALUES FOR SIZES 0 THRU 4-0
Values for size 2-0 sutures ($gm\text{-}mm^2$) appear in parenthesis.

1.	Silicone Silk	(0.65)	8.	Nurolon	(1.06)
2.	Surgilon	(0.68)	9.	Silk	(1.76)
3.	Wax Silk	(0.68)	10.	Cotton	(7.04)
4.	Mersilene	(0.94)	11.	Polypropylene	(44.9)
5.	Black Braid Nylon	(0.73)	12.	Monofilament Nylon	(49.5)
6.	Tycron	(1.18)	13.	Multifilament Steel	(71.2)
7.	Silky Polydex II	(1.67)	14.	Monofilament Steel	(385)

TABLE II COMPARISON OF RANKINGS BASED ON HANDLING CHARACTERISTICS AND FLEXURAL RIGIDITY VALUES (r=0.93)

Rank by Handling Characteristics	Rank by Flexural Rigidity
1. Silk	1. Silk
2. Cotton	2. Dacron
3. Dacron	3. Cotton
4. Polypropylene	4. Polypropylene
5. Multifilament Steel	5. Monofilament Nylon
6. Monofilament Nylon	6. Multifilament Steel
7. Monofilament Steel	7. Monofilament Steel

The interchange in ranking between cotton and dacron may be due to the fact that cotton is used by surgeons in a moist environment while the flexural rigidity tests were performed on dry cotton. Moisture will reduce the flexural rigidity of cotton. The interchange in ranking between monofilament nylon and multifilament stainless steel may be due to the lower resilience of the stainless steel. The survey also rank correlated the lack of resilience with the handling characteristics of sutures. In the case of these two materials the surgeon ranking may be strongly influenced by the resiliency of the materials.

CONCLUSIONS

The mechanical property flexural rigidity has been shown to be strongly correlated with a suture's handling characteristics. It has been demonstrated that the ring loop method is an effective means of determining this property. Although the viscoelastic phenomenon of stress relaxation is clearly present, it has little effect on the results of this study. Material resiliency appears to also influence the evaluation of the handling characteristics of surgical sutures.

BIBLIOGRAPHY

1. Herrmann, J.B., Unpublished Data.
2. Peirce, F.T., J. Textile Institute, Vol. 21, pp. T377, (1930).
3. Carlene, P.W., J. Textile Institute, Vol. 41, pp. T159, (1950).
4. A detailed derivation is available from the authors.
5. Cokelet, G.R. and Smith, J.H., Biorheology, Vol. 10, pp. 51, (1973).

COMPOSITE MATERIALS FOR LIGHTWEIGHT ORTHOTIC DEVICES

S. A. T. Long
National Aeronautics and Space Administration
Langley Research Center, Hampton, Virginia 23665

Graphite reinforced epoxy, graphite reinforced polysulfone, and carbon reinforced nylon composite materials have been investigated for lightweight orthotic devices. These composite materials are stronger and stiffer, yet lighter and thinner than the stainless steel, aluminum, plastic, and wood used in conventional orthotic-device constructions. Pelvic braces with composite waistbands and lower-leg struts, long leg braces with composite structural members, composite drop-foot braces, composite universal knee joints, and composite/metal-honeycomb shoe extenders have been fabricated. Clinical evaluations show that the composite orthotic devices fabricated not only have greatly-increased load-carrying capability, but also give greatly-increased patient mobility.

BACKGROUND INFORMATION

The NASA Langley Research Center Materials Division has developed and applied composite materials, composed of fibers held together in matrix materials, for constructing stronger and lighter aircraft and spacecraft. In Table I properties of the graphite reinforced epoxy, graphite reinforced polysulfone, and carbon reinforced nylon composite materials are compared to those of the stainless steel, aluminum, and neat polypropylene materials used in conventional orthotic-device constructions. The tensile strength is the load-carrying capability; the tensile modulus is a measure of the stiffness. It is seen that the composite materials are stronger and stiffer, while affording lighter and thinner designs, than the conventional materials.

FABRICATED DEVICES

Pelvic braces with composite waistbands and lower-leg struts and also long leg braces with composite thigh and calf bands and upper-leg and lower-leg struts were fabricated[1] from graphite reinforced epoxy. Layers of the thin composite prepreg material were laminated in metal or wooden molds and then cured for 1 hour in a 350°F oven, as shown in Fig. 1. For the pelvic brace the composite parts weigh 56% less than the replaced metal parts, resulting in a 1.4-lb reduction in brace weight. The long leg braces with composite structural members are shown in Fig. 2. For these braces the composite parts weigh 51% less than the original metal parts, accomplishing a 2.1-lb weight reduction per brace. Brace components were also successfully fabricated from graphite reinforced polysulfone thermoplastic bars using simple forming techniques along with heat and pressure. The formed brace members made of either graphite-epoxy or graphite-polysulfone can be sawed, filed, and drilled like metals. The universal knee joints for the composite braces were made by injection molding chopped-carbon-fiber reinforced nylon. The knee joints fabricated in this manner were produced very rapidly and inexpensively.

To fabricate the complex, curved drop-foot brace,[2] first woven graphite cloth was draped over a plaster cast of the leg contour. A solution of polysulfone resin in chloroform was then painted onto the cloth. The chloroform solvent evaporated quickly, leaving a solid polysulfone matrix around the graphite cloth. This easy procedure was repeated until the desired brace thickness was reached. The graphite-polysulfone drop-foot brace is much, much thinner than the conventional neat-polypropylene drop-foot brace. It allows the patient requiring a brace for one leg only to wear the same size shoe on each foot.

The composite-material/metal-honeycomb technology developed also for stronger, lighter aerospace constructions was transferred to orthotic shoe extenders. These new shoe extenders are much lighter and last much longer than the traditional wooden-block shoe fillers.

CLINICAL EVALUATIONS

The orthotic devices fabricated were clinically evaluated at The Mississippi Methodist Hospital and Rehabilitation Center in Jackson, Mississippi, and The Coastal Center for Mental Retardation in Ladson, South Carolina. The clinical evaluations showed that the lightweight, composite-materials devices resulted in increased comfort and aesthetics and greatly-increased load-carrying capability and mobility for the patients.

FUTURE PLANS

A pelvic brace with composite waistband, lower-leg struts, and upper-leg struts will be fabricated. The technologies from the composite drop-foot-brace and universal-knee-joint efforts will be combined to develop a long leg brace, by connecting leg-contour-formed struts with universal knee joints. Nylon mesh will be investigated to replace the heavier leather padding and straps on the pelvic and long leg braces. To assure the clinical applications of the basic composite-materials technologies to the improved designs of orthotic devices, the development of clinical-setting fabrication techniques will soon begin.

[1] R. M. Baucom, H. E. Johns, and R. C. Evans, NASA Tech Brief LAR-11894 (1975).
[2] R. M. Baucom and T. L. St. Clair, NASA Tech Brief LAR-11918 (1976).

TABLE I. Properties of materials for orthotic devices.

Material	Tensile strength, lb/in.^{2}	Yield strength, lb/in.^{2}	Tensile modulus, lb/in.^{2}	Density, lb/in.^{3}	Tensile strength/density, in.	Tensile modulus/density, in.	Cost, $/lb (1975)
Stainless steel	75.0×10^3	27.0×10^3	29.0×10^6	0.286	4.2×10^5	1.0×10^8	4.00
Aluminum	55.0	32.0	10.0	0.101	5.4	1.0	2.00
Neat polypropylene	5.1	5.0	0.16	0.033	1.55	0.15	0.50
Graphite-epoxy	211.0	171.0	18.5	0.055	38.4	3.4	65.00
Graphite-polysulfone	168.0	136.0	16.0	0.055	30.1	2.9	65.00
Carbon-nylon	40.0	53.3	5.0	0.050	8.0	1.0	25.00

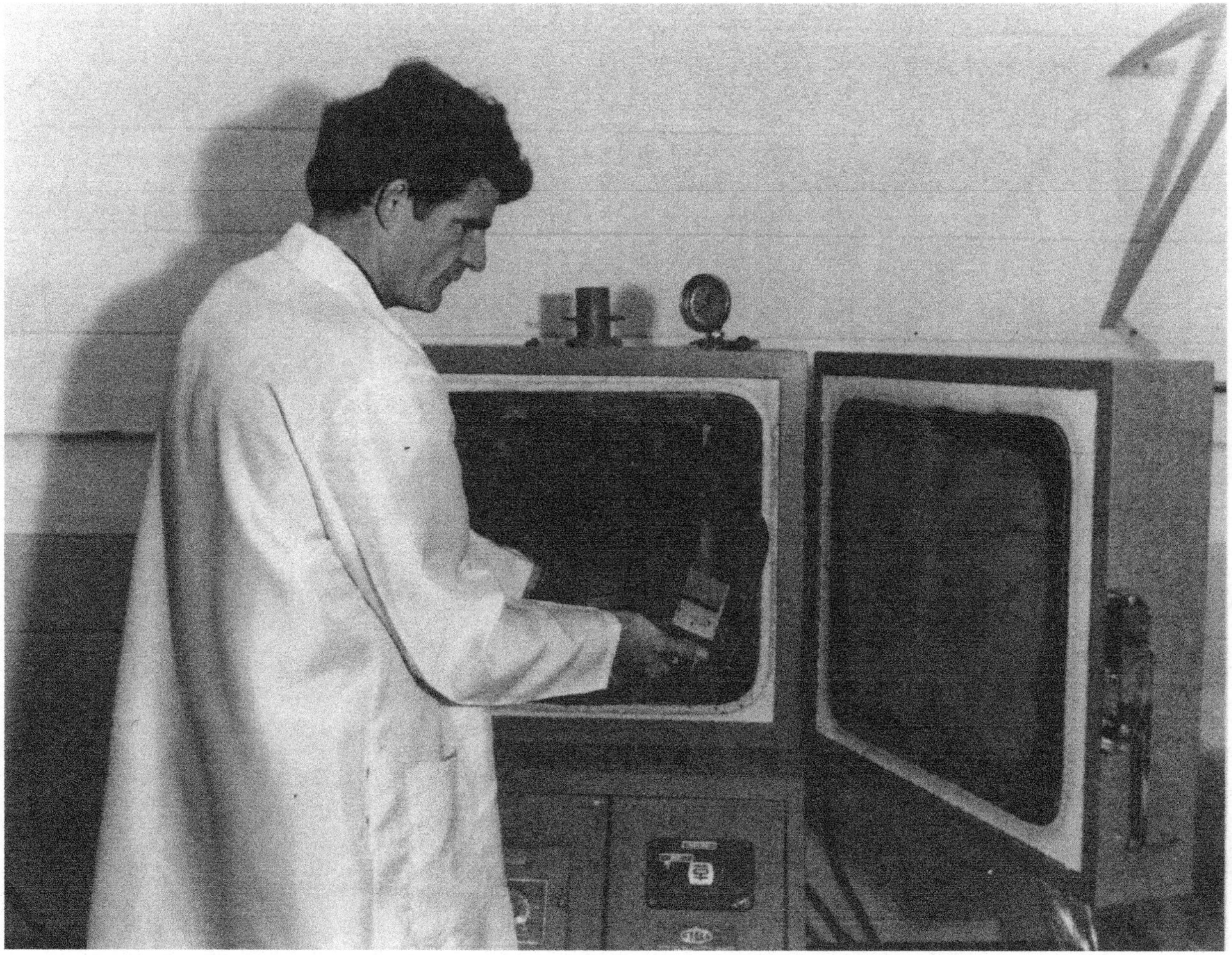

FIG. 1. Graphite-epoxy brace strut is cured in $350^{\circ}F$ oven.

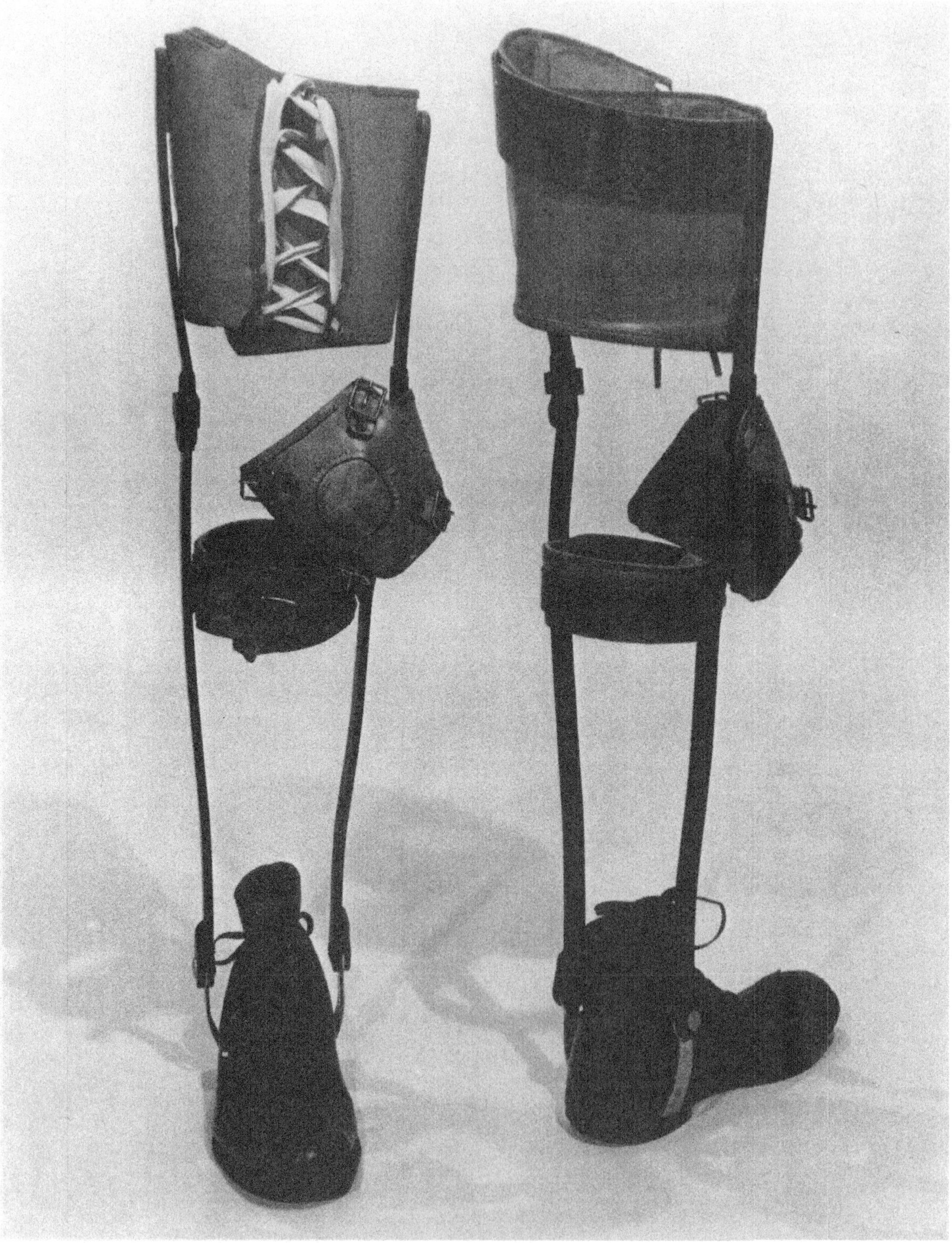

FIG. 2. Long leg braces with composite structural members.

"THE EFFECT OF FLUORIDE ON THE PHYSICAL PROPERTIES OF BONE"

James A. Albright, M.D.
Yale. Univ. Sch. of Med.
Section of Ortho.Surg.
333 Cedar Street
New Haven, Ct. 06510

Subrata Saha, Ph.D.
Yale Univ. Sch. of Med.
Dept. of Eng. & Appl. Sc.
15 Prospect Street
New Haven, Ct. 06520

Murray Ruggiero
359 Silver Sands
East Haven, Ct. 06512

Supported by USPHS Grant AM11461 and TI AM5416-07

Introduction

The specific patho-physiologic role of fluoride continues to be a subject of great interest and extensive investigation. Much of this work has been focused on the protective effect of fluoride on dental acries, an action which has been demonstrated repeatedly and is now generally accepted. Whether or not there might be an equally beneficial action on bone remains uncertain in spite of increasing attention to the subject (over 16,000 papers have been published on the biological effects of fluoride) [2].

The recognition and description of skeletal fluorosis and reports[4] of osteosclerosis in patients living in endemic areas having high concentrations of fluoride in the water supply, especially in India[6] generated the impression that fluoride stimulates bone formation and that this action might be of benefit to patients with osteoporosis. This concept was reinforced by studies showing a lower incidence of osteoporosis in areas having high levels of fluoride inthe water supply, and by metabolic balance studies which documented increased calcium retention. Other findings tend to support the view that fluoride has a beneficial effect on bone: The physico-chemical effect of fluoride on apatite results in larger and less soluble crystals; fluoride can increase bone formation and decrease bone resorption in tissue culture; bone density may increase, and the microscopic appearance of bone may improve. However, a number of epidemiological studies on the relationship between fluoride intake and the incidence of fracture have produced conflicting results.

Direct measurements of the mechanical properties of bone and their alteration by fluoride offers many advantages since the ultimate objective of treatment is an increas in bone strength. The purpose of this paper is to report the results of studies designed to determine the effect of fluoride on the physical properties of bone.

Methods

The bone specimens for mechanical testing were obtained from growing rabbits and sheep which had been given sodium fluoride in the drinking water at concentrations of 1 ppm, 10 ppm, 100 ppm and 500 ppm, plus controls (distilled water). The animals were fed processed commercial pellets (Purina Ralston Co.). The various studies included two groups of rabbits and one of sheep. In one group of rabbits 13 animals: control, 1, 100, and 500 ppm fluoride), sacrificed after 6 months, the metatarsals were loaded in torsion. The metatarsals from the other group of rabbits (8 animals: control and 100 ppm) sacrificed in 4 months, were subjected to bending. 4 sheep were treated with fluoride (control, 1, 10, 100 ppm) for 10 months. Density (specific gravity) and resistence to compression were measured on cubes of cancellous bone from the thoracic and lumbar vertebrae. Uniform slabs of cortical bone from the tibia were loaded in tension.

Torsion was measured on an Instron Testing Machine at a slow rate of loading. The metatarsals were cleaned, allowed to dry at room temperature, and mounted in a miniature lathe; the force was applied thru a perpendicular lead from a pulley on the lathe to the crosshead of the testing machine. For bending studies, a 3-point bending force was applied at 0.5 inches/min. using an Instron Testing Machine; the distance between the supporting blocks equalled 1/2 inch. Machined cubes of cancellous bone and slabs of cortical bone were produced on a Unimat miniature lathe and the dimensions measured with vernier calipers. Compression at a crosshead speed of 0.1 inches/min. was measured on wet bone. Slabs of cortical bone were parallel to the long axis of the tibia. Strain gages were attached to both sides of the specimens and connected to an independent Wheatstone bridge which controlled the Y axis on the Instron Testing Machine. Readings could be made from either gage alone or both together, and the results were accepted if the curve from one side of the bone was similar to that on the other side. Tension was applied at a crosshead speed of 0.05 inches/min.

For density determinations, the volume was calculated by direct measurement on the cubes of cancellous bone and by water displacement (wt. in air minus wt. under water) on the specimens of cortical bone, since they were irregular.

Fluoride determinations were performed on rabbit bone by a microdistalla-tion method of analysis,[1] and on sheep bone with a specific ion electrode.[3]

Results

All animals receiving 100 ppm or 500 ppm fluoride in the drinking water showed some retardation of growth and many of these animals did not appear as vigorus as those on lower levels. Grossly, there was little visible difference between the bones from the various groups, except those from animals on 500 ppm fluoride contained numerous exostoses and the outer diameter was increased; in addition the cortex was proous and had an abnormal chalky appearance.

The torsional force required to fracture intact rabbit metatarsal bones was similar in the various groups except for animals receiving 500 ppm fluoride (Table 1); these bones were less dense and failed at smaller laods. Rabbit metatarsals from the 100 ppm group were less resistant ot a bending force (ave.: 92.1 $\pm$ 24.0N), than were the controls (ave.: 120.1 $\pm$ 39.1N).

The average density of cancellous bone from the thoracic vertebrae and cortical bone from the tibiæ of sheep increased with increasing fluoride content, but the dry wt./ash wt. ratio decreased in the tibiae and increased in the vertebrae (Tables 2 & 3). The latter may have been due to an irregular pattern of mineralization which was noted in vertebrae containing high levels of fluoride, but no in the tibiae.

The ultimate compressive stress of the vertebrae, and the ultimate tensile stress and modulus of elasticity of the tibiae showed an inverse relation to fluoride content, except in the 1 ppm group. The strength tended to increase as the density increased, but this relationship was altered by fluoride. Within each fluoride group bone strength was directly related to density, but bones containing high levels of fluoride were weaker than bones of equal density but lower fluoride content (figure 1.).

Discussion

Although recent evidence indicates that fluoride may be essential for optimum growth in animals[5], it is still not known if a similar effect occurs in bone and little of the available information is applicable. Most studies of fluoride and bone have either determined the effect of fluoride on animals given deficient diets or have used high doses for short periods of time. Studies of teeth have repeatedly demonstrated a major effect from small changes in dosage when maintained over long periods of time, and differences have been reported between 2 ppm and 1 ppm fluoride in the drinking water. A similar relatioship might exist for bone.

It is now apparent that high levels of fluoride, whether given over long or short periods of time, are detrimental to bone. Our studies of growing animals treated over relatively long periods of time provide insufficient information to determine the effect of low doses; longer treatment periods and larger numbers of animals would be required.

The relationship detected between fluoride, bone density, and bone strength has significant clinical implications. The direct correlation between density and strength found in normal bone is altered by fluoride in such a way that knowledge of calcium metabolism or of bone density is not only meaningless, but may be misleading, if used to evaluate changes in the quality of bone. This can only be determined by direct study of bone itself.

Conclusion

The results of these studies indicate that treatment using doses of fluoride sufficient to produce x-ray changes in a relatively short period of time might be expected to have a detrimental effect on the quality of bone. Furthermore, since the quality of bone does not appear to be directly related to the amount of mineral present, the determination of mineral content by densitometry would provide insufficient information to monitor the effect of fluoride treatment clinically.

Although high levels of fluoride appear to have a detrimental effect on the quality of bone, at lower levels there may be a small range within which the effect is beneficial. More information is needed concerning the dose-related effect of lower levels of intake.

1. Armstrong, W.D.: Fluoride Determination. Anal.Biochem. 10:495, 1965.
2. Faccini, J.M.: Fluoride and Bone. Calc.Tiss.Res. 3:1, 1969.
3. McCann, H.G.: Determination of Fluoride in Mineralized Tissues using the Fluoride Ion Electrode. Arch. Oral Biol. 13:475, 1968.
4. Rohlm, K.: Fluorine Intoxication; A Clinical Hygienic Study, 1937, H.K. Lewis and Co., London.
5. Schwartz, K.; and Milne, D.B.: Fluorine Requirement for Growth in the Rat. Bioinorg.Chem. 1:331, 1972.
6. Singh, A.; Jolly, S.S.; Bansal, B.C.; and Mathur, C.C.: Endemic Fluorosis. Epidemiological, Clinical and Biochemical Study of Chronic Fluorine Intoxication in Panjab (India). Medicine 42:229, 1963.

Animal Group	Animal Wt. (kg) (ave)	Metatarsals (Dry)			F Conc. (femur)
		Av.Wt(gm)	Area*(mm^2)	Max.Torque (kg.Cm)	(dry Wt.) (ppm)
Control (3)	2.4	1.123±.065 (12)	5.15±.93 (23)	3.67±.89 (23)	513 (3)
10ppm[+] (4)	1.8	1.040±.098 (16)	5.40±1.11 (28)	4.26±.85 (28)	1876 (3)
100ppm[+] (3)	2.0	1.00±.097 (12)	5.11±.80 (23)	3.75±.85 (23)	4989 (3)
500ppm[+] (3)	1.8	.951±.049 (10)	7.71±1.65 (18)	2.53±.82 (18)	12,924 (3)

* Area Calculated as solid cylinder = $(\frac{\pi}{4})\, d^2$ where d was the average outer diameter.

\+ Concentration of sodium fluoride in drinking water.

Table 1. Physical properties (mean ± 1 standard deviation) of rabbit bones tested in torsion. The number of animals or bone specimens is shown within parentheses.

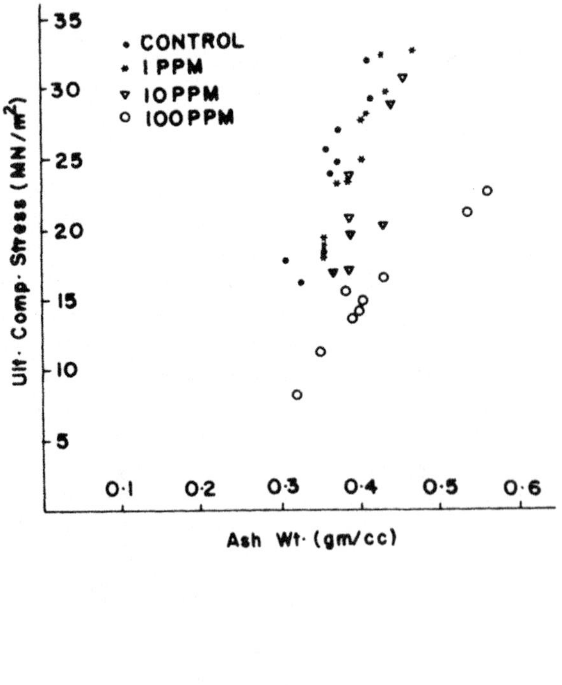

Figure 1. Ultimate compressive strength vs. ash wt. density of cubes of cancellous bone from thoracic vertebrae of sheep treated with zero, 1, 10, and 100 ppm fluoride in the drinking water.

Animal Group	Animal Wt. (kg)	Dorsal Vertebra: Dry Wt. Density (gm/cc)	Dorsal Vertebra: Ash Wt. Density (gm/cc)	Dorsal Vertebra: Comp. Strength (MN/m^2)	Lumbar Vertebra: Dry Wt. Density (gm/cc)	Lumbar Vertebra: Ash Wt. Density (gm/cc)	Lumbar Vertebra: Comp. Strength (MN/m^2)
Control	44.0	0.790±0.046 (9)	0.364±0.035 (9)	23.87±5.39 (9)	0.779±0.042 (7)	0.400±0.031 (7)	27.78±2.90 (7)
1 ppm	43.5	0.831±0.052 (11)	0.397±0.038 (11)	25.28±5.14 (11)	0.851±0.051 (7)	0.434±0.027 (7)	29.03±3.42 (5)
10 ppm	42.2	0.867±0.046 (8)	0.405±0.036 (8)	22.19±5.08 (8)	0.876±0.071 (12)	0.427±0.029 (12)	22.25±3.61 (12)
100ppm	37.6	0.959±0.178 (9)	0.416±0.079 (9)	15.40±4.41 (9)	0.783±0.163 (8)	0.365±0.077 (8)	16.45±4.76 (8)

Table 2. Physical properties (mean ± 1 standard deviation) of dry vertebral bone from sheep treated with fluoride in their drinking water. The number of specimens tested is shown within parentheses.

Animal Group	Animal Wt. (kg)	Ash Wt. Density gm/cc	Dry Wt. /Ash Wt.	Mod. of Elasticity (GN/m^2)	Ult.Tensile Stress (MN/m^2)	Fluoride: unashed	Fluoride: ashed
Control	44.0	1.350±.052 (4)	1.521±.024 (4)	25.6±3.4 (6)	94.18±6.89 (4)	399±117 (7)	890±288 (7)
1 ppm	43.5	1.340±.039 (6)	1.502±.017 (6)	24.1±2.8 (9)	83.77±27.58 (7)	551±136 (6)	1794±489 (6)
10 ppm	42.2	1.387±.035 (4)	1.493±.020 (4)	21.0±1.4 (6)	82.87±13.10 (4)	1547±793 (6)	2240±1009 (6)
100ppm	37.6	1.394±.067 (7)	1.465±.022 (7)	19.1±4.8 (8)	64.53±21.37 (6)	6207±2671 (6)	11,488±5234 (6)

Table 3. Physical properties (mean ± 1 standard deviation) of dry tibial compact bone from sheep treated with fluoride in their drinking water. The number of specimens tested is shown within parentheses.

THE EFFECT OF SOFT TISSUE ON THE NON-INVASIVE TECHNIQUES OF DETERMINING THE MECHANICAL PROPERTIES OF BONE*

by

Roderic S. Lakes and Subrata Saha
Department of Engineering and Applied Science
Yale University, New Haven, Connecticut 06520

I. Introduction. Recently several investigators have attempted to develop non-destructive techniques to evaluate the mechanical integrity of long bones in the human body. The development of a successful method of this type would be potentially very useful in the diagnosis of bone diseases like osteoporosis, in the assessment of treatment of these disorders and in the evaluation of fracture healing. Methods presently under investigation as possible diagnostic tools include impulse response measurements, determination of bone resonant frequency, and ultrasonics. These techniques are intended to provide a measure of the dynamic mechanical behavior of bones; however in the *in-vivo* condition there are other variables which can obtrude in the measurements. Among these are the properties and quantity of soft tissue overlying the bone, the forces and geometry associated with the placement of transducers, and the geometry of the bone itself. In the present study, the effect of several of these variables on the impulse response method are examined.

II. Background. Resonance methods have been used by several authors to examine diseased bones (1,2); correlations have been obtained between ulnar resonant frequencies and the degree of osteoporosis. This approach has been criticized on the grounds that rigid body modes dominated by soft tissue compliance, may have been excited (3). More recently this possiblity has been discounted on the basis of a mathematical modeling approach (4). Nevertheless, the use of resonant frequencies alone appears to be of limited value in the diagnosis of osteoporosis as a result of decreases in both bone mass and stiffness (3,5). Impedance measurements on a vibrating bone have been examined (6,7,8), however contrary to earlier expectations, soft tissue effects and the magnitude of transducer pressure affect the results (7,9). In impulse response methods, the bone is excited with an instrumented hammer and the response is measured using an accelerometer. Correlations have been obtained between data obtained in this way and the condition of a healing fracture (10,11). When transducers are placed on the skin, the results are complicated by the swelling of the soft tissue in the fractured limb (10). Direct coupling of transducers to the bone using needles (11) obviates such difficulties at the expense of using an invasive method.

III. Methods. In order to assess the effect of different variables on results obtained using the hammer-accelerometer method, the experimental configuration shown in Figure 1 was used. Tests were performed on excised specimens of human femoral bone as well as the tibia of human volunteers. The hammer impulse and the accelerometer were applied to the anterior aspect of the diaphyses of the bones examined. The hammer, instrumented with a force transducer, was impacted at a point 16 cm from the accelerometer, generating bending waves in the bone. Impacts were reproduced as closely

*This research was supported by a USPHS Research Grant No. 5R01 AM 18360.

as possible; resulting force and acceleration waveforms were recorded on a storage oscilloscope. In the <u>in-vivo</u> experiments, the impulse was delivered over the anterior portion of the tibia, where the soft tissue was judged to be thinnest. In the <u>in vitro</u> experiments, layers of rubber were used to simulate soft tissue over the bone; the force profiles of impacts on rubber and tissue were similar. Two series of <u>in vitro</u> tests were done: in one set the accelerometer was screwed directly to the bone and different thicknesses of rubber were interposed between the hammer and bone; in the second set, the hammer was impacted on a single layer of rubber over the bone and different thicknesses of rubber were interposed under the accelerometer. In the second set, the preload force was applied two ways: using a weighted arm directly (mass loading), and spring loading in which the inertia of the arm was decoupled. Further experimental details are discussed elsewhere (9).

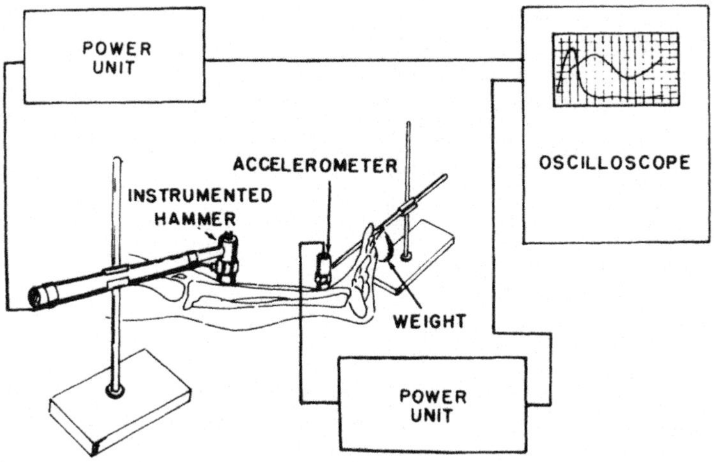

Fig. 1. Experimental configuration

IV. <u>Results</u>. Force and acceleration waveforms were obtained in an <u>in-vivo</u> experiment on two human volunteers. Peak acceleration, time delay, and peak width all depend on the static preload force and do not approach asymptotic values at any load level below the pain threshold. In addition, the curves change in shape with increasing pre-load in a different fashion even though the two volunteers were of similar age, weight and body build (9).

Six parameters P obtained from the waveforms recorded <u>in vitro</u>, were measured as functions of the preload force, and thickness and stiffness of simulated "soft tissue" as variables (x). Four parameters are shown in Fig. 2; the others are a_2/a_1 , $d_2 - d_1$. In order to compare the sensitivity of these six parameters P to variations in the experimental variables x , coefficients of sensitivity S were defined as follows:

$$S_R = [\Delta P/\overline{P}]/[\Delta x/\overline{x}] \qquad S_A = \Delta P/P_c$$

The bar $\overline{x}$ denotes the mean of the two quantities. Delta denotes difference. The relative coefficient of sensitivity, S_R , provides a measure of the relative change in a measured parameter normalized to the relative change in the experimental variable. The absolute coefficient of sensitivity, S_A , provides a comparison between parameters measured with "soft tissue" present and those measured with the accelerometer fixed directly to the bone $[P_c]$. Representative results are given below.

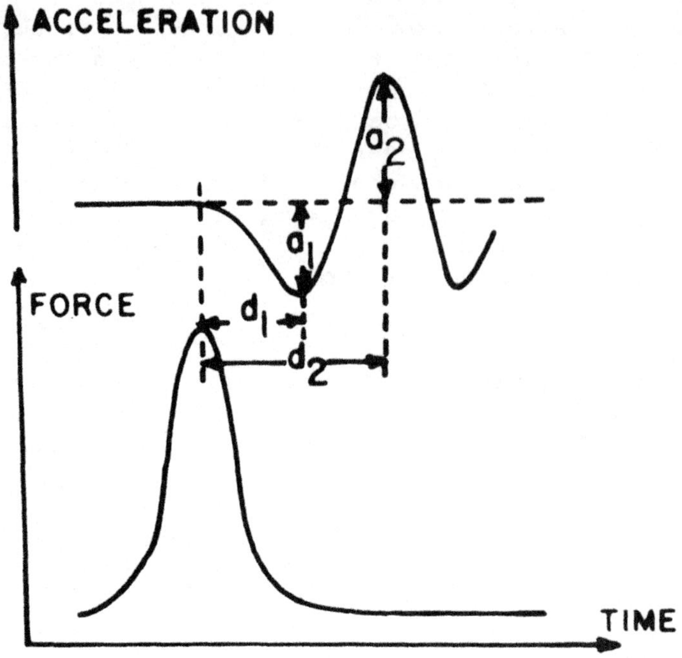

Fig. 2. Measurement of parameters from a waveform

Absolute coefficients of sensitivity, comparing parameters measured with one layer of simulated soft tissue between the accelerometer and bone, with those measured with the accelerometer directly fixed to the bone.

ACCELEROMETER PRE-LOAD [N]	TYPE OF SIMULATED TISSUE	Abs. Sensitivity (S_A) with respect to					
		DELAY			PEAK ACCELERATION		
		d_1	d_2	d_2-d_1	a_1	a_2	a_2/a_1
8.8, mass	H	1.19	1.3	1.4	-.64	-.45	.51
8.8, mass	S	2.5	1.9	1.7	-.80	-.93	-.63
8.8, spring	S	1.9	1.5	1.3	-.39	.058	.89
8.8, spring	H	.93	.36	.16	-.069	.45	.56
23.2, spring	H	.53	.22	.11	.093	.6	.47

V. Discussion and Conclusion. The impulse response technique for the monitoring of stress waves in bone, has been examined to ascertain the sensitivity of various measured parameters to variables related to the soft tissue over the bone and to transducer placement. The following observations may be extracted from the results of these experiments.

1. The application of accelerometer preload by use of a spring results in reduced relative sensitivity coefficients of the acceleration parameters and the delay difference $d_2 - d_1$, to changes in preload and tissue thickness.

2. The absolute sensitivity (variability) of all parameters except a_2/a_1 with respect to preload was less for spring loading than for mass loading.

3. For spring loading the parameters least sensitive to changes in "tissue" thickness at the accelerometer and to changes in preload force, were the delay differences $d_2 - d_1$ and the acceleration ratio a_2/a_1.

4. Nevertheless, errors introduced by "soft tissue" [i.e. the absolute sensitivity coefficient] were 16% in $d_2 - d_1$ and 56% in a_2/a_1, for one layer of "soft tissue" H, and 8.8 N spring preload. For "tissue" S or for more than one layer, significantly larger errors were seen.

5. Furthermore, these two parameters exhibited considerable sensitivity to changes in the stiffness and to a lesser extent the thickness, of the simulated soft tissue under the impactor.

6. For some combination of preload and simulated tissue thickness and stiffness at the accelerometer the second acceleration peak became an inflection point, as the higher frequency vibrations were filtered out. Higher modes of vibration are excited weakly or not at all if the "soft tissue" at the impactor is sufficiently thick.

7. In the *in-vivo* experiments, peak accelerations continue to increase as the accelerometer preload is increased, up to the pain threshold. The shapes of the acceleration curves also continue to change as preload increases. The variation in the acceleration curves as preload is changed, differs for the two human subjects examined.

The demonstration of correlations between measured quantities and disease state is necessary but not a sufficient condition for a technique to be used as a diagnostic tool. The effect of such confounding variables as soft tissue thickness and quality, on the measurement in question must be evaluated and compared with the effect resulting from the disorder which one is trying to diagnose. The dependence of measured acceleration parameters upon the state of osteoporosis and of simulated fracture healing, are being examined by the present authors. Comparison with the present results will permit an evaluation of these measurements as a potential diagnostic tool.

References

(1) Selle, W. A. and Jurist, J. M. (1966), Proc. Soc. Exp. Med. 121, 150.

(2) Jurist, J. M. (1970), Phys. Med. Biol. 15, 417-426.

(3) Doherty, W. P., Bovill, E. G., Wilson, E. L. (1974), J. Biomechanics 7, 559-561.

(4) Spiegl, P. V. and Jurist, J. M. (1975), J. Biomechanics 8, 213-217.

(5) Jurist, J. M. and Kianian, K. (1973), J. Biomechanics 6, 331-342.

(6) Thompson, G. A. (1973). Aerospace Med. Assoc. Annual Science Meeting, Abstracts, 7-10 May, Las Vegas, Nev.

(7) Orne, D. (1974), J. Biomechanics 7, 249-257.

(8) Thompson, G. A., Orne, D., and Young, D. R. (1976), Med. Biol. Eng. 14, 253-262.

(9) Saha, S. and Lakes, R. S. (1977), J. Biomechanics (in press).

(10) Lewis, J. L. and Tarr, R. R. (1975), J. Bone Jnt. Surg., 57A, 576.

(11) Sonstegard, D. and Matthew, S. (1976), J. Biomechanics 9, 689-694.

LOAD BEARING CAPACITY OF TRACTION PINS: A BIOMECHANICAL STUDY

S. Saha
Department of Engineering and Applied Science
Yale University, New Haven, Connecticut 06520

and

J. A. Albright
Section of Orthopaedic Surgery
Yale University School of Medicine
New Haven, Connecticut 06510

INTRODUCTION

Fracture of a long bone is a common problem in orthopaedic surgery. At times such a fracture must be treated non-surgically, at least temporarily, whether due to associated injuries, or due to the type of fracture itself (DePalma, 1970). In such a situation, traction on the distal fragment is commonly employed in order to overcome muscle pull and maintain length, as well as to maintain proper alignment (Fig. 1). Skin traction can be used for short periods of time, but the skin cannot tolerate more than 10-15 pounds of pull (Stewart, 1975). Therefore, when a substantial amount of weight is needed, or when prolonged traction is indicated, pin traction is generally used (Fig. 1). Traction on a pin which tranverses the bone does not cause pain, regardless of the amount of weight applied, assuming that soft tissue irritation is prevented at the points of exit on either side.

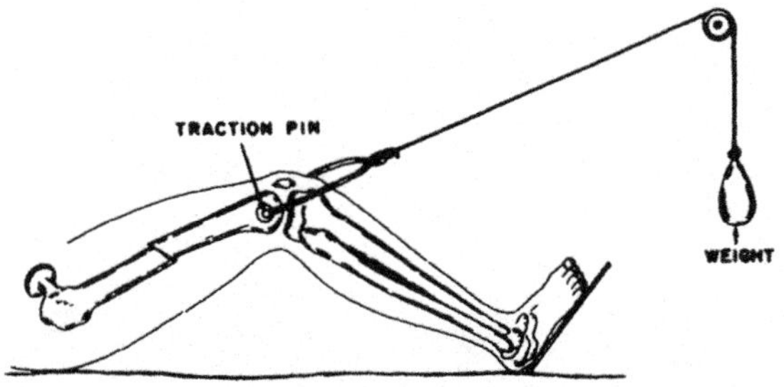

Figure 1. Skeletal traction being applied through a pin reducing a femoral fracture.

Two types of pins are commonly used in orthopaedic surgery: 1) Kirschner wires which have a small diameter (0.25 - 0.62 inches) and 2) Steinman pins which are larger (5/64" to 3/16"). Although use of a small pin is preferred as it causes less trauma, K-wires tend to bend if the traction load is large. Recently Morgan et al (1976) have suggested that the bending resistance of small pins might be improved by using pins made of stiffer materials such as tungsten-fibre reinforced composite. The bending resistance of metal pins can also be improved by applying an axial tension with the help of a Kirschner wire strainer (Stewart, 1975). However no quantitative data is available on the effect of an axial tension in reducing deflection of metal pins due to bending.

The factors limiting the amount of weight which can be safely applied to a pin include: the potential damage to joints from excessive stretching, the strength of the pin, and the holding power of the bone. In addition, pins are commonly used for the internal fixation of fractures and osteotomies, and in arthrodesis of joints; whether used for this purpose or for traction, the relationship between pin diameter and holding power in bone is often questioned.

The objective of this investigation was to study the effect of applied tension on the resistance of pins to bending due to traction load. Experiments were also carried out to determine if thin K-wires under normal traction loads produce sufficient contact stress to cut-through bone.

THEORETICAL ANALYSIS

The bending resistance of a traction pin subjected to a traction force P and an axial tension T can be analyzed by treating the pin as a cantilever beam of length L as shown in Figure 2. Using this cantilever beam model and assuming the deflection to be small and the stress-strain behavior to be linear and elastic, one can show that the differential equation governing the deflection, y, of the beam is given by (Timoshenko and Gere, 1971):

$$EI \frac{d^2y}{dx^2} = \frac{P}{2}(L-x) - T(\delta + y) \qquad (1)$$

where E is the modulus of elasticity, I is the moment of inertia of the pin and δ is the maximum deflection at $x = L$. Solving equation (1) and applying the boundary conditions at the fixed end, one gets

$$\delta = \frac{P}{2T}\left[\frac{1}{\beta}(e^{2\beta L} - 1)(e^{2\beta L} + 1)^{-1} - L\right] \qquad (2)$$

where

$$\beta = \sqrt{\frac{T}{EI}}$$

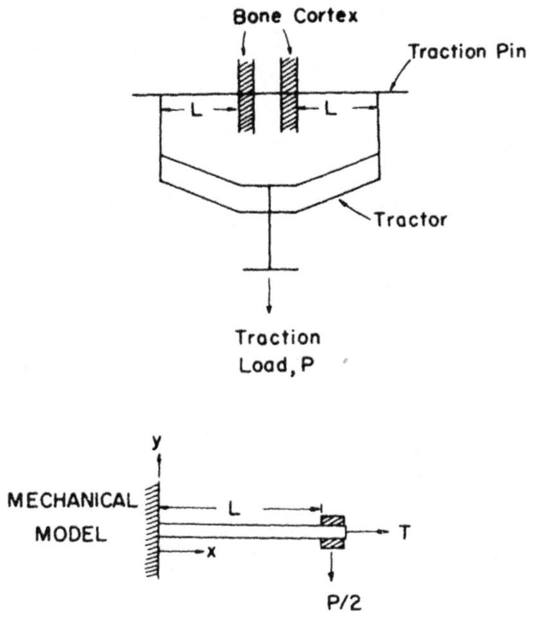

Figure 2. Cantilever beam model of the traction pin showing the loads and the coordinate system.

The numerical results from this analysis are shown in Figures 3 and 4. Figure 3 shows the maximum deflection (δ) of the pin per unit traction force (P) as a function of applied tension (T) in the pin. It is clear that for large diameter pins (diameter 0.125 inch and above) deflection remains almost unaffected by the tension T in the pin. However for small diameter pins (diameter 0.045 inch) the deflection is significantly reduced as the tension in the wire is increased.

The maximum deflection δ per unit traction force P has also been plotted as a function of the pin diameter D for two traction force T = 0 and T = 50 lbs (Figure 4). Figure 4 shows that for small diameter pins (say $D < 0.125$ in.), the maximum deflection without a tensile force T is significantly larger than when an axial tensile force T = 50 lbs. is applied. However this tensile force does not change the bending resistance of pins of diameter $D \geq 0.2$ inch.

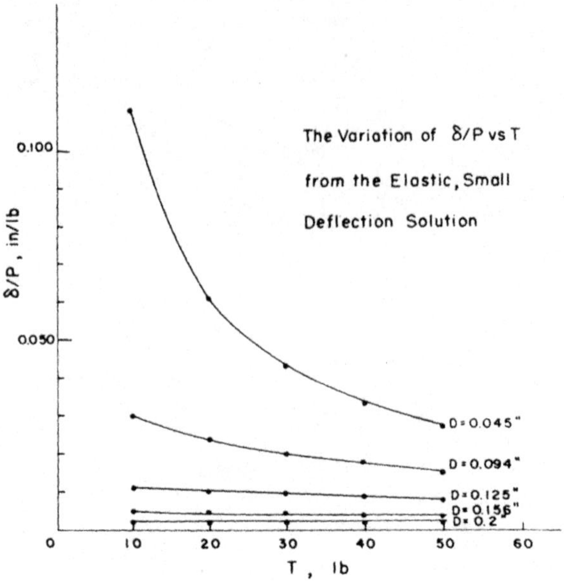

Figure 3. Maximum deflection (δ) of different diameter traction pins as a function of the tension in the pin (T).

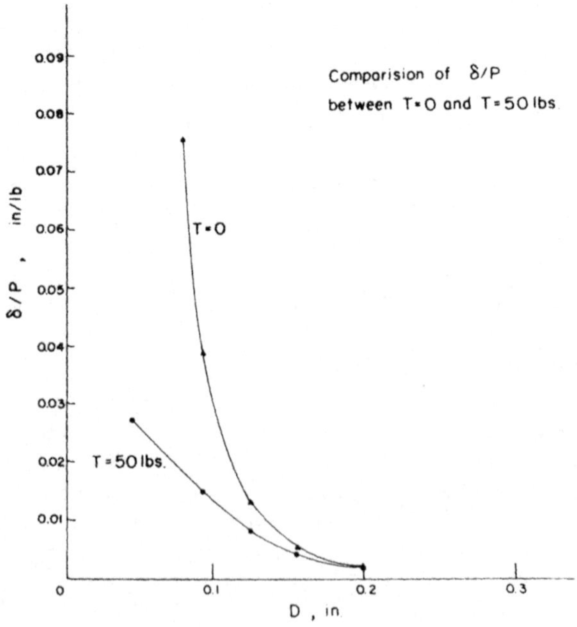

Figure 4. Relationship between the maximum deflection (δ) and the pin diameter D for two tensile load levels, T = 0 and T = 50 lbs.

EXPERIMENTAL WORK

Experiments were also carried out on the bending behavior of a small diameter pin (D = 0.062 inch) inserted into the mid-shaft of an embalmed human femur and subjected to a maximum axial tensile force which could be applied manually by a Kirschner wire strainer. Figure 5 shows the maximum deflection of this wire per unit traction force as a function of increasing traction force P. It clearly shows that due to the axial tension, the deflection per unit traction force decreased as the traction load is increased, and for a moderate traction load, the deflection is small.

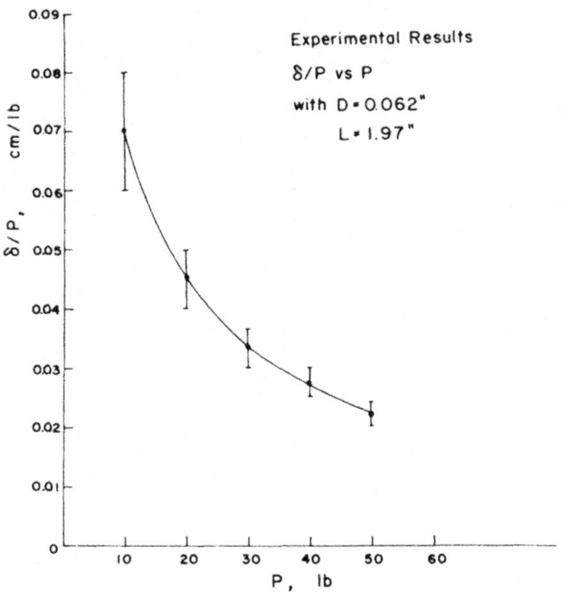

Figure 5. Variation of maximum deflection (δ) per unit traction load P as a function of the applied traction, P. Axial tension was applied by means of a Kirschner wire strainer.

To examine the possibility of small diameter traction pins under normal traction load cutting through the bone, laboratory tests were conducted using embalmed human bones. Kirschner wires (0.062 inch diameter) were inserted through the diaphysial cortex and the wire was subjected to tension by using a traction bow. Then traction loads were applied and maintained for about 12 hours. The traction load was increased in steps of 10 lbs up to a maximum of 50 lbs. There was no sign of the wire cutting through the bone.

When similar test was conducted on K-wires inserted through the distal femoral epiphyses, the wire still sustained the traction load without cutting through the bone so long as the thin shell of cortical epiphysial bone was intact. However, for those specimens in which the cortical bone covering the trabecular bone structure was damaged, the wire did show evidence of cutting through the spongy bone.

CONCLUSION

The results of this study indicate that thin K-wires could be used successfully as traction pins. Excessive bending of these wires could be prevented by applying tension. The possibility of a pin cutting through bone is minimal if the cortical bone structure of the patient is normal. However, further in-vivo study is necessary to ascertain if necrosis might occur due to excessive contract stress between the pin and the bone.

ACKNOWLEDGMENT

The authors are grateful to Mr. A. Mak for his technical help.

REFERENCES

DePalma, A. F. (1970) The Management of Fractures and Dislocations, W. B. Saunders Co., Vol. 2.

Morgan, H. G., Ryder, D. A. and Wheble, V. H. (1976) Tungsten-fibre reinforced composite pin with potential applications in orthopaedic surgery. Medical and Biological Engineering, Vol. 14, pp. 528-531.

Stewart, J. D. M. (1975) Traction and Orthopaedic Appliances. Churchill Livingstone, N. Y., p. 2.

Timoshenko, S. P. and Gere, J. M. (1972) Mechanics of Materials, D. Van Nostrand Co., New York.

EXPERIMENTAL INVESTIGATION OF FEMORAL FRACTURE

David C. Viano
Biomedical Science Department
General Motors Research Laboratories
GM Technical Center
Warren, MI 48090

Abstract: The objective of this investigation was to analyze mechanisms of femoral condyle and shaft fracture. Three mechanisms have been postulated for femoral condyle fractures: 1) splitting, 2) shearing, and 3) bending effects acting on individual condyles, while shaft failures are hypothesized to occur under induced bending from longitudinal loadings that do not act along the central axis of the femur.

Macerated human femurs were supported in a test fixture that simulated the effective pelvic mass. Six longitudinal impact environments were investigated representing rigid, lightly padded and thickly padded interfaces acting directly on the patella or condyles. A 23 kg impactor impinged at a 7 m/s velocity against the supported femur with or without the patella aligned "in situ". Failure mechanisms were identified by the use of high speed photographic coverage of the lateral and anterior aspects of the femur. The photographic coverage also permitted an evaluation of the influence of the effective induced shaft and condylar bending. Also, dynamic contact loads and structural accelerations were recorded in an attempt to correlate changes in load with fracture occurrence, and to compare responses with the performance of the similarly tested Part 572 dummy femur.

The loading environment was significant in affecting the distribution of the load over the distal aspect of the femur and, therefore, the dynamic response and failure. Changes in the loading environment also affected the location and mechanism of the failure. Rigid impacts produced a dramatic shattering failure of the condyles while padded impacts produced a more distributed pattern of segmental condylar and shaft buckling failures.

Methodology: Ten macerated human femurs, half with a patella positioned "in situ", were supported at the proximal end (femoral head, Fig. 1) and subjected to a blunt axial impact at the distal end (condyles or patella) in an initial phase series for an unembalmed cadaver impact program. The applied load and the resultant femur-pelvic mass acceleration were monitored and the sequence of events resulting in a femoral or patella fracture were photographed for three impact interface conditions: 1) rigid, 2) padded (one inch ensolite), and 3) padded + (one inch ensolite + three inch hexcell). The femur and patella specimens were anthropometrically defined (Table 1) prior to establishing reference planes and laying out photographic grid patterns on the femoral surfaces.

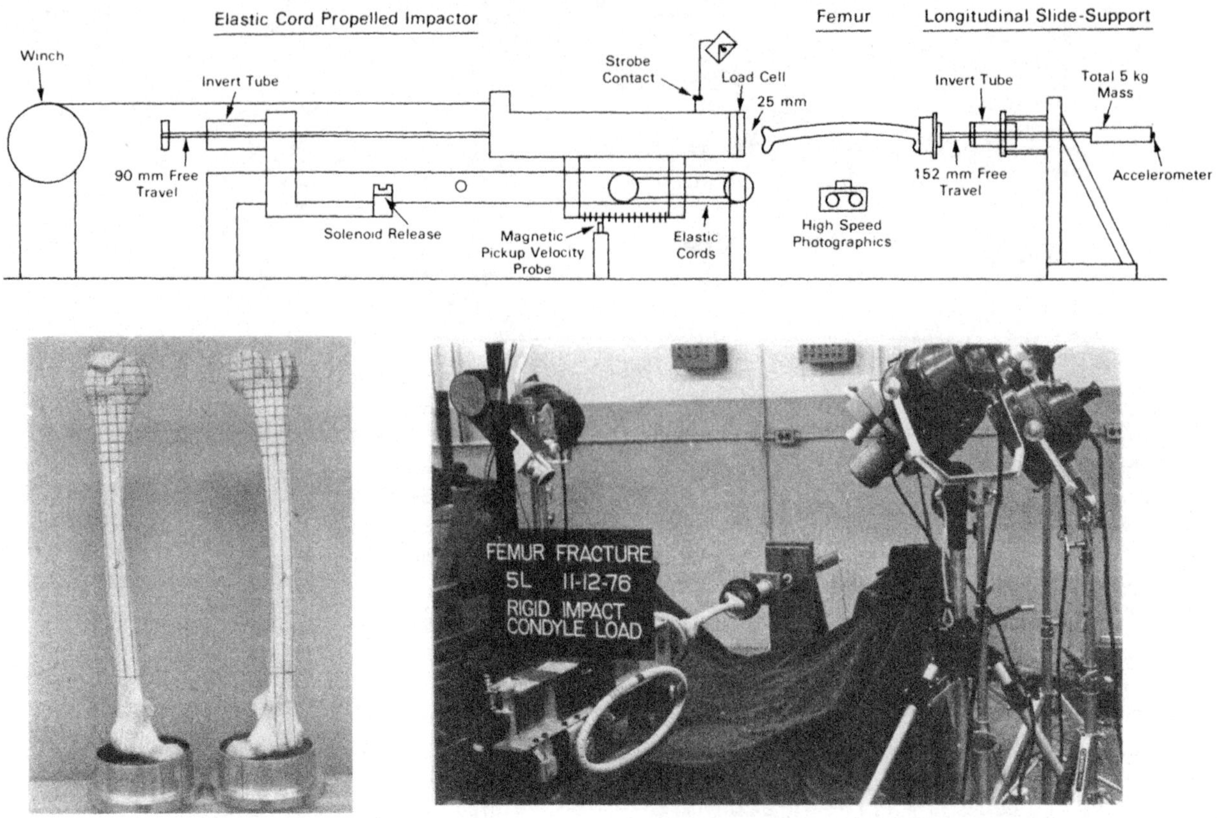

Fig. 1. Test setup for the axial femoral impacts.

The femur head was subsequently potted in a support cup (Fig. 1) so that the axis of the femur shaft (defined through the centroids of the shaft cross section at a position one third the distance from the femur ends) could be aligned with the axis of the impactor travel. The impactor employs stretched elastic cords for energy to accelerate the impacting mass (23 kg) to constant velocity (7 m/s) prior to femoral contact. Records of applied force and resultant femur-slide structure acceleration (typical records Fig. 2) were obtained for the various contact interfaces (Table 1). Two orthogonal views of the femur impact and subsequent fracture process were photographed using 1000 fps movie coverage. The femoral structure from the Part 572 test dummy was similarly tested for the padded impact so that a comparison could be established with the macerated femoral impact response.

Results and Discussion: The ten macerated (i.e. dry, defatted and blanched bone) human femurs averaged 252 gm (±44 gm) in dry weight and 42.9 cm (±2.3 cm) in overall length. The metallic Part 572 dummy femur weighed 4.7 kg (i.e., 18.7 times the dry human femur weight). The five macerated human patellas averaged 7.1 gm (±1.0 gm) in dry weight.

Table 1

Geometric Characteristics, Test Conditions, Impact Responses and Fracture Data for the Femur-Patella Longitudinal Impacts

Femur - Patella Specimen

	1L-P	1R	2L-P	2R	Part 572 Dummy Femur	3L	3R-P	4L	4R-P	5L	5R-P
Anthropometry											
Femur											
Weight (gm)	268	296	310	219	4700*	202	304	225	203	210	278
Length (cm)	46.3	46.4	43.7	42.3	49.5	43.1	43.4	39.7	40.2	40.5	43.3
Cordyle Breadth (cm)	7.6	7.7	8.1	7.4	5.1	7.1	7.8	6.5	6.8	6.5	7.8
Midshaft Breadth (cm)	2.5	2.6	2.4	2.5	4.5	2.3	2.6	2.4	2.3	2.2	2.5
Midshaft Depth (cm)	2.6	2.7	2.5	2.7	4.5	2.4	2.4	2.3	2.5	2.4	2.8
Patella											
Weight (gm)	8.0		7.6				7.5		5.5		7.0
Length (cm)	3.8		3.9				4.0		3.7		3.7
Breadth (cm)	4.1		4.1				4.1		4.0		4.2
Depth (cm)	1.8		1.9				2.0		1.9		1.9
Test Condition											
Interface	Padded		Padded+		Padded	Padded		Rigid		Rigid	
Mechanical Responses											
Load											
Peak (kN)	2.63	3.46	2.80	7.23	43.9		8.10	1.93	1.31	1.40	2.01
Occurrence (ms)	6.8	6.7	7.1	8.3	6.9		8.3	3.2	2.0	2.9	1.5
Duration (ms)	3.7	4.2	4.5	5.4	3.8		6.7	2.9	2.9	3.1	3.0
Acceleration											
Peak (g)	47.3	67.8	49.3	137.9	647		163.1	44.1	45.3		41.4
Occurrence (ms)	6.6	6.6	7.1	8:3	6.9		8.3	2.9	2.4	3.4	1.5
Duration (ms)	3.7	3.7	3.4	4.4	3.8		5.9	2.9	2.0		2.7
Fractures											
Femur											
Condyles	BS+8	BS+8	CBS+8	BS+8		BS+		CBS+4	CBS+3	CBS+2	CBS+2
Shaft				B+8[a]/B+9[b]			B+5[c]				
Head			B+12	B+10		B+	B+6	B+5	B+4		B+5
Patella			C+				C+		C+		C+

B - Bending
S - Shear
C - Compression

L - Left
R - Right
P - Patella

[a]Fracture 14.4 cm from condyle
[b]Fracture 28.9 cm from condyle
[c]Fracture 20.9 cm from Patella

+ ( ) - Time of fracture occurrence after contact, ms

* - Supported by a 2 kg pelvic mass

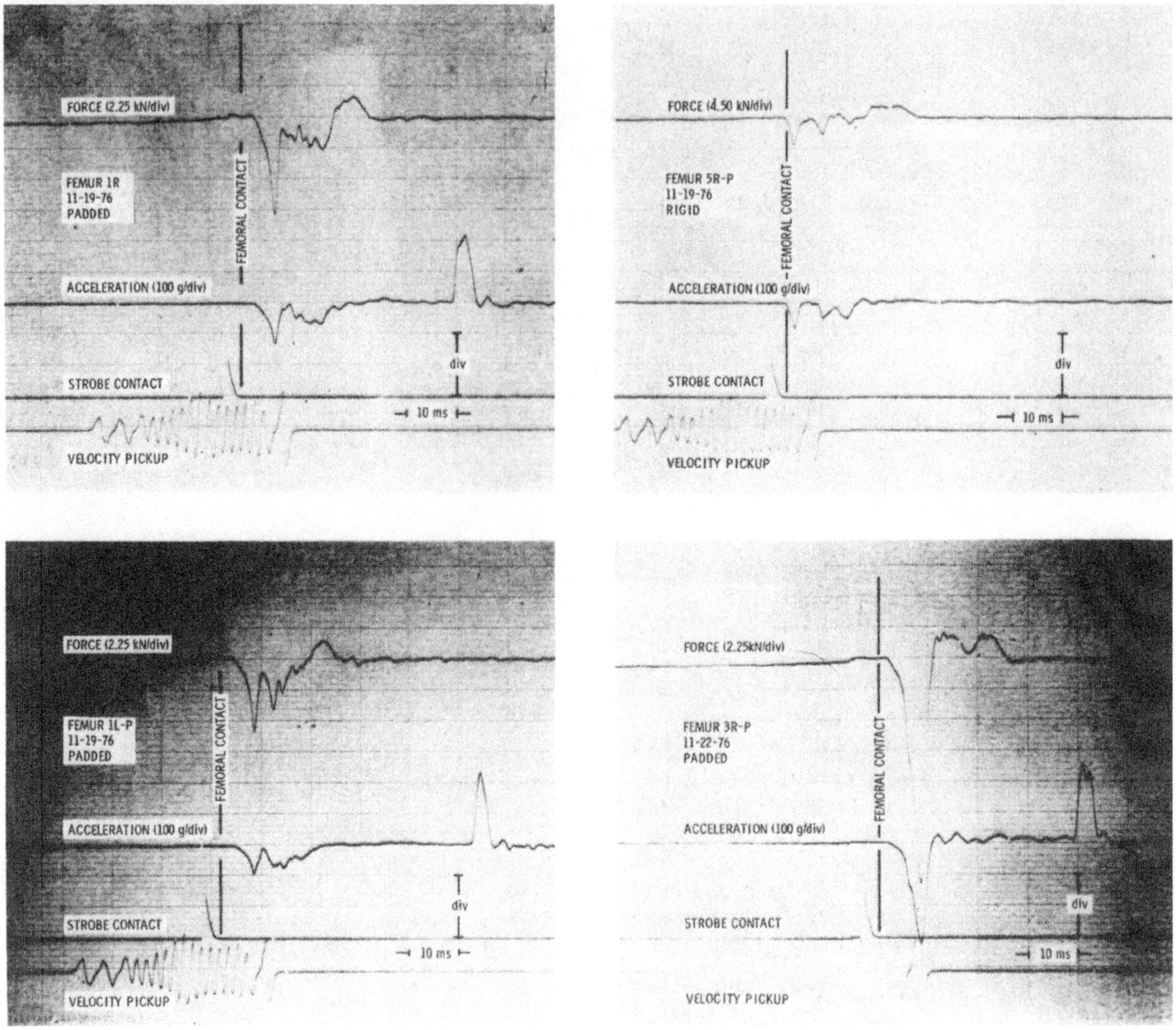

Fig. 2. Typical mechanical response characteristics.

The average peak fracture load for a rigid impact (Table 1), which is expected to be significantly lower than for fresh bones, was 1.7 kN (±0.4 kN) and occurred 2.4 ms (±0.8 ms) after femoral contact. Peak acceleration of the femur-pelvic mass averaged 44 g (±2 g) and occurred in synchronous with the peak load. The duration of the primary force for the rigid impact exposures averaged 3.0 ms (±0.1 ms). The average peak fracture load for a padded impact (i.e., padded and padded +, since both test conditions produced the same femoral response) was 4.8 kN (±2.6 kN) and occurred 7.4 ms (±0.8 ms) after femoral contact. The average fracture load was significantly higher for a shaft failure (2.5 times) than for only condylar failures. Peak acceleration of the femur-pelvic mass averaged 93 g (±54 g) and also occurred in synchronous with the peak load. The duration of the padded impact exposures averaged 4.9 ms (±1.2 ms). A padded impact test using the Part 572 dummy femur was conducted for comparison with the macerated human femoral responses. The peak load, which was 43.9 kN (i.e., 9 times greater), and resultant acceleration, which was 647 g (i.e., 7 times greater), were substantially greater in the matched dummy test for a padded impactor interface.

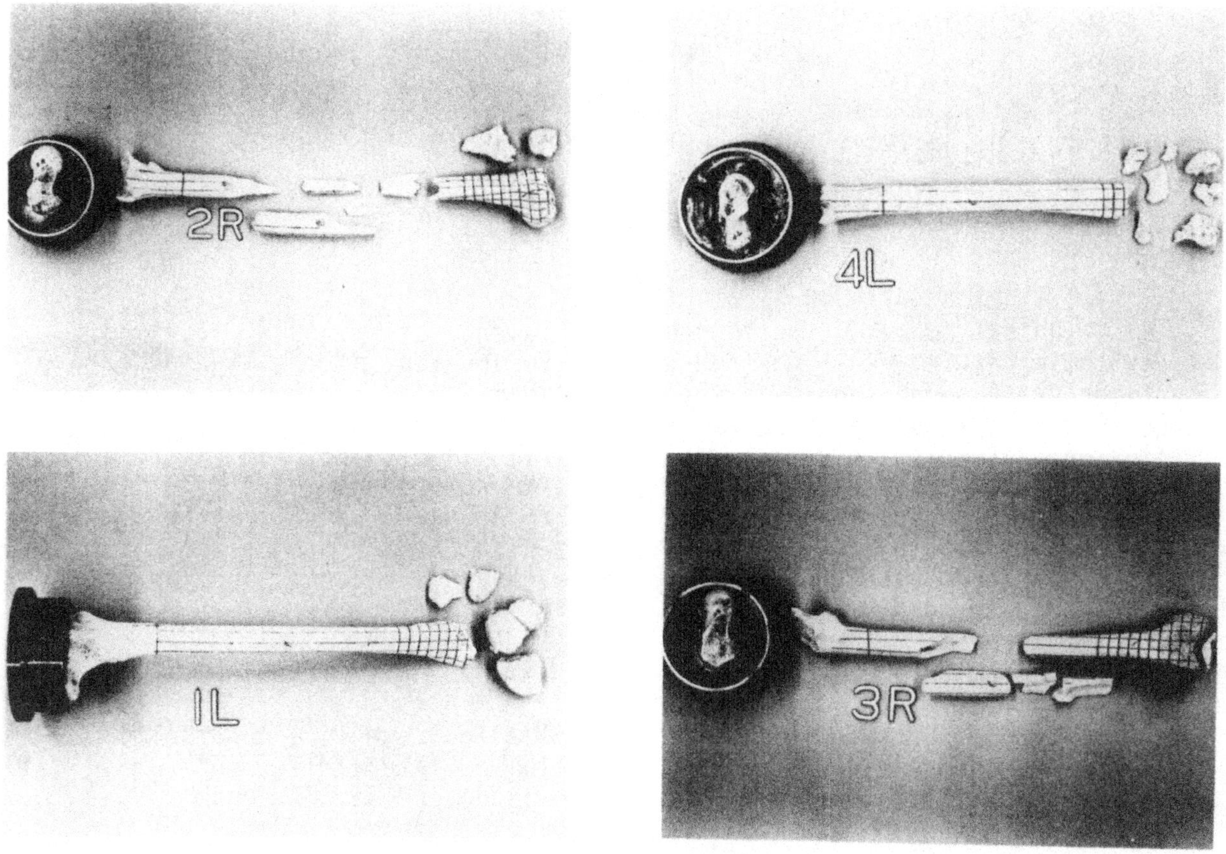

Fig. 3. Typical femoral fracture characteristics.

Mechanisms of femoral fracture can be investigated by analyzing high speed movies of the sequence of events leading to a specific fracture (i.e., condylar or shaft failure) in conjunction with the response characteristics of the applied load. A longitudinal impact of a macerated human femur produced dramatic shattering failures of the lateral and medial condyles (Fig. 3) for a flat rigid impactor interface. The failures occurred through the combined effects of compression and shear forces on individual condyles. Similar impacts with a flat padded impactor interface produced segmental failures of the condyles and occasional buckling failures of the femoral shaft. Condylar failures resulted from compression, lateral bending and shear forces on individual condyles while shaft failures were produced by excessive column bending. The more massive dummy femoral structure significantly overestimates the load and response characteristics of a longitudinal femoral impact.

The effect of the patella positioned "in situ" was more pronounced for the rigid impacts than the padded ones. The patella provided a more uniform interface and resultant load distribution over the condyles and intracondylar notch for both loading interfaces. Compression of the patella wedged the condyles laterally (lateral bending) and sheared them proximally. Without the patella interface the condyles were directly compressed and splintered by shearing actions.

KNEE IMPEDANCE TESTING

Bruce F. White, Research Engineer, M.H. Pope, Ph.D., D. Brown, M.D., R.J. Johnson, M.D., University of Vermont, Burlington, Vermont 05401.

Introduction:

Evaluation of knee ligament damage is presently performed clinically in a subjective manner. The four major ligaments are manually stressed by the clinician who then makes a subjective judgment of the ligament function (4). He considers laxity and the quality of the end point of the test for his evaluation. The medial collateral ligament is tested by inputting a varus displacement at the distal tibia while holding the femoral condyle securely. The clinician holds the knee in approximately 18° of flexion. He then displaces the lower leg in a valgus sense through the joints primary laxity, i.e. that movement which does not appreciably load the ligament (5). Then he feels the ligament as it suddenly becomes stressed. This constitutes the end point.

Impedance Method:

Our testing method makes use of a preloading technique to displace the knee to some initial stress on the medial collateral ligament. Then a sinusoidal displacement input enables us to determine the dynamic characteristics of the knee system (1,2,3). Unfortunately, the inherent difficulty of femoral fixation makes any test of the knee stiffness potentially a test of the hip in rotation. Figure 1 illustrates that the lower extremity will see a torque if the distal tibia is subjected to a valgus load. This torque causes rotation at the hip and consequently a relatively large displacement at the distal tibia, thus rendering any stiffness measurements inaccurate.

Using a test procedure which allowed us to measure the knee stiffness as a function of the angle of load application (Fig. 2), we determined that if the load were applied at angle of 20° to 40° posteriorly of the coronal plane, very little femoral rotation is observed. This is consistent with calculations made using a geometric model of the lower extremity and hip, knowing the knee flexion angle and the anterior angulation of the femoral head. It is additionally interesting to note that the clinical evaluation of the medial collateral ligament uses that portion of the load displacement curve which preceeds any marked internal rotation of the femur (Fig. 3).

The Impedance Device:

The main frame of the device is an angle iron frame attached to a massive cast iron base. The frame supports a platform for the subject which is isolated by an air bearing to insure that the body position creates no tension or compression on the knee joint. Attached to the main frame is the lower frame which can be moved about the approximate axis of flexion and extension of the subject's knee. Although this single axis of rotation does not follow the changing instant center of flexion-extension of the knee, a simple length adjustment alleviates the problem. Thus the entire lower frame provides a stiff system from which varus/valgus or rotational motion is imparted to the subject's lower leg, in any degree of flexion. The frame has various adjustments to accomodate individuals of different size. Fixation for a variety of tests is provided by the several clamping systems on the main frame.

The drive mechanisms consist of both linear and oscillating drive systems. The sinusoidal motion is provided by a simple eccentric and connecting rod assembly driving a band-type friction clutch. Cables transmit the motion from the clutches to the drive carriages under the feet. At this point, load and displacment transducers monitor the inputs to the distal tibia.

Test Method:

Our test system entails preloading the ligament to a constant tension and then imparting a 0.55mm sinusoidal displacement at the distal tibia. We effectively eliminated the problem of femoral fixation previously mentioned by rotating the femur 20° externally from the neutral position and subsquently preloading the ligament. If a decrease in preload of 20% was noted during the test, we assumed internal rotation had occurred. The test was then repeated, with 5° of additional external rotation imparted during the set up.

The sinusoidal input is swept from 0 to 12 hertz while force and velocity are monitored. Using an analog signal processing system (Fig. 4), the mobility (mobility is defined as the ratio of velocity to force and is the inverse of impedance) of the system was then calculated. We then plot the log of mobility vs the log of frequency. The resulting Bode plot (Fig. 5) may be easily analyzed for measure of dynamic stiffness, mass and damping.

Our initial reliability study was designed to evaluate our ability to repeatably obtain consistent results. We later expect to isolate variables such as fatigue, creep and anthropometric scaling factors which may be important in this type of ligament evaluation. We used two technicians who operated the machine independently. A total of five subjects was examined. Each subject was tested five separate times by each operator (see Table I).

We additionally tested several individuals with known pathology. Our results in these cases showed differences of much greater magnitude than those found with the normals of the reliability group (see Table II).

TABLE I: NORMAL CONTROLS

Results: 3

A. Examiner #1

Subject	Sample		Stiffness L	Stiffness R	∇K L-R	Damping L	Damping R	Wn L	Wn R
PL	3	$\bar{x}$	700.0	716.6	-16.6	10.3	10.2	6.6	9.3
		S	132.2	76.4	189.0	0.25	0.3	0.7	0.6
CB	5	$\bar{x}$	898.0	868.0	30.0	9.3	9.1	7.4	7.9
		S	113.2	111.2	174.2	1.1	1.9	0.4	0.6
KL	3	$\bar{x}$	783.0	733.0	-166.6	9.7	7.7	7.0	7.3
		S	202.0	236.0	125.0	0.5	1.6	0.8	1.3
MP	5	$\bar{x}$	838.0	840.0	-2.0	10.2	7.5	7.5	7.5
		S	81.9	108.4	88.14	0.3	1.8	0.9	0.8
CK	5	$\bar{x}$	790.0	900.0	110.0	10.4	9.9	7.6	7.6
		S	74.1	93.5	89.0	0.2	0.5	0.2	0.4
Group Mean			801.8	811.5	-8.8	10.0	8.68	2.2	7.9
Pooled S			116.1	126.0	133.9	0.61	1.45	0.6	

B. Examiner #2

Subject	Sample		Stiffness L	Stiffness R	∇K L-R	Damping L	Damping R	Wn L	Wn R
PL	2	$\bar{x}$	400.0	525.0	-166.0	3.6	8.6	6.4	6.3
		S	70.7	106.1	81.6	0.8	0.1	0.6	0.8
CB	5	$\bar{x}$	741.0	805.0	-62.0	8.2	10.0	6.7	7.0
		S	164.8	94.2	114.0	1.2	0.5	0.7	0.5
KL	4	$\bar{x}$	787.0	922.5	-135.0	6.7	9.9	7.6	8.3
		S	75.0	128.2	142.0	0.6	0.4	0.3	0.6
TT	5	$\bar{x}$	956.0	978.0	-18.0	7.7	8.1	7.3	7.6
		S	98.4	58.5	120.0	1.7	1.1	0.4	0.9
CK	5	$\bar{x}$	660.0	810.0	-150.0	9.2	10.4	7.5	7.9
		S	41.8	174.0	193.6	1.0	0.3	0.9	1.4
Group Mean			708.4	808.0	-106.2	8.1	9.4	7.1	7.4
Pooled S			104.7	119.9	142.7	1.2	0.64	0.64	0.93

C.

TABLE II: INJURED INDIVIDUALS

Subject	Clinical Laxity R	Clinical Laxity L	Stiffness L	Stiffness R	Stiffness ∇K
KL	Average	Loose	950	550	400
NB	Very Loose,Grade III MCL	Average to Tight	250	600	350

Experimental Error:

The pooled standard deviation calculated;

$$SP = \sqrt{\sum_{i=1}^{5} (N_i - 1) \; S_i^2 \; / \sum_{i=1}^{5} (N_i - 1)}$$

where N_i is the number of tests on an individual and S_i is the standard deviation for a given variable (i.e. left or right stiffness ∇K, damping or natural frequency Wn) gives an approximate measure of the overall experimental error. Thus for any future test with 95% confidence we may expect the actual value to equal $X \pm 2$ Sp (where X is the measured value). Where Sp has been calcualted for several variables, and has additionally been shown to be similar for different operators.

Discussion:

We are presently undertaking a study of a normal population of 40 individuals. They were selected on the basis of anthropometric variables to match a Grade II MCL injury group who we are now recalling for a clinical followup study. Table II demonstrates that the injured subjects tested so far exhibit a ∇K of 300 N/m as compared to controls with a mean of 50 N/m.

On the basis of our reliability results, we feel that this method and device provide not only accurate dynamic response data from the human knee, but may provide the basis for the development of less subjective clinical evaluation techniques. We have given particular attention to fixation methods and development of accurate testing procedures. Through this we have reduced the experimental error to a minimum. We feel this device provides accurate and repeatable results which will potentially enable the examiner to diagnose injuries to the medial collateral ligament in a less subjective manner than previously possible.

REFERENCES:

1. Pope, M., Crowninshield, R., Miller, R., & Johnson, R.: The static and dynamic behavior of the human knee *in vivo*. J. Biomech. 9:449, 1976.

2. Crowninshield,R., Pope, M., & Johnson, R.: An analytical model of the knee. J. Biomech. 9:397, 1976.

3. Crowninshield, R., Pope, M., Johnson,R., & Miller, R.: The impedance of the human knee. J. Biomech. 9:529, 1976.

4. Hughston, J., Andrew, J., Cross, M., & Moschi, A.: Classification of knee ligament instability. Part I: Medial compartmental cruciate ligament. J. Bone Joint Surg. (Am) 58:159, 1976.

5. Markolf, K., Mensch, J., Amstutz, H.: Stiffness & laxity of the knee, the contributions of the supporting structures. J. Bone Joint Surg. (Am) 58:583, 1976.

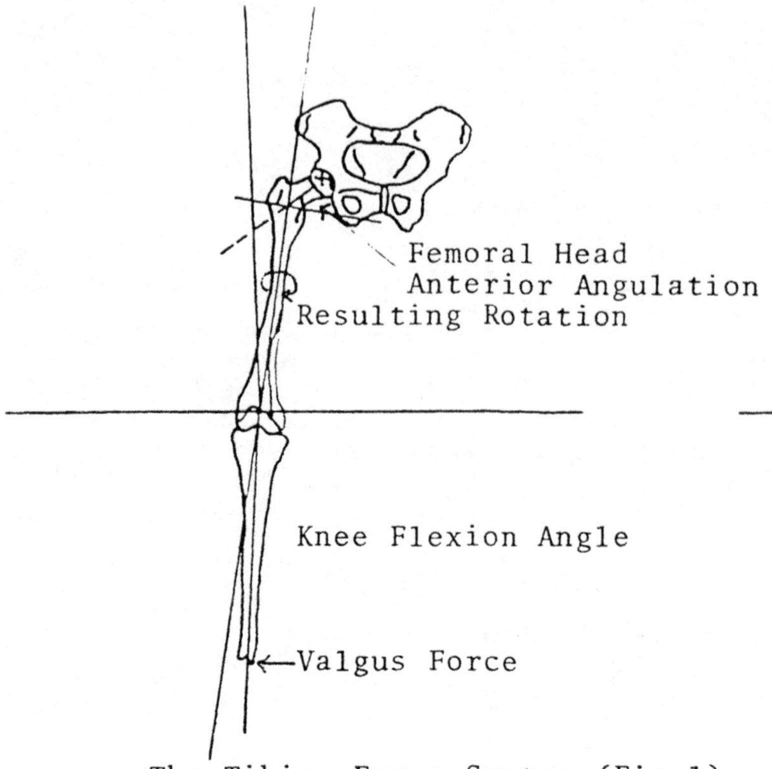

The Tibia, Femur System (Fig.1)
is Unstable in Rotation

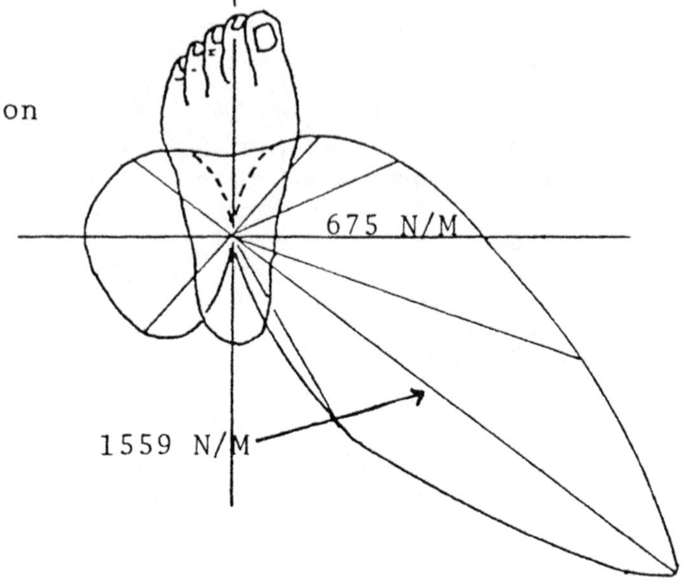

Polar Plot of Knee Stiffness
with Hip Free to Rotate(Fig.2)

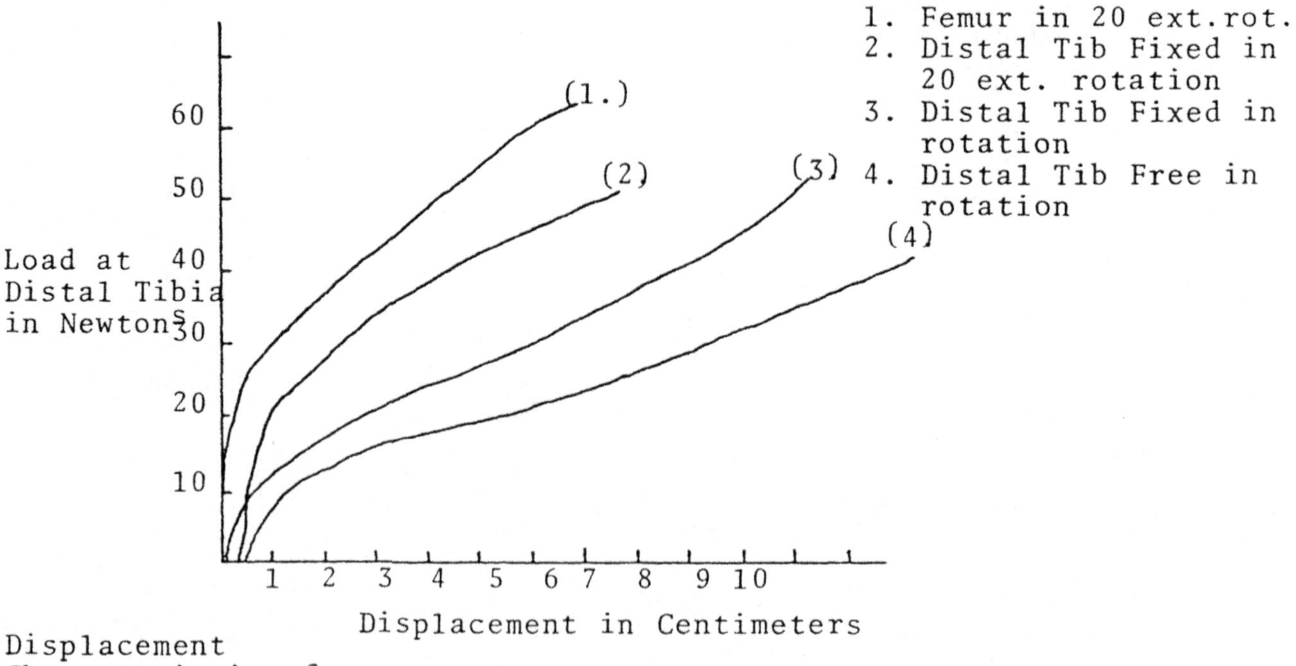

Displacement
Characteristic of
Clinical Exam

SEVERAL FIXATION METHODS FIGURE 3

ANALOG INSTRUMENTATION FIGURE 4

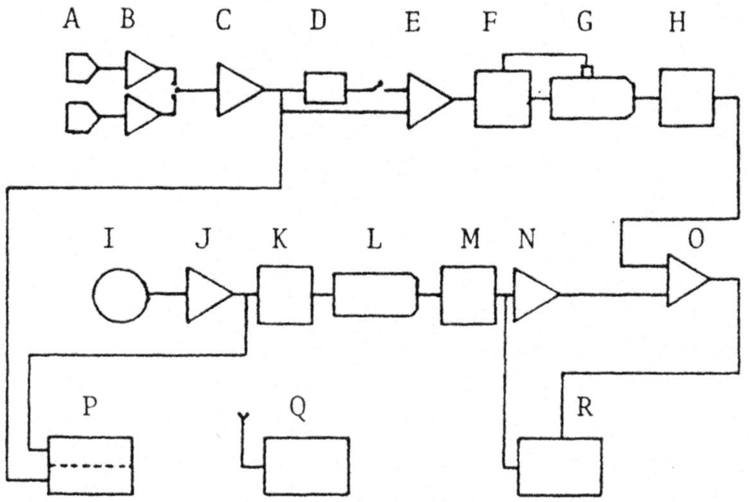

A. Load Cells	G. Track Store	M. Log Amplifier
B. Conditioning Amp.	H. Log Amplifier	N. Inverting
C. Amp & Filter	I. Velocity Trans	O. Summing
D. Averaging for Offset	J. Sig. Conditioning	P. 2 Channel Chart
E. Summing	K. Peak Detector	Q. Digital Display
F. Peak Detector	L. Track Store	R. X Y Recorder

FREQUENCY INHERTZ

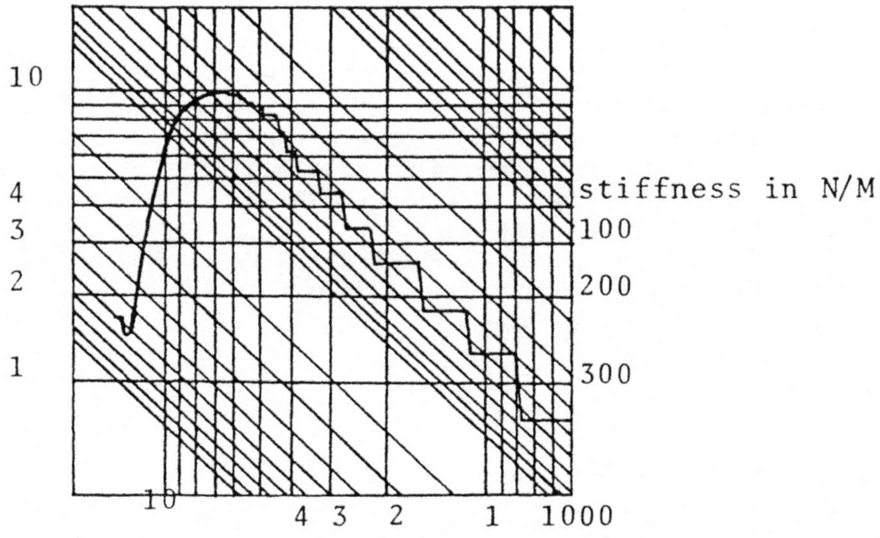

TYPICAL OUTPUT, BODECURVE FIGURE 5

DYNAMICAL CHARACTERISTICS OF A SHEET-FLOW MEMBRANE OXYGENATOR

E. Laska, P. D. Richardson, and P. M. Galletti

Division of Engineering, Brown University, Providence, RI 02912

SYNOPSIS

The performance characteristics of sheet flow membrane oxygenators depends on the transmembrane pressure because it affects the geometry of the gas and blood phases. Measurements of membrane distensibility were made for the Landé-Edwards oxygenator. The distensibility was found to be non-linear with a deflection point at high membrane distension. The mass transfer of the oxygenator was represented by means of mass transfer coefficients. The oxygen transfer coefficient was found to increase linearly with the liquid flow rate while the carbon dioxide transfer rate leveled off asymptotically at high blood flow rates. It was also found that water could condense in the gas-phase spaces, causing functional impairment similar to pulmonary edema in the natural lung.

INTRODUCTION

General characteristics of all sheet flow membrane oxygenators are: The gas and the blood phases are separated by a flat or nearly flat membrane; exchange of gases occurs across the membrane; the membranes are incorporated as flat sheets of gas permeable material with a specific exchange area (1). The Landé-Edwards oxygenator is a sheet flow type. The blood phase is made of a series of parallel envelopes of thin silicone rubber membrane. Grooved molded plastic supports are used to define the outer geometry of the gas phase. These plastic supports are inserted between the membrane envelopes. By a system of manifolding channels molded into the support plates, every envelope has 1440 capillary-like blood passages in parallel. Each passage is 30 mm long and .127 mm wide. Blood is carried to and from these capillary passages by manifolding with tapered passages, as illustrated in Fig. 1d. In this way, the oxygenator attempts to imitate architectural features of the natural lung (though with different dimensions) and, in particular, to provide a low resistance to blood flow compared with other membrane oxygenator designs. The natural lung shows distensibility of the blood vessels when the pulmonary blood pressure is increased, and this feature is imitated in the Landé-Edwards oxygenator, especially in the capillary-like passages, Fig. 1. When the pressure in the blood phase is higher than the gas phase pressure, the membranes extend into the grooves in the plastic supports and forms capillary beds. The actual micro-groove which forms the outer boundary of the gas flow channels is a 4.5 mil equilateral triangle. Most of the mass exchange occurs at these capillary beds because of their large surface-to-blood-volume ratio which facilitates exchange of gases with blood. Dry oxygen is supplied to the gas phase of each unit. Because of the architecture of the manifolding of the blood and the gas micro-channels, in one-half of the channels the blood and gas flows are countercurrent and in the other half co-current. The distension of the membranes increases the volume of the blood phase (the priming volume).

The Landé-Edwards oxygenators are built in two sizes, with 1 M^2 and 3 M^2 nominal membrane areas. The former contains 20 support plates (and 19 envelopes), and the latter has 58 plates. The rate of change in priming volume of the blood phase of a complete oxygenator with respect to the net mean internal pressure is called the oxygenator compliance, $c = \partial V/\partial(P_{blood} - P_{gas})$, where V is the blood-phase volume. There are two practical ways to measure compliance: static and dynamic.

STATIC COMPLIANCE

The oxygenator unit was put in a rigid leakproof shell, designed to collect any gas escaping from the sides of the oxygenator as well as that coming out of the main gas outlet of the device. Pure O_2 at a constant flow rate entered the gas phase. It's inlet pressure was recorded by a Stratham P23AA transducer and a HP7700 recorder. The gas phase outlet was connected to a spirometer for flow rate measurements. The liquid phase inlet and outlet were clamped tight. Known volumes of H_2O were added to and with drawn from the blood plugs by a syringe. The blood phase pressure was recorded with instrumentation similar to that for the gas phase pressure. The pressures corresponding to known changes in blood phase volume were measured and typical values are shown in Fig. 2. The static compliance is calculated as follows $C_S = [\partial V/\partial(P_b - P_g)]_{\dot{Q} = 0}$ where V = blood phase volume, P_b = arithmetic mean blood phase pressure, P_g = arithmetic mean gas phase gas phase pressure. The volume increase is large initially (40 ml for 100 mmHg) and then the oxygenator appears stiffer. An initial soft phase of the compliance is associated with deflection of the membrane into the grooves primarily by bending: for additional deflection of the membrane it becomes necessary to stretch the membrane. As the stretching increases the thickness of the membrane reduces and the elastic resistance of a sheet decreases. Constrained membranes often show weak regions when strongly stressed by pressure, and can thin down, blow bubbles and even leak through pinholes or rupture. In forming thin regions or bubbles the membrane appears to go "soft" i.e. becomes more compliant at higher pressures.

MEMBRANE DEFLECTION

It is not possible to observe directly whether the compliance as measured is associated with uniform deflection of the membrane in the oxygenator channels because the construction of the oxygenator does not permit direct observation. However, on disassembly of oxygenators it is possible to recover full-width (18.5 cm) sheets of the silicone rubber membrane. It is quite practicable to test such sheets in an open system which allows observation and measurement of deflection, and subsequently interpret the results in terms of deflections expected for the membrane in the oxygenator. A cup system was built to support a sheet of membrane, Fig. 3. The diameter of the "cup" in this device is 11.3 mm. The results of this test are shown in Fig. 4. The range of elastic strains achieved with this membrane test device was small, so that the relationship between stress and strain is linear. If we turn to consideration of the oxygenator itself and analyse the deflection of the membrane into the various channels -- the blood supply manifolds as well as the capillary channels themselves -- we can discover something of the sequence in the expansion of the blood phase. The unsupported span of membrane is much greater across the tapered manifold channels than it is across the capillary channels, and consequently when a pressure is applied to the blood phase the larger deflections are found in the manifold channels. Indeed, deflection of the membrane into manifold channels is very desirable because the flow in the gas phase can otherwise short-circuit from the gas inlet to the gas outlet without passing in contact with the membranes in the capillary channels more remote from the gas transfer ports. However, it appears that at moderate trans-membrane pressures -- even, say, 100 mmHg -- the membrane has not deflected fully into the channels provided in the plastic spacer plates because further increases in pressure allow a volume to be added which is greater than the maximum change in priming volume of the capillary channel sections themselves.

It may be noted that membrane deflections from the circular membrane tester can be used to predict the membrane deflection in the capillary channels of the oxygenator, allowing for differences in size and configuration. Let w_o be the central deflection for the circular membrane tester: w_c is the central deflection of the membrane in the capillary channel. Now, $\nabla^2 w = P/\sigma$ = const. where σ is the tangential stress. Thus, for the microchannel $d^2w/dx^2 = P/\sigma_c$, and for the test unit $d^2w/dr^2 + (1/r)dw/dr = P/\sigma_r$. The solutions to these equations under appropriate boundary conditions, with a = half-width of channel and R = radius of test cup, are $w_c = Pa^2/2\sigma_c$ for the channel and $w_o = PR^2/4\sigma_r$ for the test cup. By thin shell theory, $\sigma_c = Pr/t$ and $\sigma_r = Pr/2t$ whence $w_c = (a/R)\, w_o$.

DYNAMIC COMPLIANCE

The apparatus for measurements of the dynamic compliance was similar to that of the static compliance. In this case, however, water was circulated through the blood phase from and to a graduated cylinder in a roller pump. A variable restriction was put on the outlet tube in order to increase or decrease the pressures internal to the oxygenator. The internal increase in volume of the oxygenator blood phase plus any increase in volume in the circuit tubing corresponds directly to the volume decrease of the water in the graduated cylinder. The dynamic compliance can be calculated as follows: $C_a = \partial V/\partial[(P_{b\ in} - P_{b\ our})/2 - P_g]$. We use the arithmetic mean pressure in the blood phase $(P_{b\ in} - P_{b\ our})/2$ as an estimation of the blood phase pressure. The results in Fig. 5 show the relationship between the change in internal volume of the blood compartment (and tubing circuit) and the liquid flow rate, and they are directly proportional. Estimation of the blood phase resistance can be made when plotting pressure drop across the blood phase **vs.** flow rate. When the transmembrane pressure $(P_b - P_g)$ increases with increase of liquid flow rate the gap thickness increases and the blood phase resistance $(\Delta P/\dot{Q})$ falls. This effect has been noted before with flat-sheet membrane oxygenators of rather different configuration (3).

MASS TRANSFER

Degasses water (2) enriched with carbon dioxide was passed, at various flow rates, through the blood phase and pure dry $O2$ was supplied at constant flow rate to the gas phase. Entrance and exit pressures were measured at the blood and the gas phases, and inlet and outlet water samples were taken or partial pressure measurements of O_2 and CO_2. The mass transfer was re-presented by calculating the mass transfer coefficients (U) using the log mean partial pressure difference (ΔP_{LM}). $U = \dot{V}/A \cdot \Delta P_{LM}$ in L/m^2 mmHg min where $\dot{V}$ is the volume mass transfer of the gas across the membrane in liters per minute and A is the effective transfer area in m^2. The results show a linear increase of O_2 transfer rate with liquid flow rate while the CO_2 transfer rate levels off asymptotically with flow rate due to membrane resistance to CO_2 transfer. At disassembly of oxygenators after testing, water was found to have condensed in the channels on the gas side of the membranes.

DISCUSSION

In this design, a small transmembrane pressure has the effect that the membrane is not fully distended into the channels molded for it to provide the blood manifolding, and the capillary channels will be relatively flat. The latter effect is good in promoting efficient oxygen transfer to blood, but the lack of complete filling of the manifold channels allows gas to shunt without properly ventilating all the membrane surface. This results in reduced CO_2 removal and condensation of water in the gas space. The latter effect does not seem easily corrected once it has occurred.

Lande-Edwards oxygenators of both the 1 M^2 and 3 M^2 sizes have been used successfully clinically for many hours of support of patients (4,5). If the oxygenator membrane should tend to shrink with aging, the compliance could reduce also with age, and membrane deflections under a given transmembrane pressure would be reduced.

ACKNOWLEDGEMENTS

We wish to thank Dr. Leonard Trudell and Mr. Leonard Boudreau of the Division of Biological and Medical Sciences, Brown University, for their assistance in the experiment.

REFERENCES

1. Galletti, P. M. and Brecher, G. A. "Heart-Lung Bypass" Grime and Stratton, 108-120, 1962.
2. Galletti, P. M., Richardson, P. D. and Snider, M.T. "Blood oxygenator testing and evaluation, Part I" NIH Annual Rept. #69-2047-1, 1971.
3. Richardson, P. D. and Galletti, P. M. "Correlation of effects of blood flow rate viscosity and design features on artificial lung performance" in "Physiological and Clinical Aspects of Oxygenator Design", S. G. Davids and H. C. Engell, Eds, 29-44, Elsevier/North Holland, 1976.
4. Landé, A. J. et al. "Prolonged cardio-pulmonary support with a practical membrane oxygenator" Trans. Amer. Soc. Artif. Internal Organs 16, 352, 1970.

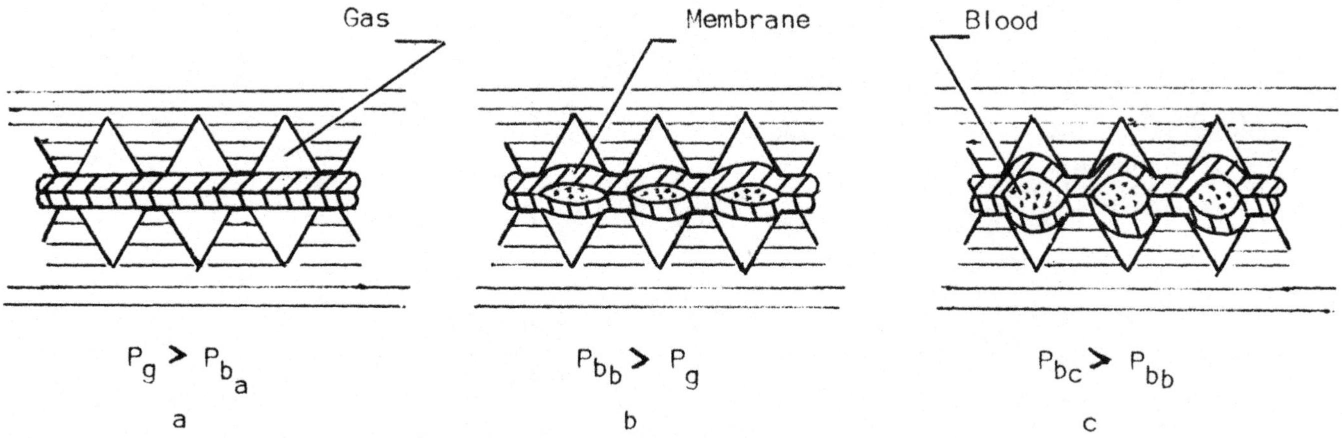

Figure 1

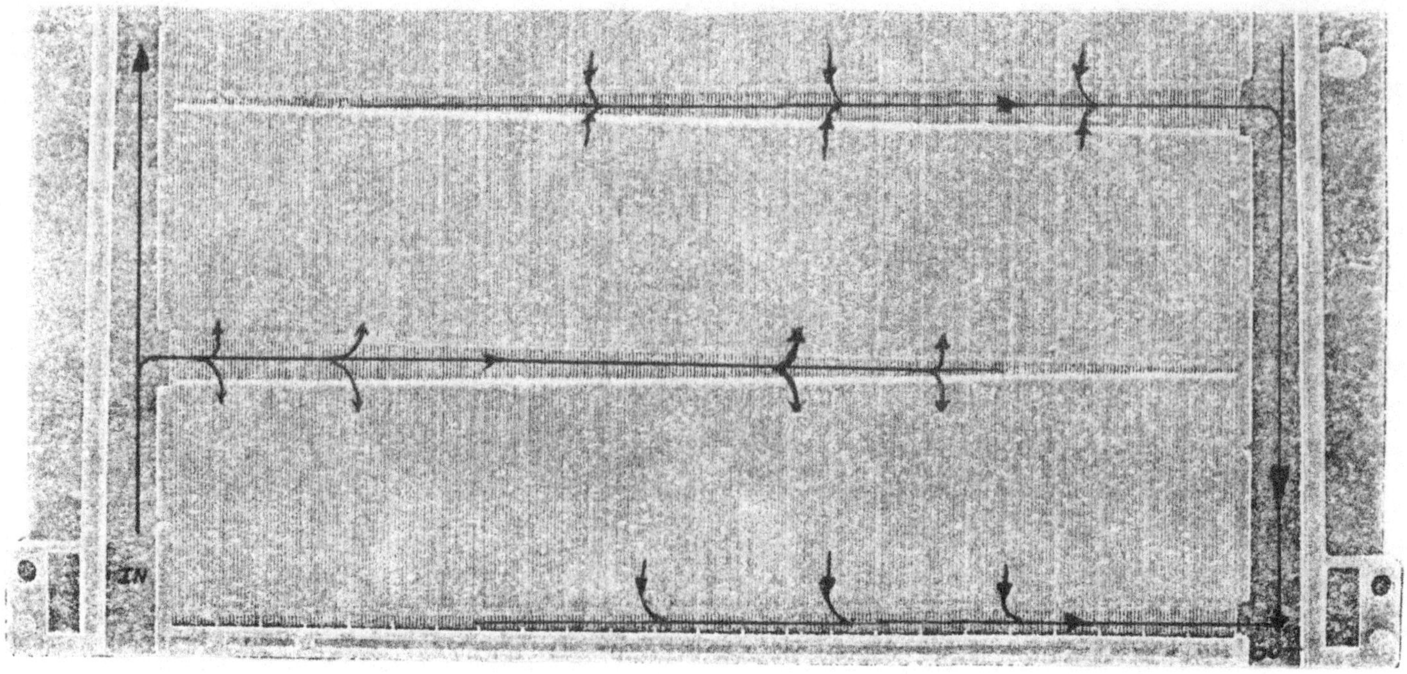

Figure 1d: showing 1/2 of the molded support plate. Blood path is shown.

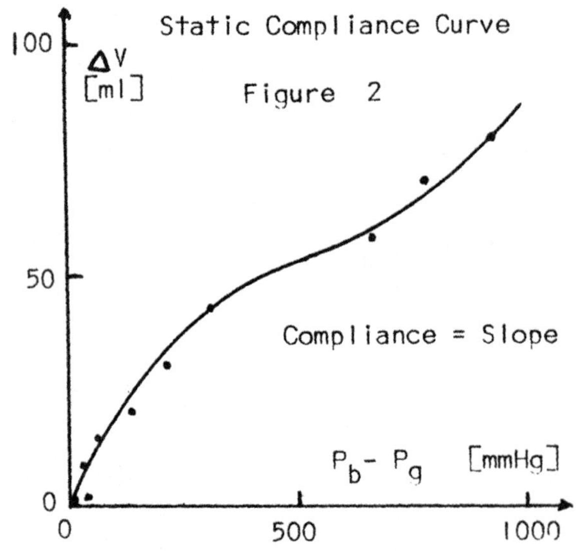

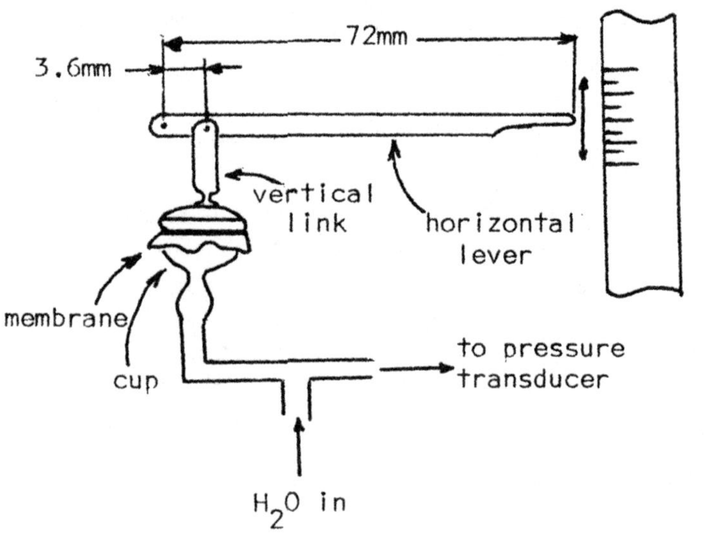

Figure 3
Apparatus for Membrane Deflection Measurements

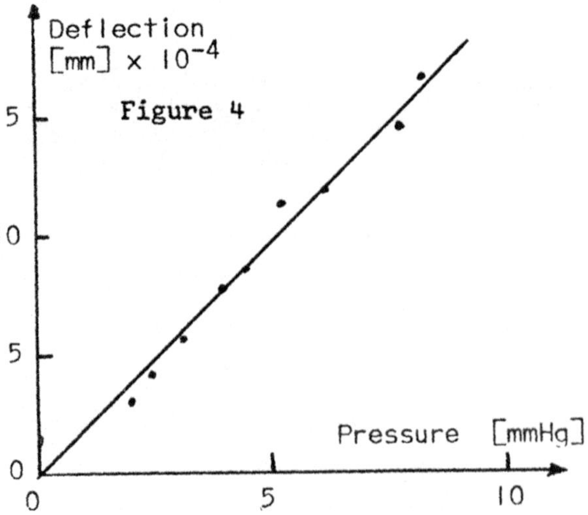

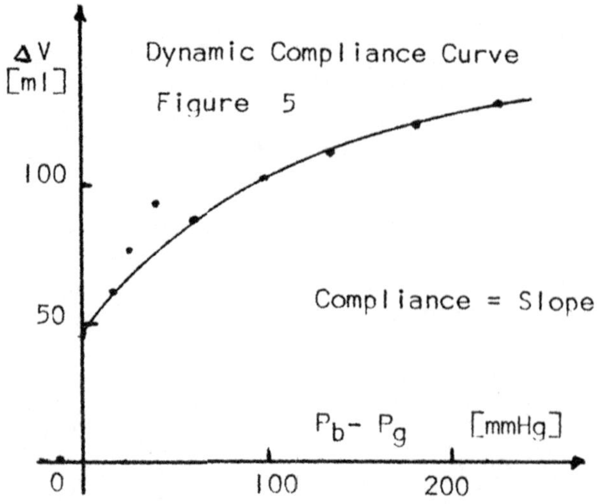

A CONTOURING METHOD FOR THE BACK

M. H. Pope, Ph. D., D. G. Wilder, B. S. M. E., J. W. Frymoyer, M. D., E. Buturla, Ph. D.

Biomechanics Lab, The Department of Orthopaedic Surgery, The University of Vermont, Burlington, Vermont 05401.

Work on the development of a convenient, accurate technique for three dimensional measurements of the human body form began some time ago. In 1440, the Italian sculptor Alberti used a rig called a "definer" to make such measurements using cylindrical coordinates and in 1841, Wheatstone's stereopticon was developed. About five years ago, Herron at Baylor College of Medicine began applying the civil engineer's stereophotogrammetric techniques to medical problems.[1]

In our laboratory, we are making multi-directional load-deflection curves of the spines of live subjects. We have previously established the motion of given vertebral bodies in space in response to given loads,[2] but are now interested in the effect of such motions on the overall topography of the back. In addition in our scoliosis clinic, we were interested in a method for assessing both the clinical progression of the disease and the condition pre and post surgery.

Herron's stereophotogrammetry method was ruled out for the following reasons: (1) equipment too expensive; (2) digitizing the number of reference points needed, too time consuming; (3) an immediate result was desired. For these reasons, we developed our Moiré fringe method.

MOIRÉ METHOD

Theory

The principle of the technique is that the object is illuminated with a point light source through a plane grating and the object is observed through the grating.[3] The small areas on the object which are illuminated through the grating and also seen through it appear bright and form Moiré fringes. The basis of our method is a 50 cm. by 50 cm. glass screen with a series of 1 mm. lines alternately clear and dark. The glass grid is mounted on a frame that can slide in a direction perpendicular to the grating lines. A polaroid camera is mounted perpendicular to the grating and a light source is mounted in line with and above the lens of the camera. The higher the light source is placed above the camera, the greater will be the number of moire fringes on the object, however, if the light source is too high above the camera, the fringes produced will be too numerous to be easily distinguished. The distance between the lens and grid and between the light source and lens axis are recorded for each test.

When using a coarser grating with dark lines 1.6 mm. wide and clear lines 2.4 mm. wide, we found it necessary to oscillate the screen during the exposure in order to keep the lines blurred, therefore, invisible. However, with fine grid lines, the problem was obviated.

Using Speight's[4] equation: (see fig. 1)

$$h_r = \frac{(r-\frac{1}{2})\,Ld}{1 - (r-\frac{1}{2})d}$$

d = grid line repeat distance
L = camera lens to grid distance
1 = camera lens axis to point light source
r = dark fringe number
h = distance from grid to r^{th} dark fringe

The distance from the grid to any fringe on the object can be computed. In our work, we found that when the values of L and 1 were large, the distance between h_r and h_{r+1} was essentially constant. This allowed us to label the first fringe encountered on an object as fringe #1 and the successive fringes as 2, 3, ..., r, r+1, instead of counting the number of fringes in the region from the grid to the object.

Method

The subject is placed in front of the grid (fig. 1) and is positioned carefully against three pair of adjustable location pads which touch the subject at the mid-clavicles, anterior superior iliac spines and at the patella. When the subject is in position, a continuous point light source is turned on, and two polaroid photographs are taken of the subject's back in rapid succession. For purposes of calibration, indicators are placed over each of the posterior superior iliac spines and the distance between them is recorded.

One photo is placed in the subject's permanent record, so the clinician can monitor any changes. The moiré fringes on the second photo (fig. 2) are digitized by means of a Summagraphics data tablet and Infoton CRT type terminal on line with a Xerox Sigma-6 computer. The three-dimensional data obtained from the fringes is stored and converted into a graphical form (fig. 3) for easy comprehension. The graphical image can be rotated about 2 axes and the height dimension can be amplified in order to exaggerate any abnormalities.

Data on scoliotic backs is being related to the bony landmarks as defined by biplanar radiographs.

Results:

The following figures represent typical results. Figure 2 is a photograph of the Moiré fringe projected on the patient's back (minor scoliotic deformity). Figure 3 is the computer plotted output in perspective of the information digitized from Figure 2.

Conclusions:

1. We have established a method of inexpensively and rapidly monitoring contours of the back using a polaroid camera and a moiré fringe pattern.

2. We have developed the software for - (1) digitizing information from a polaroid print and (2) processing the information so that a perspective view of back contours is produced.

BIBLIOGRAPHY

1. Herron, R. E.: Exploring the third dimension with camera and computer. Kodak Studio Light. Issue No. 1, 1975, pp. 26-29.

2. Pope, M. H., Wilder, D. G., Matteri, R. E., and Frymoyer, J. W.: Experimental measurements of vertebral motion under load. Accepted for publication by Orthopaedic Clinics of North America. In Proc. of 4th Annual New England Bioengineering Conference, Pergamon Press, May, 1976, pp. 49-55.

3. Takasaki, H.: Moiré topography. Applied Optics. Vol. 9, No. 6, June, 1970, pp. 1467-1472.

4. Speight, B. S., Miles, C. A., and Moledina, K.: Recording carcass shape by a moiré method. Medical and Biological Engineering, March, 1974, pp. 221-226.

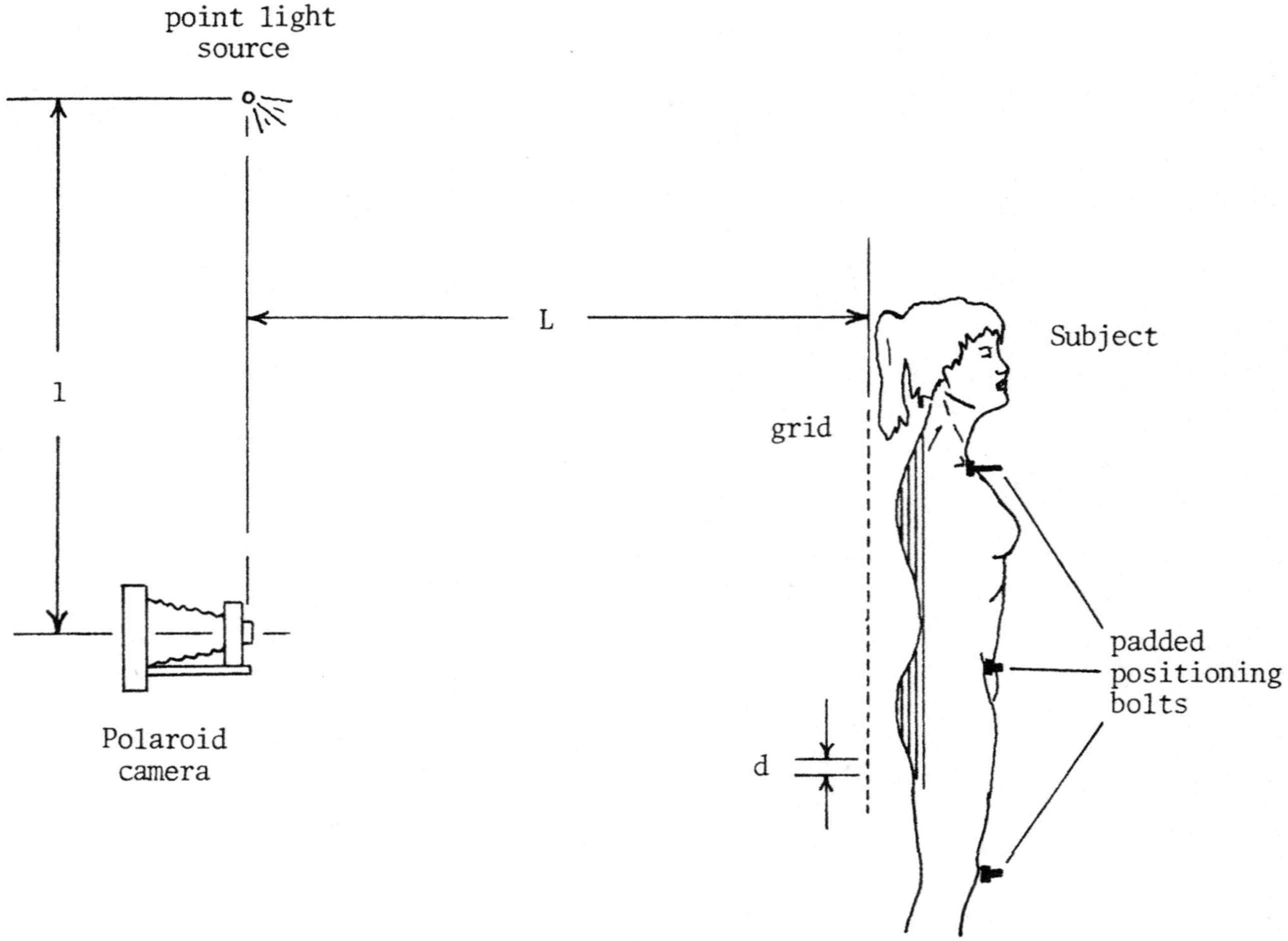
point light
source
l
L
Polaroid
camera
Subject
grid
d
padded
positioning
bolts

Figure 1

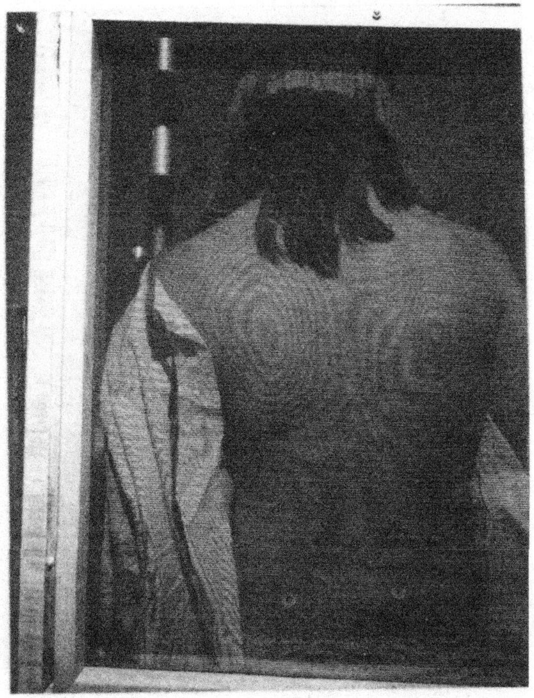

Figure 2

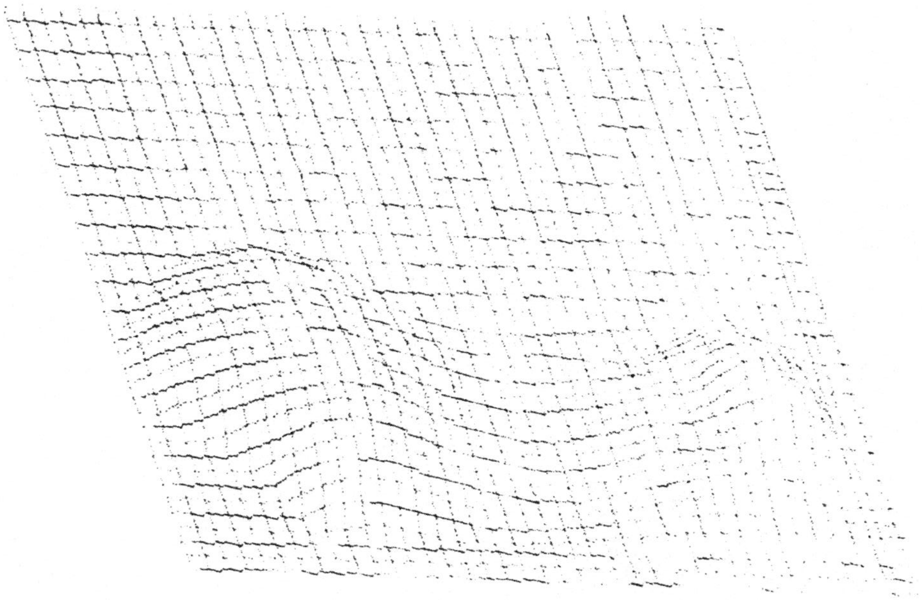

Figure 3

AN IMPROVED INEXPENSIVE FORCE TRANSDUCER MODELED AFTER THE ARKANS TRANSDUCER, AND ITS APPLICATIONS

Ira S. Tackel, W. Caldara, and J. Tiongson

Department of Rehabilitative Medicine, Veterans Administration Hospital, Albany, New York 12208 and Center for Biomedical Engineering, Rensselaer Polytechnic Institute, Troy, New York 12181.

To date, the evaluation of the effectiveness of prosthetic or orthotic devices has been subjective analysis at best, and only after the patient has used the device for a period of time. The device described here uses the same principle as the Arkans Transducer (1) but with modifications and refinements.

Use of this device in assessing normal gait patterns is the next step. By determining a normal gait range from objective data, we can then more easily identify and begin to correct the pathologic foot. As does the Arkans Transducer, the improved version provides a rapid and objective analysis of gait which will allow the prescription of a more effective corrective device.

<u>Design Objective</u>: By improving the Arkans Transducer several criteria were sought beyond those stated originally for the initial design. The transducer and associated electronics package were designed with the following in mind:
1) to provide a multi-channel ability for weight bearing discrimination,
2) to be inexpensive and easy to fabricate, (1)
3) to be non-restrictive to foot motion,
4) ease of connection to the electronics package,
5) to be able to record not only the total force exerted by the limb, but forces exerted over irregular areas,
6) easy application of the transducer yet ability to maintain it's critical positioning.

<u>Construction and modification of past design</u>: The transducer, which is capacitive in nature, is composed of four layers of material. (Fig. 1) It is specifically cut to fit the inside of a patients shoe. Instead of cutting each foil plate for use as the active pad, we have a solid piece of mylar backed copper and etched away those areas of copper which will not be used as the active transducers. This alleviates a major problem of the active pads shorting to the shield layer at the edge of the pad itself. By the etching process we can terminate the lead of each active pad at the edge of the transducer and eliminate the need for the use of the enamel-coated wire lead which was needed for each pad. Connection from the transducer to the signal-carrying cable is also simplified by using a flat-conductor mass termination connector. This is attached by crimping onto the active layer which has been terminated into a standard width flexable circuit board.

The placement of the active pads can be determined in many ways. One technique we have found quite promising is to use a Harris Mat to first locate the principle areas of force at the sole of the foot. We then transfer that diagram onto the copper to be etched into the proper foil pattern.

Electronics: The electronics package is virtually the same as described by Arkans (1) with some modifications to specific values of components. (Fig. 2) The capacitance of the transducer determines the frequency of the astable multi-vibrator (555 timer intergrated circuit). This frequency is demodulated quite linearly by a phase-lock-loop (PLL) intergrated circuit. A six channel device has been constructed to allow monitoring of six channels of information simultaneously on a chart recorder.

Applications: By using six transducers in a single package and the associated electronics, we can monitor any six areas of interest on the sole of the foot. These six signals can be displayed on a conventionsl chart recorder or multi-channel oscilloscope for analysis simultaneously.

Currently under consideration is the design of an electronics packege capable of summing vectorially the six channels of information present and displaying this information as a moving vector in real time on an oscilloscope. The information derived would be very much analogous to the type of information attained from a vector cardiograph. This technique would allow for quick, objective, and concise evaluation of an individual's gait.

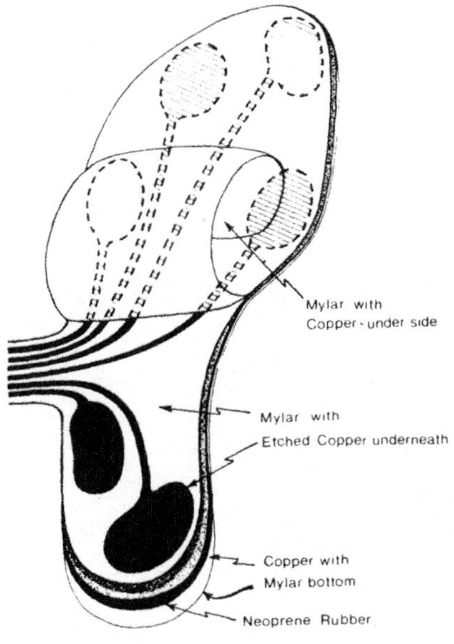

Fig. 1 Representation of the four layer foot transducer showing six active sites.

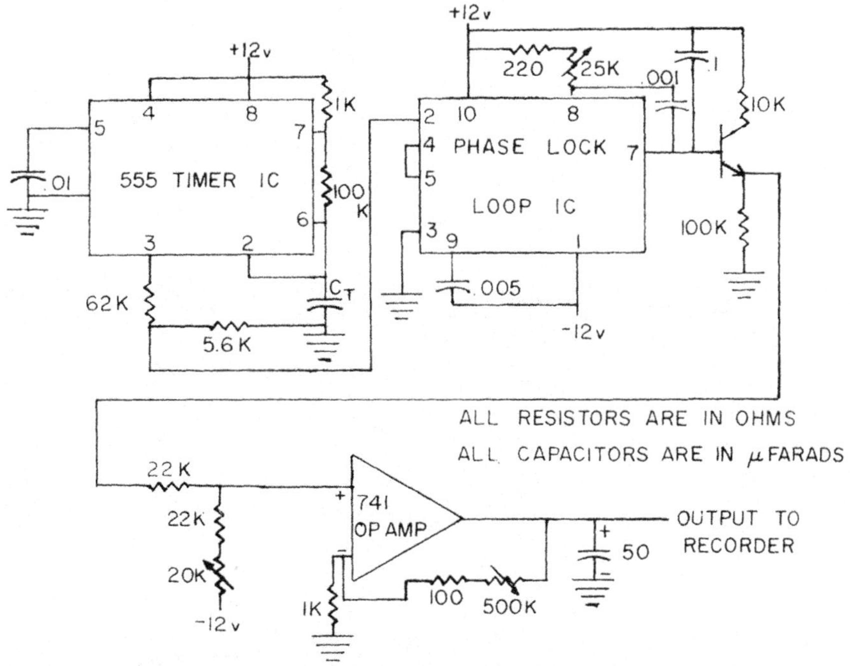

Fig. 2 Schematic diagram of the electronics package of one of the six channels

Acknowledgements

The authors are indebted to Daniel Maloney for his helpful discussions during the analysis of biomedical and electronic problems.

References

(1) E. Arkans and J. Tiongson, "A Versitile, Inexpensive Force Transducer for the Sole of the Foot," Proceedings of the Fourth New England Bioengineering Conference, May 8, 1976, 279-282

(2) R. P. Schwartz, "A Quantitative Analysis of Recorded Variables in the Walking Pattern of "Normal" Adults," March, 1964, J. Bone and Joint Surgery, 46-A(2): 324-334

A more complete list of references is available upon request to the authors.

A MOVING PLATFORM SYSTEM FOR THE STUDY OF INDUCED BODY SWAY

Robert O. Andres and David J. Anderson
University of Michigan
Kresge Hearing Research Institute
1301 East Ann St.
Ann Arbor, Mi. 48109

The evolution of the upright stance has reached its fulfillment in homo sapiens, freeing the forelimbs for specialized tasks. But standing upright does not necessarily mean standing still, and since Romberg's [1] observations of postural sway the oscillations of the body's long axis about the vertical have enjoyed considerable attention. Current knowledge indicates that this body sway is controlled neurologically, with input from the vestibular, visual, proprioceptive and exteroceptive systems being integrated at the spinal cord or higher levels [2]. The resultant effect is muscular activity intended to keep the center of gravity (CG) projection within the base of support, thereby assuring stability. With acceleratory, positional, and righting reflexes all being involved, the investigation of the neurophysiology of postural control rapidly gets out of hand in animals, not to mention human subjects.

Despite the confusion about the underlying mechanisms, clinical neurologists have been drawn to the study of postural sway because it represents a chance to develop a sensitive, repeatable, non-invasive neurological test that involves many neural subsystems. Most studies to date have been concerned with quiet stance on the part of the subject [3], [4], [5], and since Thomas and Whitney [6] many experimentors have used force platforms to project CG movements onto the subject's base of support [7], [8], [9]. The disadvantage of force platforms is the lack of information they provide about the actual position of body links. They derive the CG movements from the torques exerted on the platform by the feet, based on a static inverted pendulum body model. Of course the body is not static during quiet stance, nor does it remain rigid, either [10], [11]. Many other methods of assessing postural function have been tried (see Terekhov [5] for a review), but remarkably few involve any controlled disturbance of the standing human. Honjo and Furukawa [12] had a motor-driven platform capable of tilting at angular velocities of 1-20°/sec, Litventsev [9] designed a force platform capable of small, quick linear motion in an anteroposterior (AP) direction, and Begbie [13] used an inherently unstable platform. Gantchev, Dunev, and Draganova [14] tilted their platform around the ankle axis of the subject sinusoidally, but it was not until later that they analyzed the frequency response of the body during induced sway, in this case using horizontal linear sinusoids [15]. The results were reported for only one amplitude (36 mm peak-to-peak)and seven frequencies in a range from .2 to 1.4 Hz.

A moving force platform with two degrees of freedom (around the ankle joints, and linear AP) provided Nashner [7] with the capability to selectively stimulate the vestibular portion of the postural control system. By applying quick tilts of several small angles to the body while keeping the ankle-leg angle constant through feedback from an electric potentiometer independently sensing trunk inclination, he derived a model for the vestibular control of posture [8]. Other investigators have selectively stimulated components of the postural control system [16], but an in-depth study of the behavior of the whole body and each of its links in response to controlled postural disturbances still remains to be done. The dynamics of this response have been mathematically modelled [17], [18], with contributions from both reflex muscle activation and the elasto-viscous properties of the muscles included. But these models have not been justified with sufficient descriptive data, and they are based on an inverted pendulum body model. Our interest lies in the total description of whole-body response to a controlled disturbance of both the reflex and mechanical portions of the postural control system. For this purpose we have designed and constructed a specialized moving platform system.

The first design consideration was the nature of the motion stimulus to the base of support to be used. The choice was limited to either linear motion or a tilting motion around the ankles, because of cost. Linear motion provides the same stimulation which is experienced in normal gait and closely simulates the movement of the base of support which occurs in a slip, so we decided on a one degree of freedom, linear motion platform.

Postural control has been thought of as a complex servomechanism for many years [19], and the classical method of describing a servomechanism is through its frequency response characteristics. Stark [20] used not only the standard sinusoidal stimulus in an eye-hand tracking task, but also pseudorandom noise inputs because he found that his subjects could predict and correct for sinusoidal disturbances. The pseudorandom stimuli resulted in different frequency response characteristics than pure sinusoids because the subject could not predict the next movement of the stimulus. Thus, our second design criteria was that the linear motion of the platform must be precisely controllable, to follow input waveforms of the sinusoidal, pseudorandom noise, and impulse types.

The final purely mechanical specifications for our platform were the amplitude and frequency ranges of linear motion that we desired. Gantchev and Popov [15] could move their platform 50 mm linearly, but we wanted to provide more inertial disturbance so we decided on two feet peak-to-peak. We wished to expand the frequency band of investigation beyond any of those reported, so we specified a range of .03-3 Hz, with the amplitude attenuated at the higher frequencies because of a peak velocity limitation of 1 ft/sec on the torque motor-drive mechanism.

The intended use of the platform system was not purely for research into dynamic postural control, but also ultimately for the clinical assessment of postural function. Therefore the system had to be designed to ensure patient safety. Mechanical safety precautions must eliminate the danger of patient falls and foot slippage into moving parts. The patient must be free from the threat of electrical shock, and sudden stops or starts of the platform. An observer must always be present in the clinical environment, not only to instruct and encourage the patient, but also to act as a backup to stop platform motion to prevent accidents.

A block diagram of the platform control system is found in Figure 1. We selected a DC torque motor to drive the platform over a hydraulic system, for reasons of cost and mechanical simplicity. The torque motor (Electrocraft E703-01) is controlled by a servo amplifier (Electrocraft 6200 AP) which receives input commands from our laboratory computer. Velocity feedback is provided by a tachometer (Electrocraft 110), while any drift in position is fed back through an electric potentiometer geared to the platform.

The platform itself is 2 feet square, a sufficient size for a man to stand erect with a comfortable stance. It is mounted on two parallel stainless steel rails with ball bushings which allow it to slide two feet from stop to stop.

The drive mechanism is a .25 inch per revolution lead screw with a preloaded ball bearing nut. The platform with its drive mechanism is enclosed in a two foot by five foot aluminum box, one foot high.

Figure 2 shows the platform in the clinical setting. The experimentor sits beside the platform at an operator terminal, designating the input stimulus for each trial while monitoring the accumulation of data from the subject. Not only can the experimentor stop platform motion, the subject himself is capable of stopping motion by grabbing the surrounding railing or lifting the heel or toe of either foot off the platform, in both cases breaking microswitch contact. Instead of holding the railing, the subject's arms are folded across the chest during testing. All openings on the top surface of the platform enclosure are closed by sliding panels which collapse into the frame as the platform moves. Patients are thus protected, and clinical investigations can proceed.

The moving platform system described above will be used in conjunction with a set of two line scan cameras. Algorithms to follow the motion of the subject's silhouette are presented by Anderson, Homick, and Jones [21]. Experiments designed to determine the frequency response characteristics of all body links during whole-body oscillation will be done, in an effort to describe the types of postural corrections made by normal subjects in response. Specifically, we will investigate the extent to which an inverted pendulum body model is apropos under dynamic conditions.

The total measurement system will also be used for clinical assessment of neurological patients. Comparisons will be made between the response of patients with known pathologies and the response of normal subjects. The motivating application behind the design of this platform/camera system was the need for better quantitative tests of neurological function of both pre- and post-flight crew members in the space program, wherein lies our biggest challenge.

ACKNOWLEDGMENTS

The work was supported by the National Aeronautics and Space Administration through contract NAS-9-15106 and the 1974 Life Scientist Program and by the National Institutes of Health through Program Project Grant NS-05785.

REFERENCES

[1] M.H. Romberg, A Manual of the Nervous Diseases of Man, Sydenham Society, London, p. 39, 1853.

[2] T.D.M. Roberts, Neurophysiology of Postural Mechanisms, Butterworths, London, 1967.

[3] W.R. Miles, "Static equilibrium as a useful test of motor control," J. Indust. Hygiene 3: 316-31, 1922.

[4] F.S. Fearing, "Factors influencing static equilibrium. I," J. Comp. Psychol. 4: 91-121, 1924a.

[5] Y. Terekhov, "Stabilometry and some aspects of its applications - a review," Biomed. Eng. 11(1): 12-15, 1976.

[6] D.P. Thomas and R.J. Whitney, "Postural movements during normal standing in man," J. Anat. London 93: 524-541, 1959.

[7] L.M. Nashner, "Sensory feedback in human posture control," Sc. D Thesis, MVT-70-3, Massachusetts Institute of Technology, 1970.

[8] L.M. Nashner, "A model describing vestibular detection of body sway motion," Acta Otolaryngol (Stock) 72: 429-36, 1971.

[9] A.I. Litventsev, "Mechanisms of man's vertical posture control," Agressologie 14(B): 17-21, 1973.

[10] T.D.M. Roberts and G. Stenhouse, "The nature of postural sway," Agressologie 17(A): 11-14, 1976.

[11] T. Valk-Fai, "Analysis of the dynamical behaviour of the body whilst "standing still"," Agressologie 14(C): 21-25, 1973.

[12] S. Honjo and R. Furukawa, "The goniometer test," Ann. Otol. Rhinol. Laryngol. 66: 440-458, 1957.

[13] G.H. Begbie, "Some problems of postural sway, "CIBA Foundation: Symposium on Myotatic, Kinesthetic, and Vestibular Mechanisms, Little, Brown & Co., Boston, pp. 80-104, 1967.

[14] G-N. Gantchev, S. Dunev, and N. Draganova, "On the problem of the induced oscillations of the body," Agressologie 13(B): 51-54, 1972.

[15] G-N. Gantchev, and V. Popov, "Quantitative evaluation of induced body oscillations in man," Agressologie 14(C): 91-94, 1973.

[16] G-N. Gantchev, S. Dunev, and N. Draganova, "On the spontaneous and induced body oscillations," in 2d Int'l Symposium on Motor Control, ed. by Gydikov, Fankov, and Kosarov, Zlatni Pyasŭtsi, Bulgaria, pp. 179-194, 1972.

[17] R.V. Aggashyan and Ye. I. Pal'tsev, "Reproduction of certain special aspects of the dynamics of the maintenance of the vertical posture by man using a mathematical model," Biofizika 20(1): 137-42, 1975.

[18] V.S. Gurfinkel and S.M. Osovets, "Dymanics of equilibrium of the vertical posture in man," Biofizika 17(3): 478-85, 1972.

[19] P.A. Merton, "Speculations on the servo-control of movement," In CIBA Foundation Symposium on The Spinal Cord, Little, Brown & Co., Boston, pp. 247-260, 1953.

[20] L. Stark, Neurological Control Systems, Studies in Bioengineering, Plenum Press, New York, 1968.

[21] D.J. Anderson, J.L. Homick, and K.W. Jones, "Line scan cameras applied to posturography," in Proceedings of the 1977 San Diego Biomedical Symposium, San Diego, 1977 (in press).

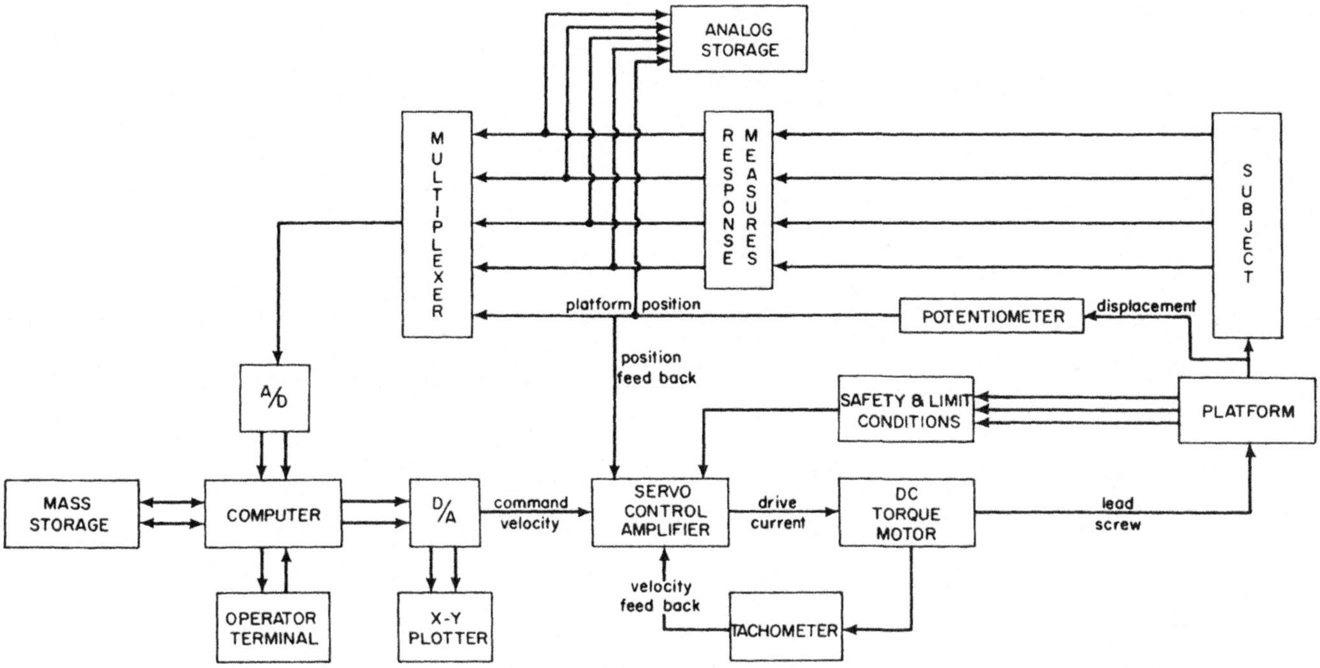

Figure 1: Block diagram of the moving platform control system.

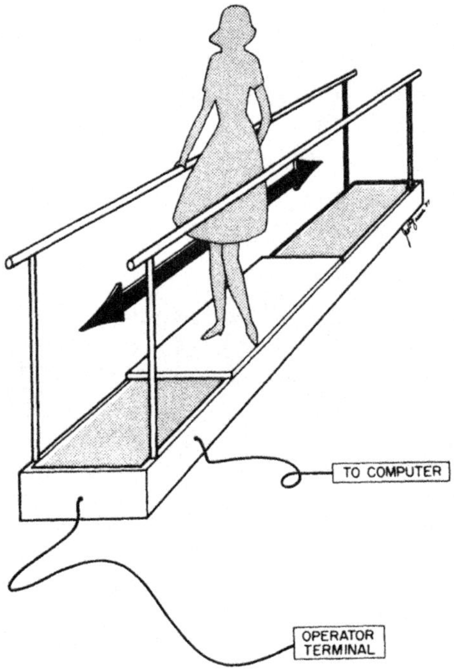

Figure 2: Scale drawing of platform in the clinical setting. The railing is 30" high. The subject holds the railing to stop platform motion.

FOREARM ROTATION

Y. Youm, Ph.D., A. E. Flatt, M.D., and B. L. Sprague, M.D.
Biomechanics Laboratory of Upper Extremity, Department of Orthopaedics
University of Iowa, Iowa City, Iowa 52242

INTRODUCTION

The motion of the forearm bones has evoked considerable interest among many investigators. Anatomically the forearm consists of the radius whose distal end is connected to the hand, and the ulna whose proximal end forms the elbow joint with the humerus. Pronation-supination occurs between the radius and ulna, and elbow flexion-extension occurs between the ulna and humerus.

A review of the literature reveals an interest in this subject that dates back to the middle of the 19th century. At that time, while most researchers agreed that the radius rotated in an arc during pronation-supination, considerable controversy rages even to this day on the role of the ulna in this movement. The path of movement of the ulna as described by earlier writers [3,4,5] has been one of an arc of a circle. Some [2,6] had hinted translatory motion of the ulna during pronation-supination movement. However, the above mentioned investigators carried out their works by either observation of specimen movement or crude experimental methods.

In recent years, elbow prostheses have been developed and widely used for joints destroyed by disease or trauma. This has focused attention on biomechanical analyses of forearm and elbow motions, an area in which little work has been done.

The purpose of our recent studies of supination-pronation of forearm and elbow motion using analytical technique as well as experimental methods are: 1) To determine the range of motion in supination-pronation, elbow flexion-extension and the carrying angle. 2) To determine ulnar movement. 3) To determine the longitudinal axis about which the radius rotates around the ulna during pronation and supination. 4) To determine the transverse axis about which the ulna rotates during flexion-extension motion. 5) To establish the design requirements for an elbow prosthesis.

MATERIAL AND METHOD

Fresh, frozen cadaver upper limbs divided at the middle of the upper arm were used. Two Steinmann pins were driven transversely through the humerus to eliminate humeral movement, and fixed into upper arm holding device. Also, two Kirschner wires were drilled oppositely into the radius and ulna without fixing the two bones.

Analytical Investigation. In the analytical investigation a 3-dimensional joint kinematic analysis of two rigid body link system [9] was applied. In the supination-pronation study the radius and ulna were assumed to be rigid bodies, and, arbitrarily, this motion was studied at 90° elbow flexion with the humerus fixed. Similarly, the ulna and humerus were assumed to be rigid bodies in the elbow flexion-extension study, and mid pronation-supination of the forearm position was used for elbow motion study.

Three non-collinear spark gaps were attached to the Steinmann pins drilled into the radius and ulna. With the ulna fixed, the spatial coordinates of three non-collinear points of the radius are obtained as the radius moves around the ulna. After the data collection procedure is completed, the instrumented specimen is carried to the x-ray room for bi-planar x-rays which are used to relate the analytical results to bony landmarks later on.

In data analysis, the right handed orthogonal coordinates are systematically constructed in the radius, ulna and humerus using the three non-collinear spark gap information, while three orthogonally oriented microphones serve as the inertial coordinates (Fig. 1). For a moving rigid body in general, there is always an axis called the instantaneous screw displacement axis in such manner that the direction of the linear velocity of any point on that axis is instantaneously coincident with that axis, and this axis is parallel to the angular velocity vector of that body at that instant.

From Figure 2, the angular velocity vector can be obtained as

$$\bar{\omega} = [\alpha(\bar{a} \times \dot{\bar{c}}) + \beta(\bar{d} \times \dot{\bar{c}}) + \gamma(\dot{\bar{d}} \times \bar{a})] \; / \; (\alpha^2 + \beta^2 + \gamma^2)$$

in which

$$\bar{d} = \bar{a} \times \bar{c}, \quad \alpha = \dot{\bar{a}} \cdot \bar{c}, \quad \beta = \dot{\bar{d}} \cdot \bar{c}, \quad \text{and } \gamma = \dot{\bar{d}} \cdot \bar{a}$$

The location of $\bar{\omega}$ in the x-ray films is obtained by the following procedure. In Figure 2, since $\bar{v}_P$ and $\bar{\omega}$ have the same direction

$$\bar{v}_P = \kappa\bar{\omega} = \bar{v}_G - \bar{\omega} \times (\bar{s} - \bar{r})$$

where κ and $\bar{r}$ are unknowns.

To solve the four unknown scalar quantities, κ and $\bar{r}$, a fourth condition is made by restricting the point P to lying on a fixed plane. For convenience, the plane is taken normal to a coordinate axis of the reference frame in the body depending upon the direction of $\bar{\omega}$. In cases when the plane is taken perpendicular to the x_i axis, $|\omega_x| >> |\omega_y| \omega_x|$ and $r_x = 0$.

Therefore,

$$r_y = s_y - (\omega_y/\omega_x)s_x + \kappa(\omega_z/\omega_x) - (v_{Gz}/\omega_x)$$

$$r_z = s_z - (\omega_z/\omega_x)s_x - \kappa(\omega_z/\omega_x) + (v_{Gy}/\omega_x)$$

where,

$$\kappa = (v_{Gx}\omega_x + v_{Gy}\omega_y + v_{Gz}\omega_z) \; / \; (\omega_x{}^2 + \omega_y{}^2 + \omega_z{}^2)$$

In the similar manner, we obtained three points on $\bar{\omega}$ and thus obtained the rotating axis.

Once these kinematic quantities are obtained, the results are related to bony landmarks through the x-ray film analysis [9] taken earlier. This x-ray film analysis was followed by the analytical indirect x-ray surveying method.

Experimental Investigation. Three methods were used to experimentally obtain the movement of the ulna, the axis of forearm pronation-supination and elbow flexion axis. Most experimental methods require x-rays of the instrumented specimen to obtain bony landmark relationships.

First, a photographic technique using L.E.D. (light emitting diode) was employed. One set of L.E.D.s was mounted on each of the K-wires implanted in the radius and the ulna. A camera was mounted perpendicular to the plane of L.E.D. motion, and open shutter shots in a darkened room taken of the L.E.D.s as the arm was passively rotated through forearm pronation-supination.

In the ulna movement study, a K-wire was drilled through the second to fifth finger metacarpals to create a different axis of rotation and the experiment was repeated. To find the longitudinal axis of supination-pronation motion, the ulna was immobilized while the radius was passively rotated around the ulna. The complete procedure was repeated twice, moving the K-wire in the radius to a more proximal point each time.

For the transverse axis of the elbow, a similar procedure was performed by attaching the L.E.D.s to the ulna and the humerus. The photographs were taken during passive flexion-extension motion of the elbow. Graphic analysis was used on these outputs to find various mechanical quantities.

Secondly, bi-planar x-ray technique was used by embedding metal markers and pieces of wire into the radius and the ulna. Serial x-rays were taken during passive motion. The x-rays were also analyzed graphically.

Lastly, the Graf/Pen system was employed in a setup similar to that of the L.E.D. method except that Graf/Pen spark gaps replaced the diodes. Computer graphic was used for the data analysis.

RESULTS

In vivo, from the study of a number of live subjects, the normal range of motion for elbow flexion-extension averages 140° ± 5°, and 130° ± 10° during supination-pronation.

The carrying angle, which is defined as the deviation between the axis of the humerus and axis of the ulna, was measured from cadaver experiment. This angle is plotted against elbow flexion angle in Fig. 3. The curve is not linear whereas Morrey, et al [7] found it to be linear in their analytical study.

The movement of the ulna was recorded during supination-pronation by the L.E.D. method. The ulna moves mainly translatory as shown in Fig. 4. The range of motion of the ulna is different depending upon the axis selected. The maximum range of motion of the ulna is obtained when the index finger is used as the rotating axis and minimum when the forearm is rotated along the little finger axis.

The axis of supination-pronation runs proximally from the distal end of the ulna to approximately the center of the capitulum, as shown in Fig. 5. This axis is oblique to the longitudinal axis of both the radius and the ulna.

The axis of rotation during elbow flexion-extension runs through the center of the trochlea.

A differential motion was observed between ulna and radius in flexion-extension motion. The radius shifts distally during elbow extension.

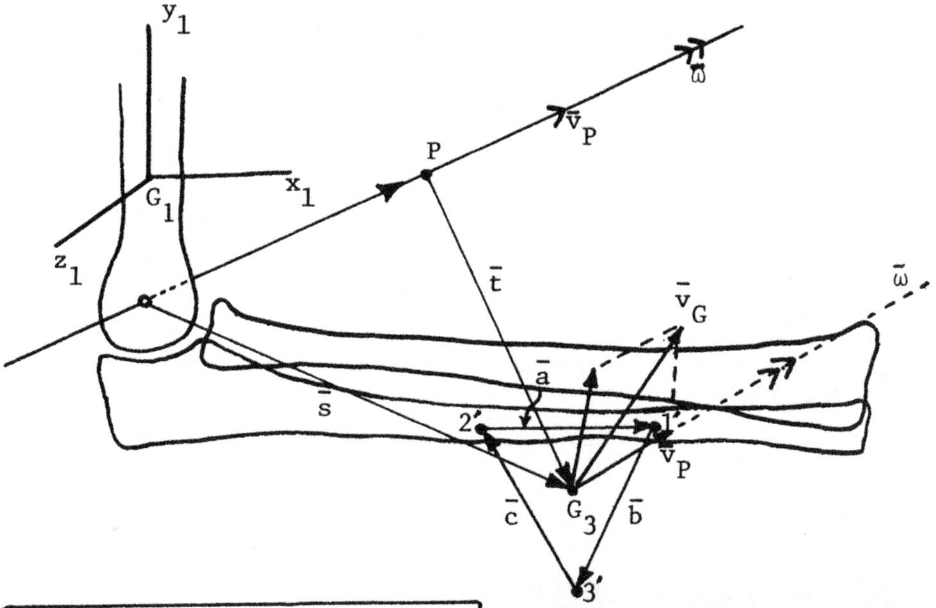

Fig. 2. Instantaneous screw displacement axis in humerus coordinate system during elbow flexion-extension.

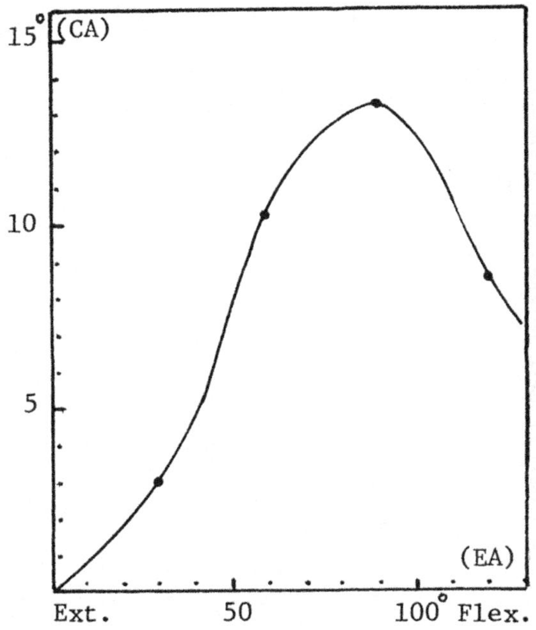

Fig. 3. Carrying angle (CA) vs. elbow flexion angle (EA).

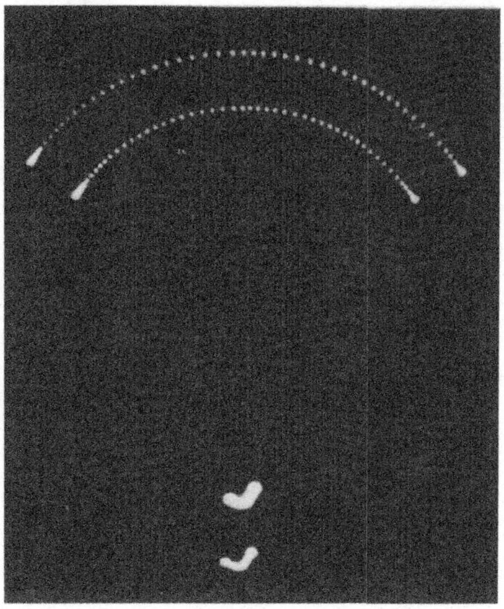

Fig. 4. Translatory ulnar movement and circular motion of radius during pronation-supination of forearm.

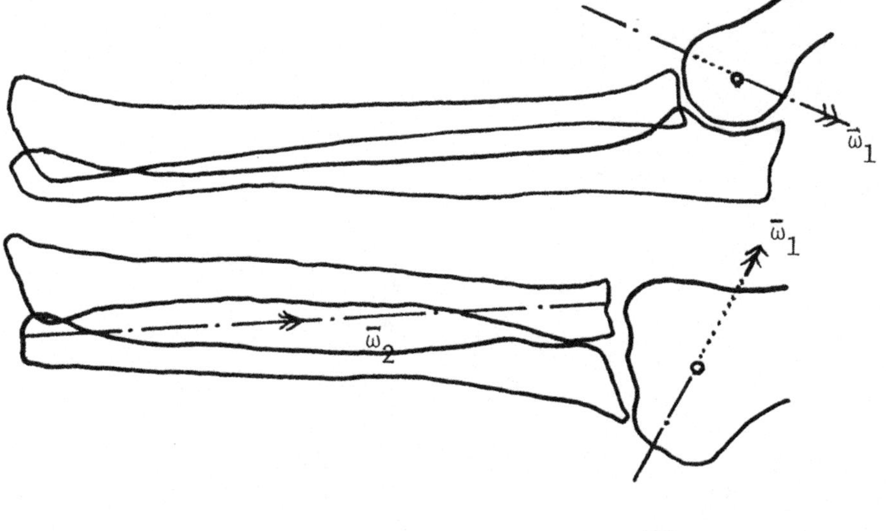

Fig. 5. Forearm axes of rotation during elbow flexion-extension (ω_1) and pronation-supination (ω_2).

REFERENCES

1. Anderson, R. J., "Rotation of the Forearm," Lancet, 2, 1901, pp. 1333-1334.

2. Capner, N., "The Hand in Surgery," J. Bone and Joint Surg. 38B, 1956, pp. 128-151.

3. Duchenne, G. B. A., Physiology of Motion, Demonstrated by Means of Electrical Simulation and Clinical Observation and Applied to the Study of Paralysis and Deformities, translated by E. B. Kaplan, J. B. Lippincott Co., Philadelphia, 1949.

4. Dwight, T., "The Movements of the Ulna in Rotation of the Forearm," J. Anat. and Physiol., 1884, pp. 186-189.

5. Hilberg, J., "The Movement of the Ulna in Rotation of the Fore-arm," J. Anat. and Physiol., 1884, pp. 237-240.

6. Kapandji, I. A., The Physiology of Joint, Vol. 1, 2nd Ed., E. & S. Livingstone, London, 1970.

7. Morrey, B. F., and Chao, E. Y. S., "Passive Motion of the Elbow Joint - A Biomechanical analysis," J. Bone and Joint Surg., 58A, 1976, pp. 501-508.

8. Ray, R. D., Johnson, R. J., and Jameson, R. M., "Rotation of the Forearm, an Experimental Study of Pronation Supination," J. Bone and Joint Surg. 33A, 1951, pp. 993-996.

9. Youm, Y., and Yoon, Y., "Kinematic Investigation of Human Wrist - General Analytical Development," to be submitted to Journal of Biomechanics.

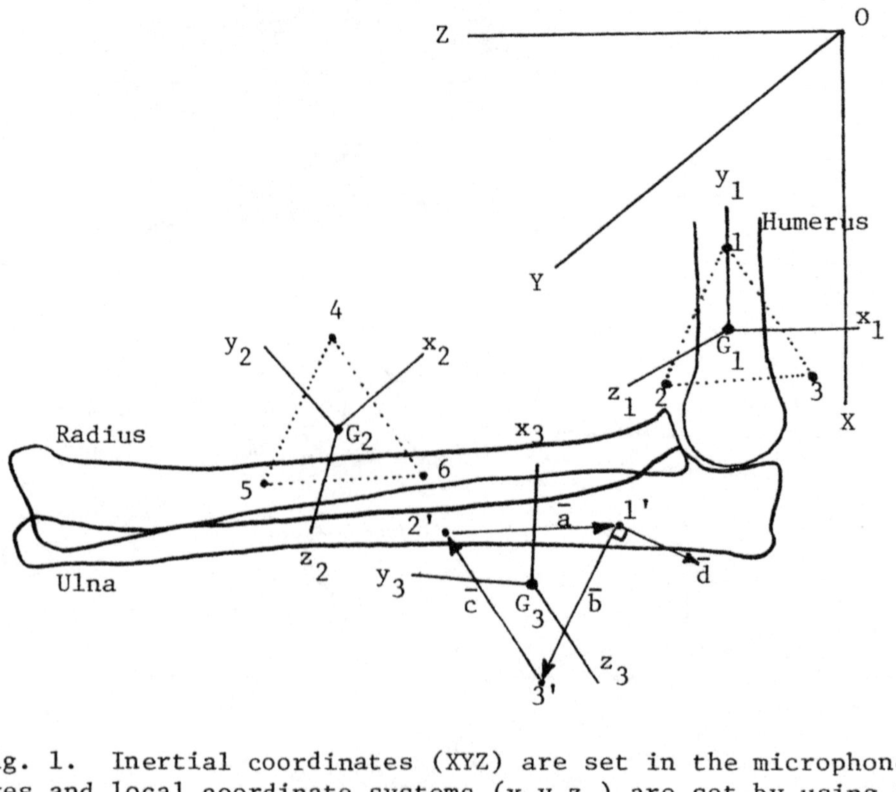

Fig. 1. Inertial coordinates (XYZ) are set in the microphone axes and local coordinate systems ($x_i y_i z_i$) are set by using 3 non-collinear spark gap information in each segment.

RATIONALE FOR KNEE PROSTHETICS FAILURE AND GUIDELINES FOR THE FUTURE

Murray R. Berkowitz
New York College of Podiatric Medicine
53 East 124th Street
New York, N.Y. 10035

Deborah M. Berkowitz
Yale University
Graduate School of Nursing
New Haven, Conn. 06520

Although improvements have been made in human knee prostheses, we are still plagued by a significant incidence of failure under normal conditions. These are usually attributed to material failures. While implant prostheses are not as successful as knee joints for amputation prosthetics, it is notable that both types of knee-joint replacements attempt to duplicate the human knee.

The knee is the largest and most exposed to observation of human joints. But it is not possible to define a "normal" knee. A wide variety of conformations are considered to be within the bounds of normality. Are all such joints "normal" in terms of mechanics, liability to injury, or wear? It is important to recognize that the gross anatomical form of the joint may determine not only vulnerability to injury but also the nature of the lesions incurred.

The knee is a movable hinge joint serving to alter the length of the lower extremity and lessen the muscle forces required to position the limb in space. Anatomically, it is seen that the hamstrings try to pull and bow the tibia backwards on the femur, and to bow the femur backwards. The two gastrocnemius attachments also tend to bow the femur posteriorly. The infrapatellar tendon, because of the pull of the quadriceps muscle on it, exerts a forward pull on the tibia which balances out the posterior force of the hamstrings. The quadriceps muscle also tends to bow the femur anteriorly by virtue of its broad attachment on this bone. This cancels most of the posterior bowing tendency of the gastrocnemius. Consequently these forces tend to balance out in any position of the joint, provided they operate at the same time.

The action of the quadriceps in producing extension of the knee joint is the basic factor in maintaining the erect position in man. The function of the patella is to improve leverage and thus to increase the power of the muscle. It is significant that this bone reaches the maximum development in modern man. These features are characteristic of man and a necessity for the erect posture. To maintain knee extension for the erect posture new muscles have not evolved; orthograde functions have been superimposed on muscles which in other mammals perform platigrade functions.

The action of the quadriceps, although similarly located anatomically, is vastly different in lower mammals than in man. In the platypus, for example, it responds to the test of extension of the flexed knee, but that is a different function from a quadriceps which supports orthograde posture. The quadriceps of the ape has a more complex function than that of the platypus, and so has that of man in comparison with the anthropoid. Even the anthropoids dispense with the support of the fore limbs reluctantly. The ape prefers the knees slightly flexed, with the aid of balance afforded by the fore limbs.

The bracing of the knee joints by the quadriceps which has enabled man to stand, walk, and run, and the ability to dispense with the fore limbs for support, is a late acquisition in evolution. It is unstable. It is little wonder that this recently acquired biological mechanism is easily deranged by an injury to the knee joint of a degree of severity sufficient to curtail normal activity. The loss of tone, volume, and control thus implied does itself constitute a disability.

The menisci of the human knee are relatively larger than those of lower vertebrates. The menisci which most clearly resemble the human are found in the anthropoids, but even in the chimpanzee, an animal capable of assuming a semi-erect posture and whose structure bears remarkable similarity to the human, these structures are comparatively narrow. It thus seems probable that the relatively large human meniscus is closely associated with the biologically recent acquisition of weight-bearing on the fully extended knee. This association with an attribute, common only to man (of recent evolutionary origin), may render the menisci structurally unstable and thus determine their extreme vulnerability to injury. The recovery of function, like the loss from disease or injury, follows an ancestral or evolutionary sequence.

The important joint properties are smooth contour and distribution of flexibility; a graduated flexibilty, maximal at the joint surface, minimal at midshaft; a mechanism which repairs fatigue and fatigue-like damage in the backing material; chemical turnover of a non-Newtonian lubricant, which is confined in a closed cavity and pumped by several means into the bearing area. These are altered by the surgical implantation of the prosthetic knee.

Kinematic analysis of the human knee joint shows that the instant center (of rotation) does not lie on the contact point, hence a relative velocity is seen to exist between the two contact surfaces, the femur and tibia. These surfaces are said to be in sliding contact. The kinematic analysis also indicates that excessive joint wear and trauma may occur because of the irregular kniematics induced into the knee by internal derangement.

It has been well documented that normal motion involves a moving instant center of rotation. Therefore, it is unreasonable to expect that normal motion may be created by the use of a single-hinge knee. A knee which produces a moving instant center of rotation is necessary for normal motion. The science of kinematic synthesis allows the production of such mechanisms. There are several classes of mechanisms which can be constructed to produce the desired motion of the knee.

The first and probably the simplest of these is the mechanism produced by duplicating the shape of the polodes curves on a solid mounting and then mounting them on their respective bones. A constraint which permits the two surfaces to roll overeach other can then be fabricated. In this manner, desired motion of the two surfaces in the knee joint may be produced.

The second method consists of the design of a four-bar mechanism, which is a series of four links and four pivot points. This can be seen in Figure 1. Links 1 and 3 are fastened to the tibia and femur respectively. Links 2 and 4 produce the desired relative motion between links 1 and 3. The four-bar mechanism is the simplest mechanism which has a determined instant center pathway. The instant center for the position shown in Figure 1 is at E. Thus four pivot points comprise the simplest mechanism which may be produced beyond the concept of a single-hinge mechanism. Link 1 is considered to be the fixed link. The velocities of points A and B must be perpendicular to links 2 and 4 respectively, because the fixed pivot points of these links are C and D. A and B are also points on link 3. Consider the velocity of point E on link 3. The orthogonal components of the velocities of each point may be obtained. Based on the principle of the inextensibility of a rigid body, the component of the velocity of E along the direction EC is equal to the component of the velocity of A along AC. Similarly, the velocity of E in ED can be determined. Drop a perpendicular from the head of the velocity of A to the line representing direction AC to determine the component of A in AC. Since this line is coincident with the velocity of A, A has no component of velocity in AC. Similarly, B has no component of velocity in BD. Therefore, E has no component of velocity in either AC or ED. The only velocity which can satisfy the condition of having the two zero components in the indicated directions is zero velocity. Hence, C is the instant center of rotation (of link 3 with respect to link 1). This procedure is repeated in Figure 2. Links 2 and 4 are extended and a new instant center, E', is shown. This process is repeated for a sequence of positions of link 3. All of the instant centers thus obtained are connected to form a polodes curve. Note that this path of the instant center closely approximates the path of the instant center of the "normal" knee.

To summarize, knee prostheses attempt to duplicate the human knee. The susceptibility of the delicate quadriceps mechanism and menisci to injury has been explained on a developmental basis. The fatigue failure in orthopedic practice is due to both large static forces and large cyclic tension and compression stresses induced by the tendency of powerful muscles to move the knee. The structurally unstable knee is due to "evolutionary lag" resulting from the poor correlation between the anatomic design characteristics and the functional parameters. Material failures are the result of microstresses set up by the wearing and friction caused by insufficient lubrication. This is the direct result of alteration of the joint properties during surgical implantation of the prothetic knee. The prosthetic knee joint implant does not accurately duplicate the sliding contact between surfaces of the human knee. Basically, knee joint protheses fail for the same reasons and as a result of the same stresses that necessitated replacement of the natural knee. Future designs must be based upon the data obtained from more precise kinematic analyses of the knee.

References

1. American Academy of Orthopedic Surgeons: Orthopedic Appliances Atlas (Vol. 2). J. W. Edwards, Ann Arbor, Michigan, 1960.
2. Frankel, V.H., and Burstein, A.H.: Orthopedic Biomechanics. Lea & Febiger, Philadelphia, Pennsylvania, 1970.
3. Frost, H.M.: Introduction to Biomechanics. Charles C. Thomas, Springfield, Illinois, 1967.
4. Klopsteg, P.E., Wilson, P.D., et al: Human Limbs and Their Substitutes. McGraw-Hill Book Co., New York, 1954.
5. Smillie, I.S.: Injuries of the Knee Joint. E & S Livingstone, Edinburgh and London, 1970.

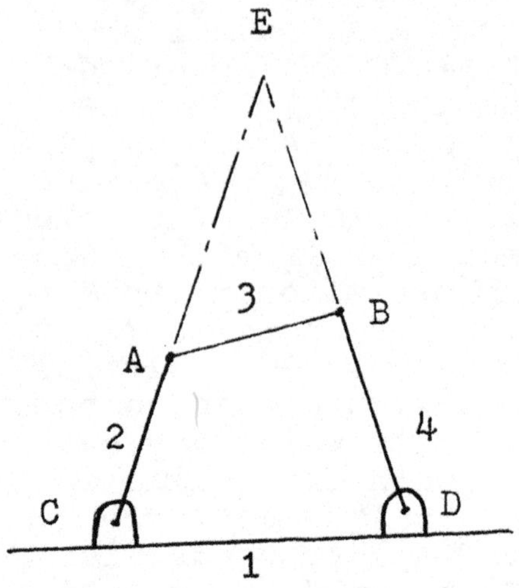

Fig. 1

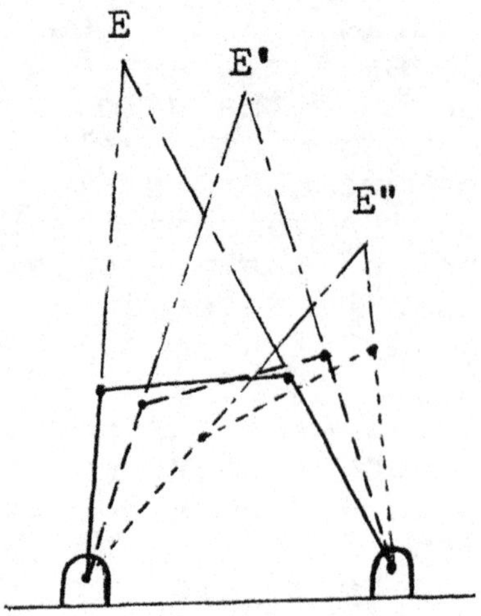

Fig. 2

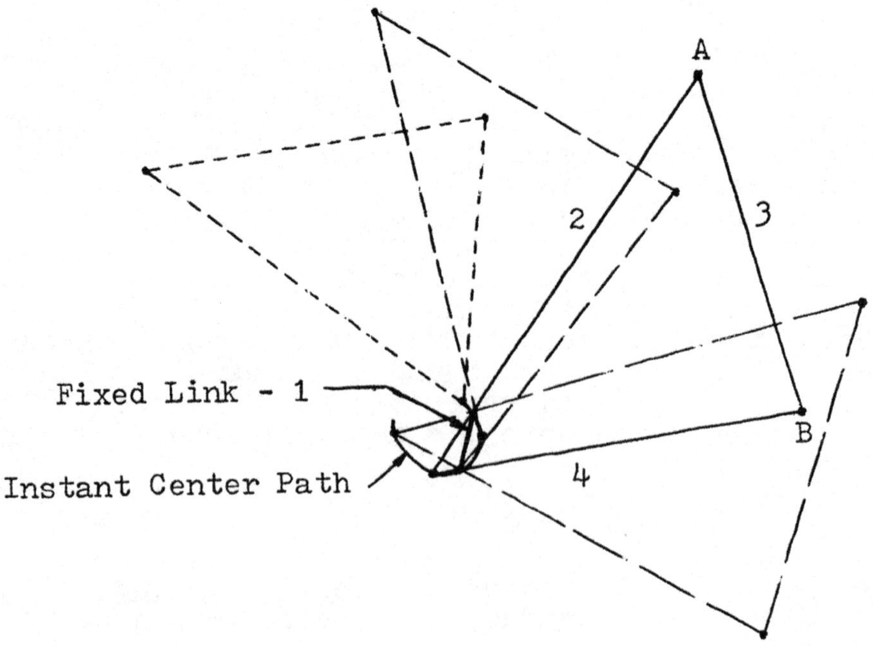

Fig. 3

COMPUTER APPLICATIONS IN BIOMEDICAL ENGINEERING

Donald A. Molony, Byoung Min, and Walter Welkowitz
Rutgers University, New Brunswick, NJ

A computer course intended to serve the needs of graduate students in biomedical engineering should be designed to fulfill several functions. It should provide every student with a reasonably complete understanding of the functioning of modern computer systems. In addition, the course should include sufficient information as to the characteristics and capabilities of generally available and special purpose peripheral devices. An introduction to programming in both low level and high level languages is essential. Finally, the use of the machines studied in a variety of practical applications of interest to bioengineering specialists should be analyzed in considerable detail. At all times the material presented should be designed to emphasize the practical aspects of the systems involved. To accomplish these goals the course at Rutgers is divided into two approximately equal parts.

The first portion of the course begins with an introduction to the basic organization and structure of existing digital machines with particular attention to minicomputers. Machine instructions are considered and elementary machine language programming is studied. The transfer of information to and from the machine is then investigated. The characteristics of devices such as analog to digital and digital to analog converters are considered in detail. Other devices for input and output of particular interest to biomedical engineers are discussed. The techniques involved in programming input and output transfers are introduced. After the students have acquired sufficient knowledge of the functioning of the machine, higher level languages are studied in reasonable detail. At the present time the Basic language proves to be of greatest utility considering the particular computer systems available at Rutgers University.

Throughout this portion of the course the student is expected to spend a considerable amount of time in contact with the available minicomputers. Laboratory practice is considered to be essential.

It should be noted that the diversity of backgrounds found in a typical class is so broad that individual instruction may be required to aid the students who have had little or no previous contact with computing machines. Students with extensive previous work in the computer field may be encouraged to undertake individual study in depth.

Following the period devoted to the digital machine the student is introduced to analog and hybrid techniques with particular emphasis on the use of such systems in modelling of biological systems.

As has been stated, the first half of the course is intended to introduce the student to computer systems in general. The second half of the course is devoted to a study of presently operating computer systems in biomedical engineering applications. Currently the areas being covered include:

1. Patient monitoring systems

2. Research laboratory systems
   (experimental and clinical)

3. Image processing systems

4. Clinical diagnosis systems
   (i.e. E.C.G. evaluation, etc.)

5. Patient data management systems

6. Clinical-chemistry laboratory systems

7. Computer simulations and modeling in biomedical research.

Where possible, reference material from available journals is provided to the students.

In each case, the material is covered by the use of a combination of lecture techniques with individual study. The instructor prepares and delivers a comprehensive lecture on each subject. Each student is required to select one subject that he or she finds to be of interest for further independent investigation. This independent study may take the form of a review of the available literature and the preparation of a paper on the subject in question. Alternatively the student may select an area in which computer related activity is currently taking place as part of one of the projects being conducted within the department. In this case, the student can participate in the work in conjunction with the research assistants involved.

It is the intent of the instructor in presenting the lecture material to be as complete as possible within the constraints imposed by the time available. For example, in discussing patient monitoring installations two representative systems are considered, Osborn's system for respiratory function monitoring (1) and Kirklin's system for cardiovascular function monitoring (2).

In the discussion of these systems the following subjects are introduced:

1. Function of the systems

2. The data acquisition methods

   a. physiological monitors
   b. signal transmission

3. Software

   a. the multiprogram executive
   b. disk layout and core area utilization

4. Real-time operation

   a. interrupt controlled program
   b. timer controller program

5. Application programs

   a. analysis programs
   b. alarm programs.

The discussion of the clinical research laboratory system is based on the system of N.I.H. cardiovascular surgical suite (3). The following subjects are discussed.

1. System description, operation and specifications

2. Signal calibration and information content (E.C.G., pressures, flows, power spectra)

3. Analog to digital conversion set-up.

Then, in the second class, the emphasis is on the design and implementation of analog-to-digital conversion for the physiological signal monitoring. Also, the A/D system based on the LPS-11 in our laboratory is demonstrated for digitization of the hemodynamic data and computation of the physiological variables. In this class, the following subjects are covered:

1. Sampling

2. Quantization

3. Analog-digital conversion

4. Choice of sampling and quantizing equipment

5. Multiplexing and sample and hold.

In the image processing system, the discussion has been concentrated on the Mayo Clinic's Scintiscanning systems (4,5).

Concurrently with the above lecture series, the students have an informal meeting with the instructor for the discussion of their progress in their assignments. The following topics have been used for these student projects.

1. Micro-processor based pacemaker monitoring and surveillance system

2. Mini-computer based arrhythmia detection and analysis system

3. Micro-processor based pulmonary function monitoring system

4. Mini-computer assisted intra-aortic balloon pumping

5. Biological signal processing of E.E.G., E.C.G., and E.M.G. signals.

References

1. Osborn, J. J., Beaumont, J. O., Raison, J. C., Russel, J., Gerbode, F., Measurement and Monitoring of Acutely Ill Patients by Digital Computer, Surgery, 64:1057-1070, 1968.

2. Sheppard, L. C., Kouchoukos, V. T., Kirklin, J. W., The Digital Computer in Surgical Intensive Care Automation, Computer, July, 1973, pp. 29-34.

3. Gilbert, D. B., Digital Analysis of Hemodynamic Data, published by H.E.W.-N.I.H., Division of Computer Research and Technology, 1968.

4. Coulam, C. M., Dunnette, W. H., Wood, E. H., A Computer-Controlled Scintiscanning System and Associated Computer Graphic Techniques for Study of Regional Distribution of Blood Flow, Computers and Biomedical Res. 3:249-273, 1970.

5. Wood, E. H., Cardiovascular and Pulmonary Dynamics by Quantitative Imaging, Circulation Res. 38(3); 131-139, 1976.

COMPUTER-ASSISTED INSTRUCTION: A VIEW FROM THE HEALTH EDUCATOR'S PERSPECTIVE

Robert G. Votaw, Ph.D.
University of Connecticut School of Medicine

I. Introduction

We are some thirty years into the computer era. In the past six to eight years the phenomenon of computer-assisted instruction (CAI) has emerged from the obscurity of a very few research centers. This instructional use of computing is in the early stages of dissemination and development. For example, in the area of health professions education, there has been rapid initial growth in the use of computer-based education materials (CBEM). The rate at which required CBEM have been incorporated into the medical school curricula of the country is seen in Figure 1.

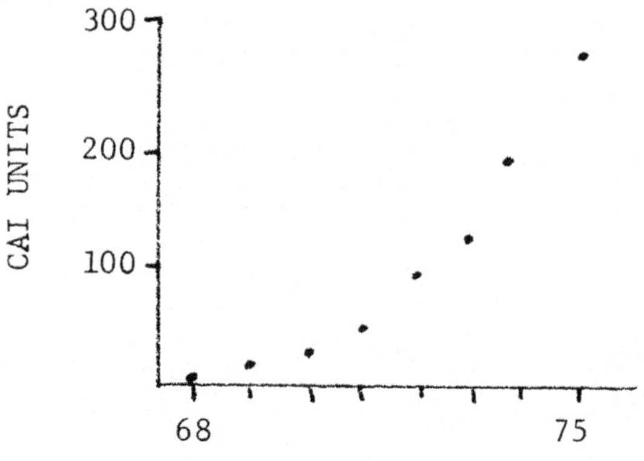

Figure 1

In 1968-69 there were essentially no units of computer-based educational materials (CBEM) required. By 1975, about three hundred units are embedded and the rate of growth is logarithmic.[1] Many additional CBEM units are used on an optional basis in these curricula. Whether or not this rate of growth continues, only time and additional analysis will tell.

In my experience, the majority of faculty and students who come into contact with computer-assisted instruction (CAI) acknowledge their perception of great potential for the medium to enhance education. Yet in any given institution, the extent of incorporation of CBEM into the curriculum is small and the proportion of faculty involved with instructional computing is small. It is my impression that a similar situation exists in engineering education.

Why is there apparently so little use of instructional computing in engineering education? With few exceptions, engineers are more oriented to systematic problem solving, mathematics, modeling, simulation and logical systems approaches than other professions. It seems that such a background should translate into ready acceptance of instructional computing and should result in faster growth in the use of CBEM in engineering education than has apparently occurred. This situation seems paradoxical.

What is rate limiting? Is it money, fear of technology, inadequate quantities of proven courseware? Have there been negative experiences with CAI? Or is it lack of interest, lack of incentives for authors and students, or lack of knowledge of what CAI is and what it can do? How can this apparent paradox be explained? What is inhibiting?

Probably, the it must be answer (e), "all of the above". The purpose of this paper is to stimulate educators to consider adopting a posture of active experimentation to find out if and how use of CAI can enhance their current educational practice.

What CAI Is (And Is Not)

In our society there is much mythology about computers, computing, and computer people. For example, we are all familiar with the layman who thinks a computer is something that answers questions posed to it on any subject. Perhaps this tradition derives from early anthropomorphic descriptions of computers as giant, electronic brains. In any event, as educators we must be careful to avoid entrapment in the pervasive myths. This is not an easy task, in part because the field is new and the early workers are prisoners of their own (limited) experience. Also, the field is characterized by very rapid change. There is, for example, a growing realization that terminal technology, i.e., graphic vs. teletype, determines two quite distinct media. Commentators whose experience does not include graphics will see a different and more limited potential. Hence, in the early adolescent literature of this field, CAI is often conceptualized narrowly and is described for others in terms unnecessarily limiting. In order to gain additional insight into what instructional computing is and what it can and cannot do, let us lay down some arbitrary dimensions in a conceptual matrix and advance along them. And in the process, let us do what we can to demythologize CAI.

It should not be surprising that computer-assisted instruction (CAI) is variously termed and defined by workers in the field. Since we are attempting insight development, we must attend to defining the bounds of what we are talking about. I use the term CAI generically to describe a situation in which students are engaged by or with computer programs in interactive instructive dialogues. This definition is very broad. A wide range of education strategies is possible. What is emphasized is an interactive process in which control of timing and sequence of lesson flow may be shared between student and program or it may be polar, i.e., reside with either.

Two common synonyms for CAI are CAL (computer-aided learning, often used in Canada, the United Kingdom and Europe) and CBE (computer-based education). Also, an instructive computer program is commonly and interchangably called a lesson. Of course, in CAI, computers and computing are means to achieve educational goals, they are not the subject of inquiry *per se* (except in computer science CAI). This is instruction with, not about, computing.

There are some very common misconceptions about CAI which probably arise from the tendency to imitation in the early use of the medium. Hence, the unfortunate legacy of programmed instruction, "It's just programmed instruction on a fancy machine." Other early efforts to re-cast books in computer code resulted in much "electronic page turning," hence, "CAI is only an efficient way of transferring information; it has no value in developing intellectual skills and critical thinking." Then there is the problem of realism in dialogues, "besides, you have to anticipate every possible student answer and program it in advance."

On the contrary, state-of-the-art CAI languages have powerful capabilities to judge student answers by concept matching in addition to the ability to check spelling and syntax. After a few rounds of trial use, a pool of real answers can be examined and the program modified to support a universe of predicted answers. Since that universe is effectively finite, a carefully tested program can respond quite idiomatically to answers or questions entered in the student's own words. This level of fluency is not always needed or even desirable but it is possible. Numeric answers can also be handled in a variety of ways.

Perhaps most important is the idea that CAI is more than automated programmed instruction. Although words cannot fully describe the nature of CAI instructional approaches and the student-computer interaction what follows is an attempt to illustrate the range of instructional approaches possible. In the following descriptions of CAI modes or instructional strategies, note the changes in the locus of control, i.e., with the student or the program.

II. CAI Instructional Modes[2]

Problem Solving

Students on their own volition use computing as a tool to solve problems which may or may not be assigned by the instructor. Often they must write programs. This means solving the problem in the general case first. Hence, the student must conceptualize the domain broadly. Such a learning experience may be more effective in developing insight than solving sequentially selected problems assigned by the instructor. Problems of greater difficulty may be assigned and the entire course may be operated at a higher level. Students need some prior familiarity with computers and programming. Computing probably adds to the cost of the course, but significant value is added. In addition to learning with computing, students learn something about computing as an everyday tool for their post graduation endeavors.

Inquiry

In this mode, the problem space is somewhat more structured. Here the computer program incorporates a data base which the student can manipulate to solve problems. Selected problem solving strategies may be emphasized by design or de facto. The learner controls the instructional sequence but may not have full freedom of choice in strategies. Often the phrase, "What do you want to do next?" identifies an inquiry lesson.

Example: COMPUTER: "This is a lesson in genetics. What do you want to do--look at the family history, obtain laboratory data or make a diagnosis?"

STUDENT: "Inspect grandparents' phenotype."

COMPUTER: "OK."

Socratic Dialogue

This is a more balanced dialogue between learner and computer; both control. The patient-physician encounter is a good example of Socratic dialogue.

Example: STUDENT: "OK, Mr. Bond, what brings you to office this morning?"

COMPUTER: "Well, Doc, I've been having this pain. . . ."

Simulation

A dynamic, "real world" environment is modeled. The learner explores the simulated universe interactively. With sophisticated techniques, very realistic simulations result. For example, actual time elapsed can be monitored to heighten realism. Control may be balanced or polar.

Example: COMPUTER: "We are going to simulate a patient resuscitation. Depending on what you do, his status may change. The next six minues are critical. Do you wish to be timed (yes/no)?"

Gaming

Gaming simply involves addition of the competition dimension. The student may compete against self (prior performance), the computer, or another student. Usually, achievement of a specific end point is sought in contrast with simulations. There are two distinct types of gaming, for fun and not for fun. Both may result in effective education. Gaming is situation dependent. To a student of physics, "Lunar Landing" is a learning exercise, to a student of English who cannot compute the velocity-time perameters, it is a "fun" game. Fun games, however, are most effective in overcoming the apprehension many newcomers feel when approaching the terminal and the pleasure/fun aspect in a competitive quiz on pharmacologic agents may be a powerful stimulant to learning.

Example: COMPUTER: "In this game of chess, the computer has been programmed to play poorly. . . ."

These five categories clearly can be used to develop intellectual skills, critical thinking and skill in problem solving. Often, computing approaches provide learning experiences possible in no other medium. We are in early infancy with respect to our knowledge of how to use effectively the unique strategies.

Tutorial

Here the computer controls lesson flow. In simplest form, a linear sequence of factual statements may be presented; student performance on questions designed to test comprehension can be monitored and feedback on answers given. In more advanced lessons, branching to enrichment material or remediation is possible. Best case exemplars are well received by students and seem to be valuable aids to learning. But all too frequently, tutorials are nothing more than electronic page turning and are not worth computer presentation. At worst, the mediocre, unimaginative tutorial is even less interactive than a good, programmed instruction text. The moral of this paragraph is that there is good tutorial and bad and it is useful to be able to tell the difference. The simplest guage probably is whether your students like it.

Example: COMPUTER: "This is a tutorial on the central nervous system for first year medical students. After presentation of some basic facts, you will be asked questions to test your understanding before proceeding to the next section. . . ."

Drill and Practice

Of course this mode is available. A computer can add power to this technique by constructing individualized lessons and review sequences. Problem generation can provide a unique file of problems for each individual. One should not overlook the special ability of computing to motivate some learners.

Example: COMPUTER: "A review of medical terminology: you will be presented blocks of ten words drawn at random from a 5000 term pool. . .Words you do not know will be stored in your own review file. . . ."

It should be pointed out that many lessons are multimode programs. The strategies that can develop critical thinking often are found in combination. For example, a kidney simulation can be "wrapped up" in a Socratic dialogue concerning the patient with kidney disease. In this case, we see the strategies as different dimensions, attributes or organizing, principles, the elucidation of which conveys insight into the style, scope and nature of the lesson.

Given the idea that increasing productivity is a high priority for all of higher education, especially in the face of declining support and enrollments, how does CAI relate? CAI is much more than programmed instruction and a means of transferring knowledge. Use of CAI can uniquely enhance current educational practice by providing powerful new kinds of learning experiences. Other applications are labor saving, a second purpose for which we use technology. Still other uses involve extending scarce assets, i.e. our faculties. For example, clinical lecture courses that are repeated six times a year can be converted to independent study format with a large CAI component, thus releasing faculty time for work with students having special difficulty or for work with students in the clinic.

III. The Heuristic Element

Introduction of a specific computing application increases the potential for discovering unforseen applications of CAI.

Computer Managed Instruction (CMI)

This term describes a computer-based approach to managing an instruction system. The student spends less than half the total curricular time at the terminal. Computer-based self-assessment tests which provide feedback to student and course manager may be taken. Students may report on progress through self-paced instruction materials, and the course manager can determine the state of an education system that is distributed, either in time or space. In contradistinction, we should define a second, more specific sense of the word CAI in which the term refers to curriculum units in which the student spends more than half the time at the terminal. This is a convenient place to define TAI (traditionally administered instruction), the classic lecture or lab.

Communications

New means of human communication--the electronic conference and person-to-person note files--often unexpectedly accompany a CAI system. For example, a CAI system with over a thousand terminals provides a multifaceted communications facility. It is possible to talk through the terminals (by typed messages) with any given individual on the system. In this system are also a number of notes or mailbox facilities which can take on the aspect of a private conference call or a public debate conducted through time for days, weeks, or months. Tens and hundreds of individuals participate in the point/counterpoint auguments which are posted in this electronic analogue to the village bulletin board. Until one has participated in an electronic or computer conference, it is very difficult to appreciate the power of this medium for increasing the ability of humans to relate. One can think, for example of a regional computer conference network in which solo practicing family physicians in rural areas are interconnected and through which they can conduct case rounds to overcome the isolation and distance factors experienced in rural practice. One can see a spectrum of interactive instructional activities ranging from such case rounds to special curricula in biomedical frontier topic areas such as current concepts in cancer chemotherapy or current concepts in burn care.

Research In Education

Because of the individualized, interactive quality of this medium, and because of the information processing capabilities of the computer, it is possible to learn in more detail the natural history of the learning process. The potential for research on learning is enormous and should not be overlooked in assessing the impact of computing on the quality of education. Perhaps for the first time, we may gain additional insight into the natural history of learning that may cause major breakthroughs.

Generative CAI

In principle and in practice, CAI lessons can be programmed to take into account the unique prior learning history and background of each student and generate unique sequences of learning experiences based on that history. This would be an especial advantage where students of widely differing ability are present in the same class. Properly programmed, in effect, this extends the ability of the instructor to provide one-on-one counseling, remediation, acceleration, and enrichment.

Service Modules

With graphics, one can think of other uses that might be of benefit although perhaps not directly considered instruction. For example, in an academic environment, a graphing capability can be used to prepare charts and graphs for publication. It's much easier to let the computer organize the data, calculate the statistics, plot the curve and do the lettering than do it yourself manually. One could also envision programs designed to help health professions students about to go into practice design the architecture of their offices, using automated design routines.

Thus, again we can observe that CAI is not a single, discrete, describable entity. CAI is many things and it is no wonder that CAI is variously described by people within the field. CAI that is useful to one person may not be useful or cost effective to another. Without belaboring the point, again one can see that the need for CAI is situationally determined.

IV. What Is Possible: Some Practical Considerations

To use CAI, one needs:

hardware	-	the computer and terminals
software	-	including a CAI language
courseware	-	lessons
peopleware	-	faculty and staff willing to explore the use of CAI
money	-	for some or all the above and possibly for lesson acquisition and/or development, faculty development, evaluative studies, data communication

Hardware and Software

CAI has been, and can be, done with just about every conceivable computer (micro, mini, midi and maxi), terminal (teletype, graphics or graphics multimedia) and language (machine, Fortran, Basic, Pilot, APL, Coursewriter, MUMPS, TUTOR,etc.) you could imagine. Some combination of the above represent low level CAI technology, others high level. Cost may not relate directly to high and low level. Some systems are dedicated to instruction, others are shared. All things considered, if I had a choice, I would pick a system with a graphics-multimedia terminal, a powerful, easily learned language and a fast response time. If I can get speed, power and flexibility, and as well some potential advantage of size such as a library of lessons, access to consultants and communications facilities, and freedom from running the system, so much the better, especially if the cost is reasonable. Two different approaches which emphasize graphics come to mind.

For example, the approach of the Computer Physics Development Project at the University of California, Irvine[3] meets many of these criteria. They use a tektronix graphics terminal and a modified APL language on a Sigma 7 computer. The system is flexible and a number of effective lessons in different modes have been developed and are in use.

We have had three years experience with the PLATO system developed at the University of Illinois, Urbana. PLATO is now an announced product of Control Data Corporation but all our experiences to date is with Illinois PLATO.

This system uses a plasma panel, multimedia graphics terminal, large computers and is designed and used exclusively for instruction. Response time with five hundred active terminals is still 0.1 to 0.2 second. The 10" x 10" plasma panel has 250,000 individually addressable dots and the TUTOR language provides broad capability to write lines and alphanumerics in any size, density, or angle desired. Complete special character design facilities and animation increase flexibility and apparent instructional power. Input is by keyboard or finger touch. Random access color microfiche and audio (playback and record) round out terminal features.

Access to PLATO also means access to several thousand lessons in over fifty university disciplines, access to consultants and access to communications capabilities. And you do not need to buy, lease or run a computer! All courseware is written or revised at the terminals.

Control Data PLATO is just emerging from the corporate workshop and it is too soon to tell if there are significant differences. In New England, Control Data and the New England Regional Computing Program (NERComP) are seeking cooperative development of PLATO and the NERComP network. This bears careful watching because, in the forseeable future, this system may offer for many the best combination of features, if the price is right.

Courseware

Good courseware is scarce. In no one discipline of which I am aware does enough exist to cover the field, let alone in a form universally accessible. Unless lessons are on a big system like PLATO (a "star" network) or a multihost network like the Health Education Network (Ohio State College of Medicine and Massachusetts General Hospital) lessons must be transferred to your own local or regional computer and that is not necessarily so easy or desirable. Yet no one institution has the dollar and people resources to produce all the courseware needed so sharing lessons is mandatory. And so it is in the book area. No one institution writes all its texts. Authors prepare materials and publishers make them shareable.

One hears that the not-made-here syndrome prevents sharing CBEM. But experience shows that some CBEM are acceptable in many institutions. This is true in health and apparently in chemical engineering. Some forty schools share the FLOWTRAN (flowsheet translator) process design simulation from Monsanto during this academic year.[4]

FLOWTRAN allows the student to model an approach and calculate material balances, cost, profit, and return on investment. So sharing courseware is possible.

The many facets of CAI described in Parts II and III suggest a need for the ability to modify programs produced elsewhere and to produce some courseware to respond to local need. This need may include insight development for faculty who reached intellectual maturity before the advent of universal computer literacy. The large system which has these options may have advantages at this stage of the development of CAI. In ten years, standard, "published" courses of proven effectiveness and transferability will be available. Then a playback only system may suffice. Look for the intelligent stand alone terminal--a mini or a micro--combined with videodisc technology to change current educational practice profoundly.

In regard to the participation of faculty in lesson development, it must be recognized that conventional author incentives are minimal. There is no generally accepted analogue to the peer review of journal articles. No general publishing mechanism is yet in place. Hence, royalties and academic credits toward promotion and tenure are often not forthcoming. I believe the research necessary to develop this new CAI field will require as much rigor and creative scholarship as any other field. We must demonstrate to our colleagues that this is the case in order to invest workers in this field with appropriate credit.

Costs

This is also a complex topic. Presently, we are using three different system, Illinois PLATO, the Health Education Network and our own Univac systems. Direct computer and communications costs fall in the range of $2-$10 per hour of student instruction at a terminal. Lesson development costs are not included.

With regard to cost benefit analysis, a caveat is in order. Although we often use the student terminal hour as a unit of measure, there is no justification for comparing the hourly cost of CAI and TAI unless the outcomes of student learning are known.

Having said that, let us observe that for an entry level class of one hundred students, one hundred hours of CAI at $5 per hour ($500) is not cost effective compared to one hour of lecture at $20. For a medium level class of ten students, ten hours of CAI at $5 ($50) may look much more attractive. If the faculty must teach the course repeatedly, or wishes to offer more than one medium level course simultaneously, this simpleminded cost benefit analysis looks much more favorable. If one reckons with the value added because of unique CAI features, e.g., FLOWTRAN, the analysis is highly favorable. Some CAI is certainly cost effective now. It takes two to five years to become skilled in using CAI.

In five years computing costs will be much less.

Peopleware and Where to Begin

One need not be a computer expert to use CAI. To a large degree, others can help with the technology. It is much more important to be an experienced teacher and informed concerning instructional computing. It is helpful to have some easily obtained familiarity with a CAI language. This can be acquired in a few hours by most people of high school age or above. It must be recognized that no universal set of rules has yet been worked out concerning the effective use of computing in instructional environments. Applications are largely the result of local initiatives by individual faculty or small groups. Teaching style and educational need, i.e., the situation, determine how and if computing is used. The approach is also conditioned by available hardware and software and may be importantly related to academic discipline. In each discipline, there are examples of creative application of one or more CAI approaches to the instructional environment. Patterns of useful approaches are beginning to develop. One can examine the experience of others to generate ideas. Now is a good time to consider the use of CAI. As a first step, the object is to become familiar, in principle, with what computing can and cannot do in order to identify potential computing application in your own instructional situation.

REFERENCES

1a. Lefever, Dale and Jenny K. Johnson. AAMC Survey on Use of Computers in Instruction, 1976.

b. Tidball, C.S. George Washington University Medical Center, Personal Communication.

2. We have observed or used examples of all modes. However, two interesting discussions are found in:

a. Mosmann, Charles. Evaluating Instructional Computing, University of California, Irvine, 1976.

b. Levien, R., et al. The Emerging Technology: Uses of the Computer in Higher Education, New York: McGraw-Hill, 1972.

3a. Bork, Alfred. "Learning via Computer Graphics," Amer. J. Phys., 43:81-88, 1975.

b. Bork, Alfred. NATO Advanced Study Institute, Louvain-LaNeuve, Belgium, July, 1976.

4. Seider, Warren. Dept. of Chemical Engineering, University of Pennsylvania, Personal Communication.

SOME SOURCES FOR MORE INFORMATION

1. The Association for the Development of Computer-based Instructional Systems (ADCIS), c/o Joan Hayes, Secretary/Treasurer, Western Washington State College, Bond Hall, Bellingham, Washington 98225 (206-676-3360).

2. EDUCOM, Interuniversity Communications Council, Inc., P.O. Box 364, Rosedale Road, Princeton, NJ 08540 (609-921-7575).

3. New England Regional Computing Program (NERComP), c/o New England Board of Higher Education, 40 Grove Street, Wellesley, Mass. 02181.

4. SIGCUE (Special Interest Group Computer Uses in Education) through Association of Computing Machinery, 1133 Avenue of the Americas, New York, NY 10036.

5. CONDUIT, Central Office, P.O. Box 388, Iowa City, Iowa 52240.

Computer Applications In Patient Management - Its Impact Upon Medical Technologists

Joseph D. Bronzino, Ph.D.
The Vernon Roosa Professor of Applied Science
Director of Biomedical Engineering Program
Trinity College - Hartford Graduate Center
Hartford, Connecticut

The increasing availability of automated clinical laboratories and health testing centers and the resultant services provided to the physician by these facilities has stimulated the development of computer systems designed to provide permanent patient records. There is no question that there exists both a medical and legal need for permanent records of laboratory work and the health profiles provided at these testing centers. These records must be physically incorporated into the patient's chart. In the past, the need has been met either by inserting laboratory slips into the chart or by transcribing the values to a cumulative record page.

In efforts to replace this approach with a printed record which is clear, complete, accurate and inexpensive, computer-generated permanent record systems have been designed and incorporated within the clinical environment. Through the use of these new patient-management type systems medical and nursing staff have instant access to vast bodies of information and to all other hospital departments (Laboratory, Radiology, Admitting, Pharmacy, Food Service, etc.). In these institutions health professionals no longer have to rely totally on their own memory capability, and have all the patient data presented at a time and in a form to facilitate the decision making process. Such systems make it possible for the medical staff to be in continuous control of the patient's status from pre-admit to discharge. Thus there is general agreement that a computer-stored, integrated medical record is essential to the functioning of a data system for any large medical program. (Bronzino, 1977) In addition, concise records constitute teaching documents and instruments of learning, thereby allowing health professionals to be trained to deliver better medical care.

However, such a record and the system that maintains it must be so structured as to store all classes of patient-related data. It must be capable of continual updating over the lifetime of the patient, accepting the variable format of data input, responding to the need for real-time inquiry to individual patient records and it must be maintained by reliable error detection routines. Let us consider its impact upon hospital care if such a system were to be utilized.

Since about one-third of a hospital's per diem costs are for information handling, (Collen _et. al._, 1974) the greatest impact of this type of computer technology in the hospital will be on its medical information system. The establishment of such a hospital information computer system would be capable of providing the following functions:

1) Enable the health professionals providing medical care (doctors, nurses, professionals, etc.) to communicate individual patient data into the patient's own computer medical record. Once this record is established it could then be communicated to other professionals (e.g., dietitian) and hospital services (e.g., radiology) to inform them of the needs of the patient.

2) Permit patient data obtained from subsystem components (e.g., automated multiphasic screening laboratory or intensive care unit) to be inserted into the patient's computer medical record.

3) Provide a means by which clinical services (i.e., nursing stations) can communicate with auxiliary services (e.g., ECG or radiology) regarding a particular patient.

4) Establish scheduling files and communicate information for scheduling of patients, personnel and medical care services.

5) Establish a data base for administrative and business functions.

6) Establish a medical data base that can fulfill research and teaching objectives.

Such comprehensive computer hospital information systems are not yet operational anywhere in the world. However, in large medical centers some subsystem components, such as automated multiphasic health testing programs, patient monitoring systems, admission and bed census systems, outpatient scheduling systems, and pilot hospital information systems are in various stages of development.

In 1962, one of pioneering efforts to utilize available computer technology in a clinical environment began at Hartford's Institute of Living, a 400 bed, private, non profit psychiatric hospital. In the intervening years, (Glueck and Stroebel, 1969; Stroebel and Glueck, 1972; and Glueck _et. al_ 1974) an appropriate computer and terminal hardware system has been developed for the day-to-day clinical care of patients as well as furnishing support for the administrative and business operations of the institution. This system (presently consisting of two Digital Equipment Corporation PDP-15 computers capable of handling inputs from approximately 40 terminals located throughout the hospital) is available to hospital staff aroudn the clock on a reliable basis. This is due in part to the built-in redundancy of the system, since one computer completely updates the other every hour; thereby insuring that little information is lost if one computer developed problems.

This system has not only enabled the hospital to maintain a continuous file of the patients admitted, but has been invaluable in generating special reports concerning the patients within the hospital. For example, in an effort to evaluate the treatment of alcoholism at the Institute of Living, two of the psychiatrists recently requested a list of all patients in the hospital who had problems with misuse of alcohol. Since alcohol abuse may occur as a symptom in several psychiatric illnesses multiple diagnoses were given for the search. A general extractor program (written in MUMPS) was used to search the files for the specifically identified diagnoses and print a list of matches. The requested report was produced in less than 20 minutes.

One of the most useful applications of the system has been the conversion of the Nursing Notes to the automated system. By using a statistical analysis called Factor Analysis that groups and weighs the multiple statements of observed behavior available in the nursing notes, it is possible for this computerized system to help the hospital staff evaluate the patient's behavior. Behavior as seen by separate observers can be evaluated whereas this is virtually impossible using the traditional nursing procedures.

Using this statistical analysis approach, 13 categories, or factors are scored daily for both the day and evening indexes. These factors include the following: (1) Acceptable Behavior, (2) Inapropriate Appearance, (3) Social Isolation, (4) Sexual Behavior, (5) Eating Habits, (6) Resentment to Confinement (7) Antisocial Behavior, (8) Hostile-Aggressive Behavior, (9) Agitation, (10) Paranoid Behavior, (11) Depression, (12) Retardation and (13) Disorganization. These factors appear on the Nursing Behavior Report in discending order according to the amount of behavior displayed by the patient. Thus the information supplied on these reports provides nursing and medical personnel with a continuous of current information about patients with the institution.

Using these computerized nursing notes as the data source, this system can itemize unit and patient scores for any of these factors. As a result, it is possible to compare individual patients with the group and with day-to-day changes. Consider the following example, a patient who was usually outgoing and apparently cheerful received some unpleasant news. A computer-constructed graph of one of the factors, social isolation over a period of one week, demonstrated the sensitivity of this technique as well as the accuracy of the observations. The computer presentation clearly indicated that the patient exhibited behavioral patterns associated with "Social Isolation" on the day of the receipt of bad news. The patient's efforts to cover his feelings were reflected by a return to a more normal behavioral pattern on the day following his withdrawl.

The potential use of Automated Nursing Notes for research in basic behavior, for assisting the physician in making a more accurate diagnosis, and for evaluating the effect of treatment are, at the moment, being studied. However, this concept of quantifying various ranges of behavior in different psychiatric hospitals should improve when multiple, meaningful, different discrete descriptions of behavior can be shared by using this approach. (Stroebel and Glueck, 1970)

The use of the nursing observation index at the Institute of Living has resulted in a number of unexpected advantages. One is the educational value of the index. Experience has shown that by providing nursing personnel with guides to their observations, they have become better in their observation of patients. The sets of relatively unsophisticated statements enable the student nurse to record her notes in a superior manner. They have encouraged her to see and report areas of behavior which she might otherwise ignore.

One of the major medical concepts that has arisen because of the availability of this new type of record keeping system is the Problem Oriented Medical Record (POMR). One of the leaders in this new approach to patient care, Dr. Lawrence L. Weed of the University of Vermont Medical School emphasizes that the record should clearly identify: (1) what you are doing for the patient (2) how you are going to accomplish it and (3) some means of auditing the record and the accomplishments. (Weed, 1975)

This concept represents a departure from the source-oriented approach usually followed in obtaining patient information. The information initially collected in establishing the POMR is the "data base" and consists of the patient's history, a physical examination, certain laboratory tests and x-rays. When any member of the health care team decides what is wrong, a "problem list" is created. When he decides what to do about it - to develop a goal and a plan of action for management of the problems - he has reached the Plan. Finally, he carefully watches the results of the plan and in effect establishes "the progress notes". This technique is called "Problem Oriented" because each plan and progress note is identified (titled and numbered) with respect to a problem. As we have all learned from experience, new information is continually supplied to the physician as the patient reacts to a prescribed treatment. He changes the plans as needed and continues the process of evaluation and ajustment until the best possible outcome is reached. Thus each progress note has four parts: (1) subjective (e.g., the patient's reactions), (2) objective (i.e., physical findings such as laboratory data, x-rays, etc.), (3) assessment (any new interpretation by medical personnel and (4) plan (the revised plan for the problem).

With the establishment of the well documented and well organized charting system associated with POMR, the nurse's work is also facilitated. The computer can relieve the nurse from the physical task of writing requisitions, making medicine cards, etc. In addition, it eliminates the resulting opportunities for error and in the process correlates all the information in a problem-oriented way. Every diet, every medication, every laboratory test must be ordered or done for a specific patient problem (Gane, 1975).

Of all of the members of the health care team, nurses enjoy the longest sustained contact with the patient. Consequently, they are truly the essential factor for making the POMR a success or a failure and for coordinating the system. With the increasing legal and moral pressures placed upon nursing professionals today, POMR can be of great assistance in allowing these nurses to document their activities and be held accountable for them.

Since it provides a logical framework for teaching problem-solving logic, POMR also serves as a teaching document (Berni and Readey, 1974). Many educators today agree that this approach has created an effective means for providing continuing education for members of the health profession. By requiring all indisciplinary personnel to communicate with one another, POMR provides an opportunity for these individuals to enhance their own medical knowledge. By identifying problems after a careful collection of data, devising plans to control or eradicate these problems, POMR serves as a teaching document and learning tool. It enables the health professionals involved to evaluate the success of the approach taken todate, stimulates thinking in an organized orderly fashion, and may lead directly to action beneficial to the patient. Such a process could assist nurses and other health professionals cope with the startling technological breakthroughs that have an enormous impact upon medical practice.

Several institutions have adopted the POMR concept and demonstrations of this medical record-system have merited enthusiastic acceptance from both physicians and nurses. It has been shown to (1) promote a logical and structured method of gathering and recording patient data. (2) increase record efficiency by use of display system to recall information in electronic and/or paper form (that is legible) in seconds and simultaneously from different locations. (3) enhances communication between physicians, nurses and other members of the health care team; (4) acts as an excellent education vehicle and thereby can be used to further the learning process of health personnels; (5) provides a means to audit this information for correction of discripancies or systems updating; finally, (6) the physical record cannot be misplaced because the original electronic record never leaves the computer system.

POMR equates specific problems with the respective treatment programs, official orders, clinical progress and therapeutic regimine. This organized approach to all the steps in appropriate patient management can be of extreme value in simplifying the communication between health professionals about a particular patient. The system's ultimate role in the delivery of health care and the implications that it may have for nursing are as yet undefined. However, the potential of this type of system is obvious. With responsible leadership this technology can be effectively exploited so that an idealistic concern for massive medical and social problems can be turned into constructive action.

These approaches - the imaginative utilization of computer technology - can serve as an example of how computerized systems can be incorporated into the health care field and effectively affect the methods employed by the medical technologists. Thus, the medical profession and the computer industry should be encouraged to cooperate in ventures designed to achieve specific meaningful tasks that can bring health care to the people.

References

1. Berni, R. and H. Readey. Problem-Oriented Medical Record Implementation. The C. V. Mosby Co., St. Louis, 1974.

2. Bronzino, J. D. Technology and Patient Care. The C. V. Mosby Co. (in press, publication date July 1977)

3. Collen, M. F., L. S. Davis, E. E. Van Brunt, and J. F. Terdiman Functional Goals and Problems in Large-Scale Patient Management and Automated Screening. Federation Proceedings 33: 2376-2379, 1974.

4. Glueck, Fr., B. C., R. P. Erickson, and C. F. Stroebel. Use of a Psychiatric Patient Record System. Federation Proceedings 33:2379-2384, December 1974.

5. Glueck Jr., B. C. and C. F. Stroebel. The Computer and the Clinical Decision Process: II Amer. J. Psychiat. 125:2-7, January 1969

6. Gane, Dr. An Example of the Effects of Computer Usage on Nursing Practice Your Health Care and How To Manage It. Weed Essex Publishing Co. Vermont, 1975.

7. Stroebel, C. F. and B. C. Glueck Jr. Computer Derived Global Judgements Psychiatry. Amer. J. Psychiat. 126: 1057-1066, February 1970.

8. Stroebel, C. F. and B. C. Glueck Jr. Computers in Medicine. Practice of Medicine, Vol. II, Chapter 61. Medical Department Harper and Row Publishers Inc. Hagerstown, Maryland, 1972

9. Weed, L. L. Your Health Care and How to Manage It. Essex Publishing Company, 1975.

SIMULATION AND THE BIOENGINEER

Earl J. Kletsky, Assoc. Professor of Sensory Sciences
Institute for Sensory Research
Syracuse University
Syracuse, NY 13210

Introduction

We have been exploring the role of computers in the education of bioengineering students. My assignment today is to examine the area of biosimulation in this context. Since the term is widely used and interpreted, let me first try to establish those aspects of biosimulation that are of importance to bioengineers. We will then be in a position to suggest a general outline for an upper division course in biosimulation. Finally, I would like to share with you some typical examples of simulation taken from both our undergraduate program in Bioengineering and our graduate program in Sensory Sciences at Syracuse University.

Why We Simulate

Over the years there have been semantic ambiguities and differences in scientific research philosophies among those concerned with modelling and simulation of biological processes. On the one hand, Fein (1965) noted "the questionable utility of digital computer program simulations and neural net models." On the other hand, Stewart (1965) stated that "model techniques contribute most to the process of exploration and discovery in nature by providing versatile and novel means for framing hypotheses"; and Perkel and Moore (1965) describe "three instances as clear-cut cases of neural models being uniquely useful to research workers in neurophysiology." Against this diverse background of opinion Fein (1966) presented a paper entitled "Biological Investigations by Information-Processing Simulations" in which he attempted to establish logically and precisely the structure and character of a "useful investigative simulation." While one may disagree with his concept of usefulness, he has, nevertheless, examined the process of modelling quite thoughtfully and has clearly distinguished among several different categories of models. His paper should be required reading for those involved in biosimulation education.

Fein defines an Investigative model as one expressly designed to gain new knowledge. Such a model should allow its user to assert that he now knows something about a particular structure, operation, function, or other aspect of the biological system under study that he did not know before.

Educative models are those used for pedagogical purposes. They help organize thinking, display the complex interactions between parameters, and in general enable users to gain a better understanding of what they already know.

Emulative models are designed to perform functions similar to the biological functions of interest to the modeller. The internal operations need not bear any relationship to the internal operation of the functions being modelled. Prosthetic devices fall in this category of model.

Imitative models are designed to imitate the past behavior or response of the system under study with the expectation that future behavior will be equivalent to that of the actual system.

It is, of course, obvious that a given simulation developed to serve a specific purpose may serve other needs as well.

Bioengineers are called upon to perform simulations in all of these categories. Our bioengineering students should therefore be aware of the various roles played by biosimulation and be certain that conclusions drawn from a simulation study are consistent with the category of simulation performed. We are thus led to the first component of our course in biosimulation:

a. A discussion of the philosophy of simulation with emphasis on the nature and limitation of results.

How We Simulate

The technical and scientific expertise required to perform a meaningful biosimulation is drawn from many sources. Minimum prerequisites from the physical sciences should include facility with differential equations, linear systems analysis in both the time and frequency domain, digital computer programming, and basic instrumentation techniques. Prerequisites from the life sciences should include a solid foundation in the neurosciences, anatomy, physiology, and biochemistry. The life science background is exceedingly important since it highlights the complexity of the problems attacked by bioengineers. Furthermore, it is essential that bioengineers and their life science colleagues be able to communicate effectively with each other so that both clearly understand the system to be simulated as well as the nature and importance of simplifications and approximations.

With the prerequisites cited, the course in biosimulation can concentrate on a study of tools and techniques. These include methods of digital simulation, analog simulation, hybrid simulation, as well as simulation by means of analogous systems, closed-form analytical solutions, and numerical solutions. In addition to understanding the techniques, students must be aware of the effort involved in implementation as measured in terms of time and money. Consideration must be given to the ease of use of the simulation; the level of second-order effects (noise, round-off, dynamic range); temporal scaling; the interrelationship between repeatability, accuracy, and precision; and the available input-output formats (graphical, analog signals, tabular data, hard copy, etc.). We can summarize this component of the course as follows:

b. A study of the techniques for digital, analog, and mathematical simulation. Examination of the effectiveness of the simulation based on time, cost, and results.

The biosimulation course should introduce a number of simple exercises to enable the student to master the mechanics and technical fundamentals. This should be followed by one or more in-depth studies of actual life-science simulations drawn from the literature or from the research experience of the instructor. These studies should place primary emphasis on an evaluation of the simulation itself rather than on numerical results. Thus, the third component of the proposed course would be:

c. Laboratory exercises and case studies drawn from the literature.

Examination of the three components for the proposed course reveals no particular emphasis on the digital computer. While this may bruise the egos of many of those present and brand me as a bit of a maverick, critical examination shows that the digital computer is only one of a number of simulation tools.

What We Simulate

I would like to close by describing briefly three simulations that have been used in our biosimulation courses as exercises and examples.

The first example, used in our upper division course, is a digital computer simulation of neural firing dynamics (Kletsky, 1975). The simulation was performed in order to study the characteristics of a neural spike train produced when summated excitatory post synaptic potentials (EPSPs) interact with a time-varying threshold as shown in Fig. 1. EPSPs, each with an exponential tail, occur randomly with inter-event times controlled by a Poisson process. An output spike is generated when the summed EPSPs exceed threshold. After each spike, the threshold enters an absolute refractory period followed by a slow recovery toward its resting value. Students used a variety of computers and algorithms. A summary of results is shown in Fig. 2. The exercise resulted in a number of accomplishments. It acted as a vehicle to uncover many aspects of digital simulation. It allowed comparison of a number of high-level computing languages and programming techniques. Finally, it provided a practical seminar in oral presentation and in program documentation.

The second example is drawn from the research activity of the Institute for Sensory Research and is concerned with the evaluation of a particular integration time-constant in auditory receptor units (Kletsky and Stengrevics, 1972). The simulation was carried out using a real-time analog model of the auditory receptor and first order neuron. A block diagram is shown in Fig. 3. Physiological data indicate that neural firings are highly correlated with positive-going excursions of the auditory stimulus waveform as shown in Fig. 4 (Brugge et al., 1969). A more detailed examination of these data reveals, however, that the histograms decay more slowly than predicted from the approximating sinusoids, particularly at higher frequencies. This suggests a temporal integration or lowpass filter in the receptor unit. In the simulation study, the time constant of the implied lowpass filter was varied to obtain a best match with the physiological data. Representative runs for two different time constants are shown in Figs. 5 and 6. The simulation established that the best match is obtained with a time constant of 0.15 msec. Fig. 7 illustrates the results obtained using this time constant and a more complex two-tone stimulus.

The last example is again drawn from the research at the Institute and illustrates a simulation that has Investigative, Educative, and Emulative aspects. The simulation is a real-time electrical analog model of the peripheral auditory system. Sound waves impinging on the external ear are transformed by the mechanical and acoustic properties of the ear canal, middle ear, and the cochlea or inner ear before transduction from mechanical vibrations to neural impulses takes place. Equations can be derived that describe the transfer function for sound pressure in the ear canal as well as the transformation between sound pressure at the eardrum and motion of the stapes (stirrup) located at the entrance to the cochlea. Hydrodynamical analysis leads to a partial differential equation describing the vibrations of the basilar membrane stretched along the length of the fluid-filled cochlea.

Using the theory of analogous systems, the mechanical system was converted to an equivalent electrical system. Passive electrical networks were used to provide simulation of the ear canal and middle ear. A 96-section lumped transmission line approximates the partial differential equation describing displacement of the basilar membrane. In its Investigative role, the model has been used to study the pattern of vibration along the basilar membrane as a function of frequency and as a function of the mechanical properties of the membrane. As an Emulative simulation, it provides input to real-time models of auditory receptor cells. As an Educative simulation, it allows visual examination of the complex wave motion that exists in the inner ear. In 1972 we made a short 9-minute film* employing the simulation to be used as a teaching aid. I would like to close the formal part of this presentation by showing you the film.

*Wave Motion in the Cochlea, #1-10223, Syracuse University Film Rental Center, 1455 E. Colvin St., Syracuse, NY 13210.

Brugge, J. F., Anderson, D. J., Hind, J. E., and Rose, J. E. (1969). Time structure of discharges in single auditory nerve fibers..., J. Neurophysiol. 32, 386.

Fein, L. (1965). Appendix in Biophysics and Cybernetic Systems, Spartan Books, MacMillan & Co., Ltd., London, p. 169.

Fein, L. (1966). Biological investigations by information-processing simulations. In: Natural Automata and Useful Simulations, Spartan Books, MacMillan & Co., Ltd., London, p. 181.

Kletsky, E. J. (1975). Simulation of neural firing dynamics: a student project. (presented at the 1975 Annual Conference ASEE, Colorado State Univ., Ft. Collins, CO).

Kletsky, E. J. and Stengrevics, J. M. (1972). Estimation of the integration time-constant in auditory receptor units. J. Acoust. Soc. Am. 52, 141(A).

Perkel, D. A. and Moore, G. P. (1965). Appendix in Biophysics and Cybernetic Systems, Spartan Books, MacMillan & Co., Ltd., London, p. 176.

Stewart, R. M. (1965). Appendix in Biophysics and Cybernetic Systems, Spartan Books, MacMillan & Co., Ltd., London, P. 178.

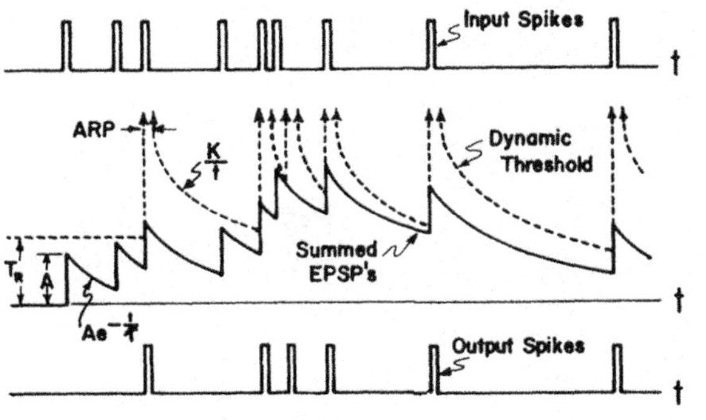

Fig. 1. (above) Simulation of neural firing.

Fig. 2. (right) Results from seven different simulation studies.

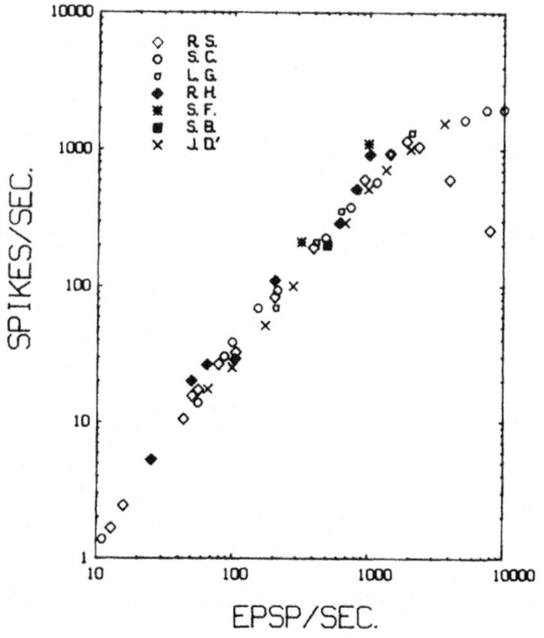

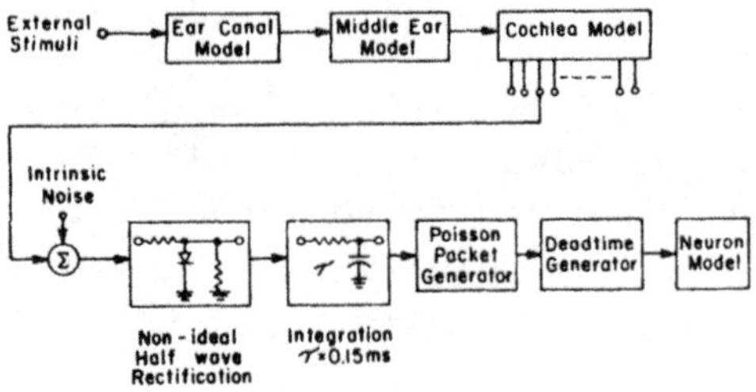

Fig. 3. Auditory transducer model.

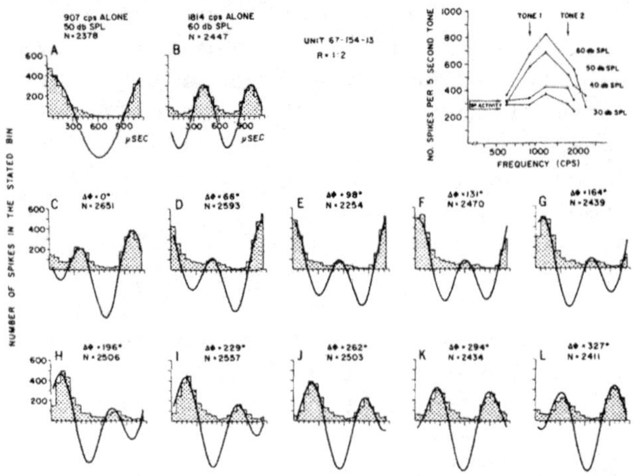

Fig. 4. Neural firing in response to low-frequency periodic tones.

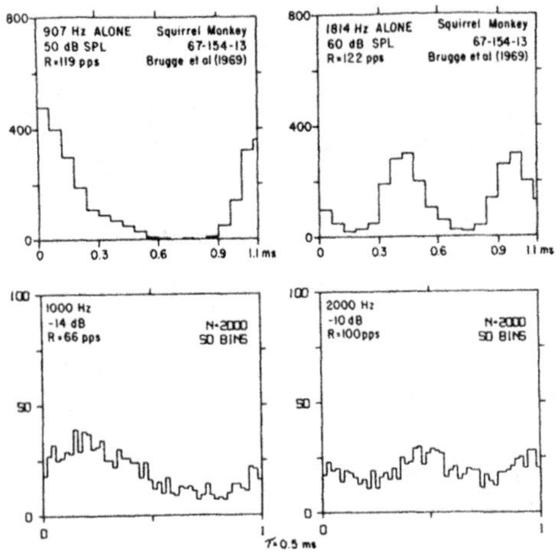

Fig. 5. (above) Upper: physiological response to 907 and 1814 Hz separately. Lower: simulated response at 1000 and 2000 Hz with τ = 0.5 msec.

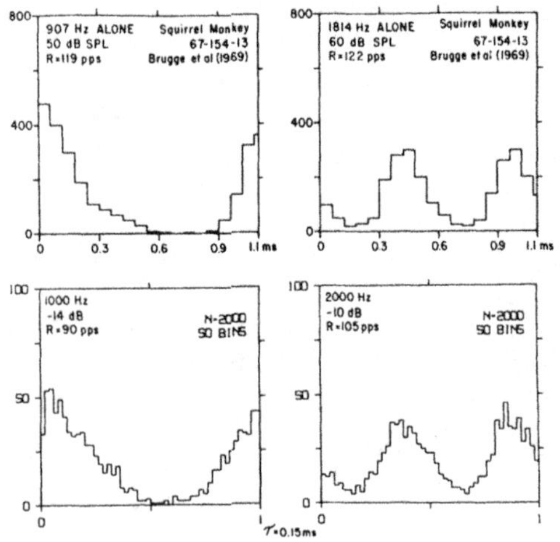

Fig. 6. Same as Fig. 5 with τ = 0.15 ms.

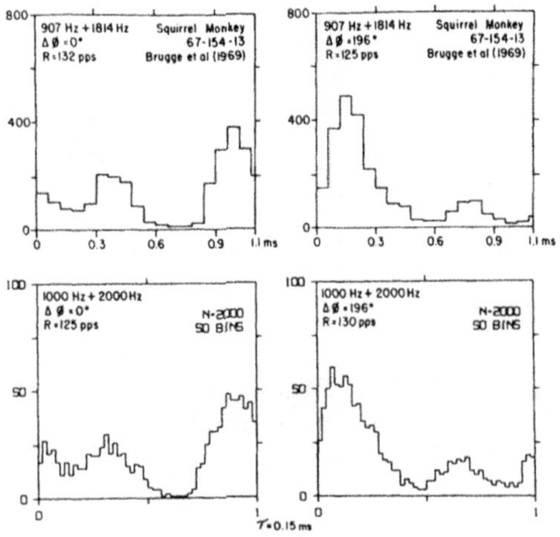

Fig. 7. (right) Upper: physiological response to 2-tone complex with different phase relations. Lower: simulated response with τ = 0.15 msec.

PERSONALITY DETERMINERS OF SUCCESSFUL PROSTHETIC AND SENSORY AID USE

Walter E. Needham, Ph.D. and William R. De l'Aune, Ph.D.

Psychology Service and Eastern Blind Rehabilitation Center
Veterans Administration Hospital, West Haven, CT 06516

It is generally acknowledged that the successful utilization of a sensory aid or prosthetic device is dependant upon both the qualities of the device and the characteristics of its user. Attempts to investigate the latter, however, have for the most part either been limited in scope, descriptions of clinical observations or nomethetic in nature and have yielded little hard data. Yet because most devices are successfully used by some clients and not by others, we are forced to conclude that personal characteristics determine to a significant degree the outcome of the device-user interaction.

Intuition alone has proved a poor guide in identifying which idiographic factors are significant. Researchers involved with the evaluation of individual differences and aids for the blind have attempted to clarify the presence of many such relationships, and have often provided results contrary to armchair speculation. For example, success of a blinded client in the comprehension of time compressed speech would seem to be strongly linked to verbal intellectual abilities and lack of auditory impairment, but De l'Aune et al. (1975) reported an absence of such relationships. The ability of a blinded veteran to use acoustic cues to identify the opened or closed corridors within a hospital setting would seem to be highly related to hearing as measured by standard audiometric threshold techniques but De l'Aune et al. (1974) found no such linkage. It also seems obvious that skill in the use of a reading machine would be very dependant upon the verbal intelligence of its potential user since the task involved is basically a verbal one. However, data from Weisgerber et al. (1974, 1975), Schoof (1974), and Gadbaw et al. (1976) have indicated no consistent relationship.

The authors have found relationships between individual differences in the above mentioned tasks and individual differences in personality functioning, as measured by scores on the Minnesota Multiphasic Personality Inventory (MMPI) and the California Psychological Inventory (CPI). A similar set of relationships were also found between the test results and ambulatory velocity and veer in the blind (De l'Aune et al., 1975a and Needham et al., 1975). These findings, viewed as a whole, underscore the importance of personality function in the use of prosthetic and sensory aids. Success with tasks thought to be essentially perceptual, motor, or a combination of the two, appears associated with subtle indicators of psychological health.

The subjects were required to take the MMPI and the CPI as part of their initial assessment upon admission to the Eastern Blind Rehabilitation Center. The MMPI consists of 566 true-false statements covering a wide range of subject matter from the physical conditions to morale and social attitudes and is designed to assay those traits that are commonly characteristic of disabling psychological abnormality. The scale scores obtained from this test consist of three validity scales: Lie (L), Frequency (F) and Correction (K); and ten clinical scales: Hypochondriasis (Hs), Depression (D), Hysteria (Hy), Psychopathic deviate (Pd), Masculinity/Femininity (Mf), Paranoia (Pa), Psychoasthenia (Pt), Schizophrenia (Sc), Hypomania (Ma) and Social introversion (Si). The CPI is also an objective personality measure consisting of 480 statements but is intended for use with non-psychiatrically disturbed subjects. Its scales are addressed to personality characteristics important for social living and social interaction. The results of this test are presented in 18 scales: Dominance (Do), Capacity for status (Cs), Sociability (Sy), Social Presence (Sp), Self-control (Sc), Tolerance (To), Good impression (Gi), Communality (Cm), Achievement via conformity (Ac), Achievement via independance (Ai), Intellectual efficiency (Ie), Psychological mindedness (Py), Flexibility (Fx), and Femininity (Fe). Once on answer sheets the tests are either handscored by means of templates in the case of the CPI or machine scored and given preliminary computer analysis for presence or absence of psychiatric diagnosis in the case of the MMPI.

The relationships between these scales and the tasks previously described can be found in the original papers. While the specific scales of the psychological tests lack consistency in their relationship with the different tasks, it was noted that the trends shown by the broad category of "psychological health" and success in all of the endeavors can provide the researcher with access to a very powerful factor in the evaluation of the less than apparent reasons for a client's acceptance of a device or his acquisition of skill in processing the information presented to him by the sensory aid (or his environment).

The results of personality testing for blinded veterans at the Eastern Blind Rehabilitation Center (1969-76) are shown in Figures 1 and 2. Especially on the MMPI one finds a shift in the direction of disturbed feelings and behavioral limitations. The scales most strongly associated with a neurotic adjustment, e.g. the "neurotic triad," consisting of the Hs, D, and Hy scales, are most elevated. At or approaching T scores of 60 are the Pd and Ma scales, elevations on the former associated with nonconformance with a traditional social role which emphasizes personal responsibility and on the latter, impulsivity and hyperactivity. On the CPI low scores (at or approaching standard scores of 40 or less) were found on the To index, which attempts to identify persons with permissive, accepting, and non-judgemental social beliefs and attitudes, and on the Ie scale, which is felt to be an indicator of the degree of personal and intellectual efficiency which the individual has obtained. It should be noted that the 50 level on both tests refers to mean scores of non-psychiatric populations.

When the MMPI data from the blind center is compared with analogous data from chronic pulmonary patients (Kass, 1971) strong similarities emerge. While the pulmonary patients appear to score higher than blinded veterans on most of the scales denoting psychopathology, the most striking thing about the comparison is the similarity of pattern with the neurotic triad being especially elevated. Possibly this similarity of psychological profile for such diverse groups could reflect the same pervasive aspects of psychological health (or its absence) as found in the blinded veterans. If such a phenomenon, which the authors have termed "dysfunctional overflow" occurs in other disability groups and affects adaptability of the patient to successfully employ a prosthetic device or a sensory aid, then many implications, both for product research and clinical use, for these other groups emerge.

It should be noted that the more than three hundred veterans sampled in this series of studies did not have extremely severe psychiatric pathology. They had been screened for the most part by a psychiatric interview and judged as free from any significant psychiatric illness which would hamper their participation in a blind rehabilitation program. However, the computer scored MMPI tests indicated that over 68% exhibited significant psychiatric symptoms. An immediate implication of this finding is that psychological state, possibly reflected most accurately by scores on psychological tests, should be considered in both the procedures for the development of aids and in the evaluations leading to the prescription of a device for the individual client. Possibly their application in product development is most important since the decision to continue or discontinue work on a specific aid potentially affects greater numbers of people. Accordingly, if one wants to get the user population with the highest probability of success in acquireing skill in the use of a device, he should ascertain that he uses people that are not only intellectually, physically, and sensorially able, but who are also in good psychological health as measuredby careful testing.

References

De l'Aune, W., Scheel, P., Needham, W., and Kevorkian, G. Evaluation of a methodology for training indoor acoustic environmental analysis in blinded veterans. Proceedings of the 1974 Conference on Engineering Devices in Rehabilitation, Tufts-New England Medical School, Boston, 1974, 26-31.

De l'Aune, W., Needham, W., Lewis, C., and Nelson, J. Speech compression and blinded veterans. Proceedings of Devices and Systems for the Disabled, Krusen Center for Research and Engineering, Philadelphia, 1975, 69-75.

De l'Aune, W., Needham, W., and Kevorkian, G. Relationships between indices of mobility and personality factors in the blind: I. MMPI. Journal of the International Research Communications Service, 3, 80, 1975a.

Gadbaw, P., Dolan, M., and De l'Aune, W. Optacon skill acquisition in a blinded veteran population. Journal of Blindness and Visual Impairment, (in press).

Kass, I. A Program to Identify the Factors Involved in the Rehabilitation of Patients with Chronic Obstructive Pulmonary Diseases. National Technical Information Service, Springfield, Virginia, 1971.

Needham, W., De l'Aune, W., and Kevorkian, G. Relationships between indices of mobility and personality factors in the blind: II. CPI. Journal of the International Research Communications Service, 3, 81, 1975.

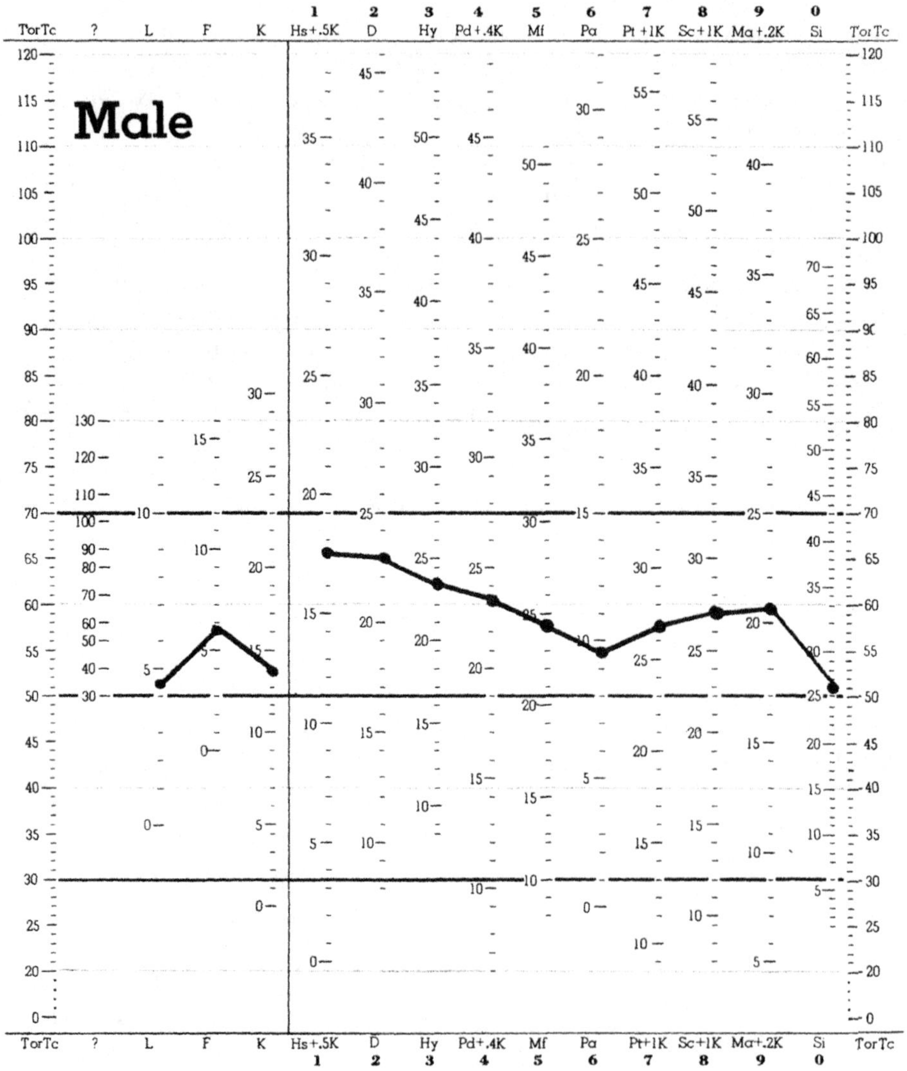

Figure 1. Mean Minnesota Multiphasic Personality Inventory profile for 244 blinded veterans.

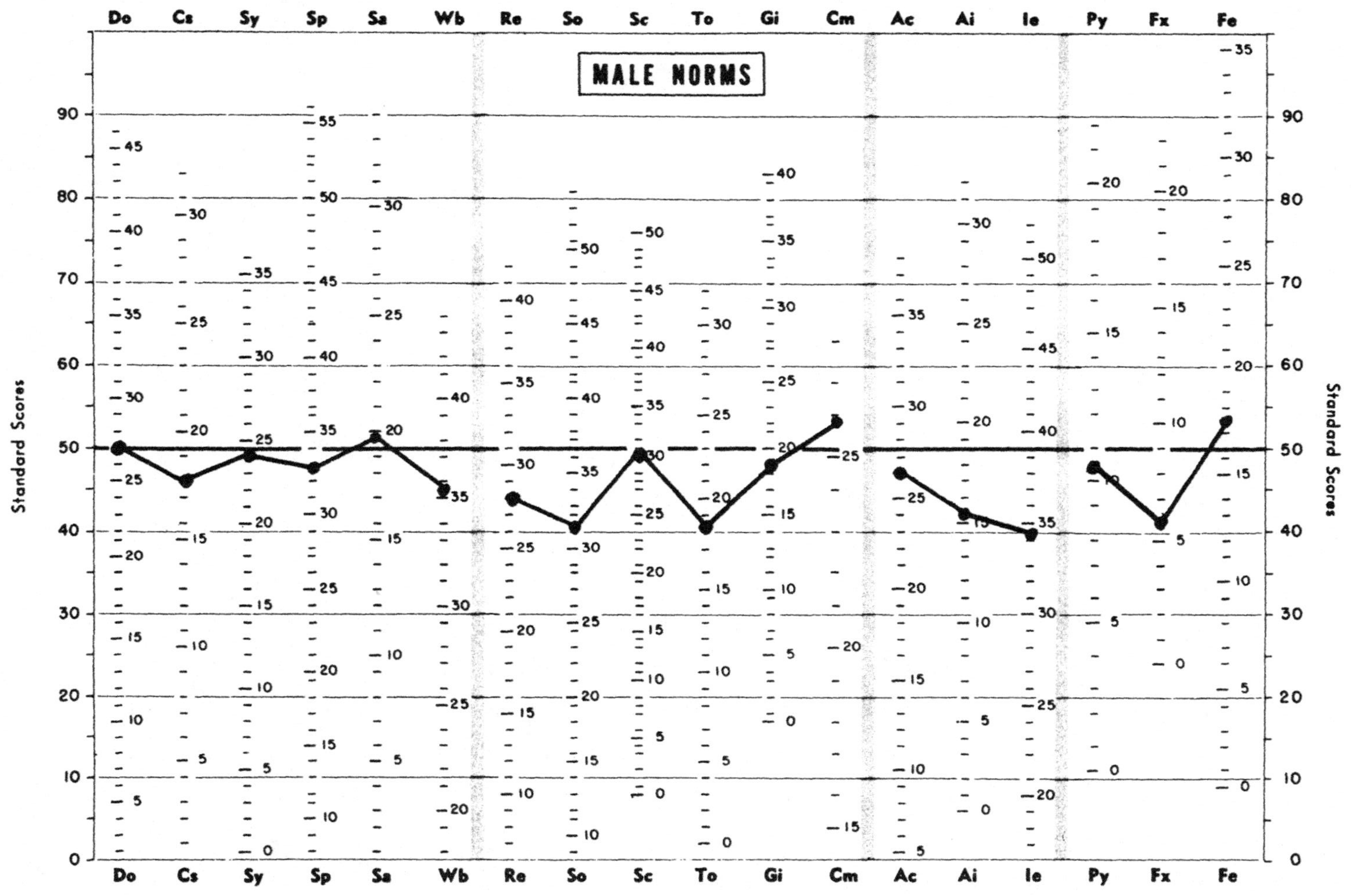

Figure 2. Mean California Psychological Inventory profile of 228 blinded veterans.

ADAPTIVE COMMUNICATIONS ENHANCEMENT FOR NONVERBAL PHYSICALLY HANDICAPPED

Louis G. Esposito, Vice President
Adaptive Therapeutic Systems, Inc.
Madison, Connecticut

An increasingly large percentage of our gross national product is being committed to the education of the handicapped. New federal legislation carrying promises of "free and appropriate education for all handicapped, in the least restrictive programs; non-discriminatory testing and evaluation; and more make special education bigger than it has ever been before. Since bigger is not necessarily better, more study is needed now than before to make certain that each special education dollar is spent producing its appropriate increment of total educational product. To accomplish this a systems point of view is necessary.

Although much of what Public Law 94-142 is bringing was foreseen by the special education hierarchy in this country, the time table for the occurance of individual events has been accelerated. This acceleration creates pressures for program initiation which may not be completely planned through with the necessary view towards the prospective students receipt of his share of educational product.

Various alternative structures designed to deliver educational services are in use across the nation. These structures range from mainstreaming in a public school setting to institutionalization. Organizationally, provisions for education can be made within a district, by service contract, regionally, or by special school district. Costs vary with the structure and organizational method chosen for delivery of service, but one factor can be considered a constant. The physically handicapped student is more expensive to educate then a non-handicapped student by a factor of 3.64. Since severely handicapped are the prime focus of new legislation, the major share of new funding will be spent here.

Whatever the program, structure or organizational set up, for the student to benefit some type of communication is necessary. The impact of a physical handicap caused by cerebral palsy (cp) goes well beyond the motor side of the problem. Perhaps the most insidious aspect of the problem for many cp victims is the inability to communicate. Mancino discussed the problem from a teachers point of view, in that, how does a teacher make contact with a child who has so many physical problems that the most basic response is impossible to perform? The largest single expenditure for state special education programs is the delivery of instruction, particularly salary cost for teachers and teachers aids. Without considering federal contributions, these costs exceed 180 millions annually in California and range down proportionately based on populations of handicapped. It adds up to this: new programs and new funding will be directed towards severely handicapped. A particular problem of many of these is lack of effective means of communication. Lack of communication will frustrate the efforts of teachers whose salary costs make up the largest single increment of special education expenditure. Much of the new federal funding will therefore be wasted unless communications channels exist between teacher and student.

By inspection, there is an expectation of physical return on invested special educational dollars. What we're looking for is a restoration of function at a minimum. For the orthopedically impaired this means that the whole person is serviced. Therapists and others work on motor deficiencies; teachers and aides work on educational format. The problem here is that the approach is of necessity tandem with therapy first and education last. Although it is difficult to imagine the active use of methods which are costly, difficult to perform, and abstract when considered in terms of the actual educational needs of the physically handicapped consumer, such is the present situation. Much of the presently relied upon educational programming for these consumers consists of interdisciplinary teams operating in tandem with special educational staffs attempting to modify the student rather than the equipment used or the task involved. There are those who argue that not only does the old method not work, it can't work in view of what is now known about the criteria for its successful application. For a special education program to be forced to wait until therapy can turn over enough in the way of motor performance to accomplish educational goals requiring this performance is a waste of valuable manpower.

Modern communications technology ranges from symbol usage to synthesized assembly of artificial language. Although some of these will be presented briefly below, it is beyond the object of this paper to discuss all of them along with their relative advantages and disadvantages. Others have done this job of research and presentation much better than any such attempt this writer can make. What will be presented is the observation that communications channels can be established between teacher and physically handicapped students; that by doing so greater cost/effect ratios can be achieved with new funding; and further that a physical return can be achieved simultaneous with the educational side of the program. The capability for all this lies not so much with the communications device used, but in the manner of interface of the device to the student.

For an integrated communications enhancement program to be operative, the consumer turns out to be not only the student, but also all those who deliver services to him. Staff needs will be considered later. Centering on the orthopedically handicapped student for now, he's severely motor-impaired; the most simple motor task is nearly impossible to perform. His handicap makes him a prisoner in his own terrain. Often he knows what he wants to do but his resultant motions, unassisted, are inappropriate to his desires. He's bright in 30% of the cases and either non or marginally verbal.

Direct approaches to the language aspect of the total communications problem have produced an array of devices from simple to sophisticated and complex. Codes and symbols, e.g. "blissymbolics" are widely used to train the nonverbal to communicate. Moving up technologically, linear scanning devices which offer selections to the user with a rotating pointer are next. The user signals the device to both start and stop.

X-Y scanners are another alternative for communication. With this method selections are offered to the user in timed sequencial patterns usually row then column. We travel from this level to the most advanced speech synthesizers by deliberate additions or more sophisticated technology.

The thread that runs through these devices is the basic need of the student and that the user must physically do something or nothing happens. The user must be mated to the device somehow or it can't be used. To accomplish this takes a least common denominator of motor skill. In at least 90% of the cases there is some gross motor opportunity with which to work. Combining that opportunity with a well designed adaptation is the key to a cost effective educational program. Earlier physical return on investment was mentioned. We know that education is the goal and that communications is the effective means by which that goal can be accomplished. We also know that devices exist that will allow communications to occur. Our objective now is to add these ingredients together in a manner which brings about physical return on investment of educational dollars at the same time as education is occurring. It takes an integration of techniques designed to produce the result of therapeutic benefit and educational gains simultaneously.

An overview of special education classroom considerations must take into account the needs of both the special education teacher and the physically handicapped student. The needs of the special education teacher are certainly no different then anyone else's needs, however, it's difficult to self-actualize when only receptive participation by the student is possible. The teacher needs a foundation upon which to build a meaningful program of education. Equipment designed to treat cases of cp have been around since 1843, and numerous isolated devices have always been available. But these devices have been largely prescriptive, and their operation and application were generally known only to their inventors and perhaps a small circle of educators surrounding.

In the face of central nervous systems damage the teacher still needs methods of establishing connections with the student which allow her to help the student develop motor skills in the direction of writing skills; improve eye-hand coordination; learn directionality; and develop an awareness of the relationships between vertical and horizontal planes. What's important here is that the teacher cannot accomplish all of this by herself, mechanical help is required.

The physically handicapped student has a need for self expression; a need to recognize himself as a functioning person. They require some satisfaction of their need for autonomy. Because of the motor side of their problem they have greater difficulty in this expression than others. The physically handicapped student cannot express joy in spontaneous movement, and when they disagree with program they cannot make the ultimate protest of running away. What is usually described as their patience is really not that at all, it's the sheer inability to express impatience or apathy born of discouragement. The student needs methods of interfacing with his environment and work areas, but they too cannot do it alone.

It would appear that the needs of teachers and students are similar; an approach to communications enhancement focusing on the idea of establishing an expressive outlet by whatever means possible, sometimes using mechanical means, would appear to satisfy the needs on both sides of the problem.

Adaptations which accomplish the above can be put to use as early in the educational scheme as the scheme itself will allow. Early intervention programs, also known as early uptake, sensory stimulation, and other terms, are vitally needed by the special student. They are the start of an educational continuum, and the need for equipment and adaptations is pervasive throughout. Tactile devices, Crawlers, and Prone Boards are actually communications enhancement devices. They allow practitioners to engage in activities aimed towards increasing muscle tone, basic reflex integration, facilitating co-contraction, and developing equilibrium.

Beyond early intervention, proper placement within the school system is essential to the progress which the student will ultimately make. Hanson, Levy, and Muller all stressed the need for adaptive headgear in testing and evaluation procedures. Muller has listed nine immediately applicable standardized psychological tests which can be taken by the severely physically handicapped, without modification of the test, simply by using an adaptive headgear. Levy stresses that success can be internalized, and that it is possible to diminish strong dependence on an adaptive headgear and transfer newfound skills to other motor areas.

Speech therapy procedures as well can benefit from adaptive devices. The combination of overhead slings which elevate the upper extremities thus diminishing the affect that gravity has upon them and certain eating devices can allow the physically handicapped student to perform some of the routines that speech therapists sometimes stress. Examples of this would include the sucking and swallowing experiences gained from the use of a lollipop. Allowing the student to do these activities himself gives the motor basics the attention that they require. If one starts off from a foundation of safe and comfortable positioning, the inhibition of drooling can also occur. Since overhead slings can increase the mobility of the upper extremities exploration of workspace area is encouraged. It is fairly well known that mobility can enhance vocalization.

The work of Held, Bach-Y-Rita, Harris, Wooldridge, Herman and others has produced general considerations regarding effective motor learning in the presence of central nervous system dysfunction. Among these considerations are that specific goal oriented activities and self initiated motor responses relating to achieving these goals are necessary. If a student has the ability to interface with his communications system in a manner which allows for frequent responses and efforts to achieve the goal where the intervals separating the efforts are as short as possible, and that further he receives an immediate awareness of performance which is continuous precise and objective, there is the possibility of motor learning (or a physical return on investment) while an educational program is taking place.

Burns and Jacobs were able to successfully demonstrate that severe and moderately involved physically handicapped students could be interfaced to self teaching machines designed for non-handicapped students using only simple headgear adaptations.

The use of adaptations is not universal throughout the various education programs already in existence. Programs which use adaptations as the central core of their program delivery system report gains unequaled by any alternative form of approach. When adaptations can be matched with what is already known by the special educator in terms of availability of software, the delivery of educational product can be highly streamlined.

One of the major blocks to the integration of new techniques to special educational classrooms is that special education suffers from the same malaise as general education. There is no effective accountability for results on behalf of those who actually deliver programs. It is generally known that as educational budgets have increased linearly over the past decade, achievement test scores have fallen off consistently during the same period. In special education this problem is compounded by a subtle downgrading of expectations considering the handicaps involved. For the severely handicapped, up until recently there was little if any expectations of educational product at all. Often the severely motor impaired student found himself in an institution because it was impossible for him to express his educational possibilities.

In 1972, Erway suggested a performance contracting system involving adaptive devices. He stressed that most of the objections previously brought up in opposition to performance contracting for general education would not be applicable to special education. Most of what would be accomplished would be done in the non-typical classroom setting. The educational programs would deal with special children in unique settings. There would be little conflict with teachers and/or unions since the children involved would not be in the regular classroom setting. Some of the most severe may have earlier been written off in terms of normal educational process. It would therefore be little infringement on the teachers territory. He assumed that the response to a performance contract would be highly favorable since no one at that time had found a way to educate the most severely handicapped student. But the bad name that performance contracting received in general education seems to have overflowed into special education as well, and this writer knows of no present contracts which are centered on payment for performance. In light of the recent pressures above cited, this situation may change.

REFERENCES:

1. Bach-Y-Rita, P.: Brain Plasticity as a Basis for Sensory Substitution. 1975 Conference on Devices and Systems for the Disabled, Krusen.
2. Erway, M.: Performance Contracting. A Report. Private Research, August, 1972.
3. Esposito, L.G.: The Systems Point of View and its Application to the Education of the Physically Handicapped. Biotelemetry, Vol. 2, #1-2 (49-50) May 1976.
4. Herman, R., Monster, W.A., Gronner, M.: Neuromotor Control Systems. . .Final Report, Res. Grant #23P-5515, 3-02 (1971) Krusen Center, Philadelphia, Pa.
5. Holt, K.S.: Facts and Fallacies about Neuromuscular Function in Cerebral Palsy as Revealed by Electromyography, Dev. Med. Child. Neur., Vol. 8 (255-267), 1966.
6. Jacobs W., M.S., Burns, N., OTR: Use of the Borg-Warner System 80 Audio/Visual Unit with Physically Handicapped Children Using Special Adaptive Headgear. Copyright 1973, Adaptive Therapeutic Systems, Inc. Madison, Conn.
7. Levy, I., Ph.D.: Handwriting Instruments for the Trainable Mentally Retarded Child. Copyright 1973, Adaptive Therapeutic Systems, Inc. Madison Conn.
8. Levy, I., Ph.D.: Psycho-Educational Summary of Headgear and Its Implications. Copyright 1973, Adaptive Therapeutic Systems, Inc. Madison, Conn.
9. Mancino, P.: The Severely Involved Cerebral Palsied Child - An Adaptive Approach in Meeting His Educational Needs. Copyright 1973, Adaptive Therapeutic Systems, Inc. Madison, Conn.
10. Muller, B.P., Ph.D.: Review and Analysis of the Mancino Adaptive Headgear with Respect to Standardized Psychological Testing. Copyright 1973, Adaptive Therapeutic Systems, Inc. Madison, Conn.
11. Raitzer: Three Basic Approaches to Communication. . .and. . .Adaptable Aids. 1976 Conference on Systems and Devices for the Disabled, Tufts.
12. Taft, L.T.: Are We Handicapping the Handicapped? Dev. Med. Child. Neurol. 15, (401-402), 1973.
13. Teachers Manual, Hand Right. Copyright 1974, Motor Skills Research, Inc. Horsham, Pa.

Computer Animation as a Tool for Teaching Reading to the Physically Handicapped

R. Daniel Bergeron
Department of Mathematics

Leo D. Geoffrion
Department of Education
University of New Hampshire
Durham, New Hampshire 03824

Abstract

When a child's communication skills are severely hampered by sensory, cognitive or physical handicaps, initial reading skills are extremely difficult to master. The Computer Animated Reading Instruction System (CARIS) is designed to introduce reading in an exploratory learning environment that capitalizes on the infinite patience of the computer, coupled with the powerful descriptive and attention holding potential of animated displays. Using CARIS, children experiment with reading and writing by creating their own sentences from a predefined vocabulary. Brief computer animated cartoons provide a visual feedback of the meaning of sentences created by the child.

Introduction

The Computer Animated Reading Instruction System (CARIS) is designed to develop reading and language skills in handicapped children through the use of computer-generated visual displays. The system provides a highly stimulating environment in which the child is motivated to explore the reading and typing of words and sentences. Children learn to read by directing the computer to make simple brief animated cartoons. The principal goal of CARIS is to provide a responsive environment in which a child with poor communication capabilities can nevertheless control the environment with relative ease. With an adequate machine/child communication system, the reading activities of CARIS should facilitate the development of better communication between the child and others as the child develops reading skills.

Communication Breakdown

Among the most fundamental needs of human beings is the ability to express one's own ideas and feelings and to understand those of others. Communication is the basis for cognitive and emotional growth.

The work described in this paper was partially supported by a grant from the UNH Central University Research Fund.

Perhaps the most universal characteristic shared by handicapped individuals is the inability to communicate effectively through conventional symbol systems. Various handicaps differ as to the locus of the communication breakdown, but once broken, the results of a lack of communication are very similar across handicaps.

The most visible result of a communication breakdown is the failure to learn the symbol systems used by other people. Without the complete feedback loop, children are slow to learn the vocabulary and syntax of English. A less obvious but equally debilitating effect of communication breakdown is the way that a handicapped child quickly loses interest in surrounding activities, particularly those involving interaction with other people. The child appears to withdraw into a private world with little interest or concern about the activities of others. Indeed Van Lint's autobiography [13] of her adaptation to paralysis suggests that this passivity is a learned response to the inability to communicate efficiently.

Another striking result of a communication breakdown is its effect on those people interacting with the child. Without communication, the handicapped child is often treated almost like a nonperson, incapable of judgement or reason. This destructive interaction is obvious not only for children who are born with communication handicaps but also for adults who lose the ability they once had.

One sort of communication breakdown arises from hearing impairments. Without the ability to hear the speech of others accurately, deaf children seldom become proficient in understanding the English language. Even though most possess normal or superior visual skills, few deaf children become proficient readers [4]. Many demonstrate the passive attitude described above, particularly in situations where manual signing is not permitted.

Another obvious breakdown in the communication loop comes from physical handicaps such as cerebral palsy. In these handicaps the child has adequate receptive skills to hear and understand speech but lacks the motor skills to complete the loop. Without the expressive component, most cerebral palsied children fail to become proficient in understanding English in spite of normal receptive skills. Furthermore many cerebral palsied children who have near-normal cognitive skills are identified as severely retarded only because of this inability to communicate effectively.

A third group of people have adequate sensory and motor development but lack the ability to interpret and manipulate the symbols involved in communication. Included in this group are people afflicted with such varied problems as mental retardation, aphasia, and autism. Their similarity is demonstrated by the fact that differential diagnosis among these conditions is often difficult [1].

The first step in the education of these handicapped individuals is the development of effective communication skills. The optimal technique to achieve this is to emphasize exploratory learning in a responsive environment. In such an environment the child has the opportunity to manipulate a system, form predictions based on that experience, and then test those predictions. Through such interaction the child learns to master the feedback loop that is fundamental to communicative competence.

Existing Systems

Edison Responsive Environment. The oldest computer-based system for promoting exploratory learning of reading has been the Edison Responsive Environment, more commonly known as the "talking typewriter" [8]. The system is based on a computer driven typewriter and audio response system that uses four stages for promoting reading and writing skills.

Studies of the effectiveness of the talking typewriter with the mentally retarded [8], autistic [6], culturally disadvantaged [7] and reading disabled children [12] have been very encouraging. The principal disadvantage of the talking typewriter is that it builds to reading words from extensive practice identifying individual letters. While this synthetic phonic approach is clearly effective for normal children [2], it is not the best approach for all children, particularly those with severely deficient language backgrounds. For these children it might be better to begin with brief meaningful messages and only later provide a transition to alphabetic manipulations.

PLATO. The PLATO system developed at the University of Illinois is a general purpose computer-based instruction system which includes a Reading Project [9,10] that combines both exploratory learning and drill-and-practice approaches. Data on the effectiveness of the Reading Project has not yet been released but preliminary results suggest considerable excitement concerning the value of their exploratory activities for helping young children to learn to read. While experiments are now being conducted in the use of PLATO with handicapped children, there has been no attempt yet to adapt the system to the special needs of these children.

LOGO. Another outstanding system designed for normal children but recently adapted to the handicapped is the LOGO system developed by the Massachusetts Institute of Technology. LOGO is unique among computer-based educational systems in that it has no curriculum *per se*. LOGO is a general purpose computer language that is simple enough to allow very young children to become engaged in the thrill and challenge of teaching the computer to perform various feats such as picture drawing, puzzle solving, and game playing [11]. Most of the learning activities within the LOGO system have focussed on developing mathematical and logical problem solving skills. There has been little emphasis on learning activities for reading and typing skills other than for words needed to control the system.

Recently the LOGO project has begun to explore the value of LOGO for handicapped children. Through special terminal adaptations, even children with severe physical handicaps have been able to operate the system successfully. Case histories of these children using LOGO provide dramatic examples of the power of exploratory learning approaches [5,14]. When presented with a system they can control and manipulate, even children with severe handicaps soon become engrossed in the fun of trying to make the computer perform a desired action. At the same time they are learning cognitive skills which are more sophisticated than many had assumed them capable of learning.

CARIS

The CARIS project is an attempt to create an exploratory learning environment with the specific goal of developing reading skills in children with reading handicaps.

The eventual goal of the CARIS project is to provide a low cost (less than $15,000) standalone computer animation system based on a microcomputer/floppy disk configuration. It is hoped that the cost can be made low enough to be within the resources of most special education centers. In order to investigate the feasibility and effectiveness of the CARIS approach, a prototype system has been developed on a PDP11/40 with a DEC VT11 graphics display.

The CARIS Prototype System

The CARIS prototype system is designed to provide a child with three distinct phases which require increasingly more reading competence.

Introductory phase. When the child is first introduced to the system he/she will be presented with a display containing about five words (nouns) printed on the screen. The child learns that he can select a word by touching a light pen to that word. When this happens, the nouns immediately disappear and the picture representing the word selected occupies the center of the screen. On the left of the screen are five new words (verbs). When the child selects one of these verbs, the figure on the screen acts out the intended meaning of the verb. After the animation is complete, the noun list reappears to allow a new selection.

As the child becomes more familiar in recognizing the words, additional words are added until the complete lexicon is available to the child. In all cases, however, the child can choose the desired word by simply touching it with the light pen.

The system is designed to be tailored to the needs and abilities of each child. Some of the system parameters that are specified independently for each child include:

- -the number and selection of words in the initial lexicon,
- -the frequency with which new words are added to the child's lexicon,
- -the maximum number of words to be displayed at one time,
- -the frequency with which the words in a list are reordered.

Sentence phase. Once the child is familiar with its use, the system will shift to sentence phase. In this mode of operation the animation is deferred until the child chooses a complete sentence consisting of subject, verb, object (optional), and punctuation. The purpose of this phase is to encourage the child to think of communicating in complete sentences. As before, however, words are chosen by merely touching them with the light pen.

Spelling phase. After the child has become proficient in forming sentences, the spelling phase is introduced. In this phase the light pen response for choosing words is no longer accepted by the computer. Instead the child must spell the word to use it. Not all words will convert to the spelling phase simultaneously in order not to frustrate the child. To provide the child with a smooth transition between sentence mode and spelling mode, the child first chooses the word desired using the light pen. The computer then requests that the child spell out the word. Since the computer knows which word is intended, it can help by ignoring incorrect letter choices or by providing other clues.

Statistics keeping. In all three phases the system records information concerning the responses made by the child as well as the response times. this information can later be analyzed to determine response patterns (if any) both individually and within or across handicaps or reading levels.

Current lexicon. The current CARIS lexicon consists of approximately 30 nouns and 20 verbs. Due to memory limitations, the active lexicon for each child is now limited to about half of that number. Both nouns and verbs are defined entirely as command tables in system files, so adding new words is a relatively simple process.

Nouns. Construction and definition of nouns is very simple. The noun table is composed of commands to draw lines or points and to move the drawing beam to a new position.

Verbs. The verb command table is considerably more complex since it defines time sequences of actions on one or more nouns. The entries in the verb command table correspond to statements in a Simple Action Definition Language (SADL). Although there does not yet exist a program to translate from symbolic SADL code to actual entries in the command table, the SADL constructs are a convenient notation for describing the capabilities of the CARIS animation.

SADL provides just five basic commands: PLACE, MOVE, ROTATE, SIZE, and LOOP. The MOVE, ROTATE and SIZE commands are normally enclosed within a LOOP and specify composite actions such that each iteration through the enclosing loop performs the next incremental move, rotation, or size change. Once the complete composite action for a command has occurred, that command is ignored during further iterations of the loop.

The ROTATE and SIZE commands are constrained to specify equal sized increments for each step of the loop. The ROTATE command includes an initial orientation, a final orientation and the number of steps desired. The SIZE command includes initial size, final size and number of steps. The MOVE command references a path definition which is composed of a list of arbitrary incremental x and y coordinates to define the path to be followed by the noun being referenced.

In order to achieve variable speeds for all types of motions, each command may also specify a time delay to occur before the execution of that command. The delay is specified in milliseconds, but is accurate only to multiples of 1/60th of a second.

Adaptation to the Physically Handicapped

As currently implemented, the prototype system functions adequately for children with normal motor facility. The system is being used by children with hearing impairments, mental retardation, and learning disabilities. However, a major goal of the CARIS project is to be able to support children with moderate to severe physical handicaps also.

Future versions of CARIS will allow children with only moderate physical impairment to operate the system with an enlarged keyboard such as the "touchpanel" communicator [3].

A more extreme modification is needed for children with severe physical handicaps. In these cases the display system used by CARIS will be modified to scan among the available student responses. The student waits until the display scans to the desired word, and then indicates his response by a simple switch closure using any available physical reflex. Scanning rate and the specific reflex movement used can be varied to fit a particular individual's limitations. For example, one child might interact most effectively with a foot pedal, another might prefer a padded switch controlled by head movements, and so forth. The scanning system will demand much slower response time than the use of an enlarged keyboard, but it also provides an extremely flexible user interface that should be adaptable to even the most severely handicapped children.

Acknowledgements

The authors would like to acknowledge their appreciation to Brian Johnson and Dennis Flynn whose programming talents were responsible for implementing the prototype CARIS system.

References

1. Bakwin, H. and R. Bakwin, Clinical Management of Behavior Disorders in Children, W.B. Saunders, Philadelphia, 1966.

2. Chall, J., Learning to Read: The Great Debate, McGraw-Hill, New York, 1967.

3. Foulds, R. et al, "The Tufts non-vocal communication program", Proceedings: 1976 Conference on Systems and Devices for the Disabled, Biomedical Engineering Center, Tufts-New England Medical Center, Boston, 1976.

4. Furth, H., "Linguistic deficiency and thinking: Research with deaf subjects 1964-1969", Psychological Bulletin, 1971, 76, 58-72.

5. Goldenberg, E.P., "A computer technology for children with communication handicaps", Unpublished memo, MIT LOGO Lab., June 12, 1976.

6. Goodwin, M.S. and T.C. Goodwin, "In a dark mirror", Mental Hygiene, 1969, 53, 550-563.

7. Logan, J. and M. Fleming, "Talking typewriter: Title I evaluation 1971-72", ERIC Document No. ED 078 390, 1973.

8. Moore, O.K., "Autotelic responsive environments and exceptional children", in O.J. Harvey (Ed.) Experience Structure and Adaptability, Springer, New York, 1966.

9. Obertino, P., "The PLATO reading project: An overview", Educational Technology, February 1974, 14, 8-13.

10. Obertino, P., "The PLATO reading project", ERIC Document No. ED 105 377, 1975.

11. Papert, S. and C. Solomon, "Twenty things to do with a computer", Educational Technology, April 1972, 12, 9-18.

12. Pines, M., "What the talking typewriter says", New York Times Magazine, May 9, 1965, 23, 74-79.

13. Van Lint, J., My New Life, Neyenesch Printers, San Diego, 1975.

14. Weir, S. and R. Emanuel, "Using LOGO to catalyse communication in an autistic child", D. A. I. Research Report No. 15, Department of Artificial Intelligence, University of Edinburgh, 1976.

THE COMPUTER AS INFORMATIONAL PROSTHETIC FOR THE CHILD WITH SEVERE COMMUNICATION HANDICAPS

E. Paul Goldenberg
Logo Group
February 1, 1977

M. I. T. Artificial Intelligence Laboratory
545 Technology Square
Cambridge, Massachusetts 02139

Computers can be used to provide a radical improvement in the scope and quality of life for a large class of people with serious communication problems such as are caused by deafness, autism or severe physical handicaps. In this paper, I will focus primarily on children with highly restricted motor capacity. Pilot work with these children has already shown the great potential for this approach and motivates a serious call for research.

The principle is simple:

1. If you have voluntary control of any muscle, there currently exists an interface that provides you the capability of operating an electric typewriter.

2. If you can control an electric typewriter, you can control a computer.

3. If you can control a computer, you have a flexible aid to communication and access to a vast range of educationally, vocationally and recreationally valuable activities.

Our work at the M.I.T. Logo Group has concentrated on the third step of the idea, especially as it relates to very young children. Part of what makes the computer exciting to the programmer is that it is not a static device with prepackaged and limited programs. If a computer is to be an exciting instrument for a person who is not a professional programmer (for example a child), the computer language must be easy to learn yet genuinely powerful, and the devices to control must be varied and flexible. In our Children's Learning Laboratory, young children easily learn to make computers draw pictures, generate music, manipulate text, speak out loud, operate robotic devices, and more. Multiply handicapped children are no different. They, too, rapidly master the computer, almost regardless of their previously observed abilities and disabilities. If only through the effect of seeing themselves as able and active learners, the children have shown enormous motivation and excitement that in several cases has carried over into daily activities not involved with the computer.

Without a powerful technology, we would be handicapped in implementing our philosophy, but it is the *philosophy* and not the technology that I wish to emphasize here.

Computer as Tutor vs. Computer as Eyeglasses

Perhaps the most common non-bookkeeping use of the computer in education is as a private tutor. A program is designed by teachers and programmers to lead a child through a sequence of steps to learn some desired behavior. Although the child may be an active member of the child-computer team, it is the child who is being programmed. The interaction is guided by the computer which generally initiates all of the transactions (poses a problem, asks a question, gives an instruction) and chooses its next course of action based on the child's response to the last one. Proponents of this use of the computer argue that the computer, unlike a person, is reliably patient, able to keep detailed and flawless records, and able to juggle those records to choose the smoothest path for the student's learning based on his past and current performance. This use of the computer might be called the *Hospital Model* of educational intervention. Within the Hospital Model, students are tested, screened and diagnosed. Weaknesses are identified and programs of remediation are prescribed. This model is prevalent even in settings where the most immediately visible and preemptive characteristic of the children is not some severe handicap.

Serious risks attend this model. Though perhaps practical for certain academic-skill goals, computerized programmed learning streamlines an aspect of teaching which properly constitutes only a tiny fraction of the business in which students and teachers ought to be engaged. Moreover it assumes, *a priori*, that the child is *not* the agent, but the patient. Thus, whenever our goal for a child is to build his autonomy and initiative, the model is counterproductive. At the best, progress toward autonomy remains superficial; at the worst, it is undermined altogether. For the handicapped person, the Hospital Model is a superfluous reminder of his own disabilities.

Unlike a tutor, eyeglasses have no agenda. They contain no information for the wearer to learn and they don't try to teach. Yet they allow the wearer to do things that would otherwise be difficult, perhaps impossible. The computer allows a child to make an animated cartoon without requiring that he have the drawing skill in his hands; to experiment with musical ideas without requiring that he *first* gain competence at an instrument. The computer extends and amplifies the abilities of the person with normal needs. Again, one sees the same (and more) vast possibilities for people who have special needs. From this point of view, the distinction between people we consider normal and those we consider handicapped becomes, to quote John Eulenberg, a "sociological and technological artifact." We know how to build a device that largely allows mildly to moderately visually impaired people to ignore that handicap -- and we have learned to accept glasses as a normal aid. We have not yet become so accepting about hearing aids, headsticks or wheelchairs, and those technologies are still far from optimal.

The important distinction between the tutor metaphor and the eyeglasses metaphor is the shift of emphasis from programming or fixing oneself, to learning how to influence one's surroundings -- from focussing inward, to focussing outward.

One Student's Experience

(This paragraph and **The Promise** are quoted from a working paper by Paul Goldenberg and Seymour Papert)

> The problems and the potential are illustrated by our work with Jay, a cerebral palsied adolescent with no effective use of hands, legs or speech. Before the experiments started, Jay had been experienced in the use of a head-stick with which he could type, though slowly and inaccurately, on a electric typewriter that was specially modified to have recessed keys to guard against accidental key-pressing. After only one day (approximately 3 hours), Jay's teachers noticed a vast improvement in his typing ability. In three days, Jay had learned enough Logo to draw some very handsome pictures. The interest of these events is greatly increased by the fact that Jay's intellectual potential had previously been assumed to be extremely low. Jay's experience with the computer demonstrated for the first time a high level of analytic and of spatial ability. Indeed, some puzzling problems for developmental psychology are raised by the fact that a person who had never manipulated objects or drawings could perform so well a task requiring quite sophisticated spatial reasoning. The case study was conducted in a very informal exploratory spirit but nevertheless makes it quite obvious that more work in this direction could be extremely interesting for theoretical psychology besides being of more immediate practical value from an educational and humanitarian point of view.

The Promise

Jay's achievements after three days were already enough to make a difference to his life not only because his own image of himself as a learner had improved through a uniquely stimulating intellectual experience but because those who take care of him had seen a new dimension of his mind. But the image we want to create here of the possible value of access to the computer is qualitatively very much more than even this. To develop it we have to ask the reader to grant that if Jay could learn any significant piece of programming skill in a few days (and on our first presumably clumsy attempt at teaching someone like him) then he could come to a high enough degree of proficiency in the course of a few years to be employed as a professional programmer. Naturally, this suggestion raises many questions about the broader education he might need to operate effectively as a computer (or other) professional and the material conditions which would make it possible for a person with his disabilities to interact with a work world. Our present state of knowledge makes these all routinely researchable questions to which one can expect with reasonable confidence to have acceptable and demonstrated solutions within a few years.

A Plan for Research

The grand vision is clear. Some of the particulars -- which environments might particularly suit which children -- are also known. But a great deal of work remains to be done. This kind of investigation leads us near the frontier of work with the handicapped. It also requires a state-of-the-art technology and new pedagogical assumptions and skills. Finally, the children themselves may furnish a wealth of surprises as they begin to engage in activities they had never dreamed of, and sensitivity to them may lead to research in directions other than those suggested below. Environments must be designed that allow us to test out the basic assumptions. The designing of these environments is a clinical task, tailoring hardware, activities and pedagogy to each individual child. At the same time, it seems important to do early responsible explorations into the directions that computer-based habilitation and education centers might go in the future.

Following are three specific areas in which work is needed.

1. Equipment Development:

The ultimate communication device must be wheelchair-installable, as available to the user as are his glasses. It must also be flexible enough for recreational, social, educational and vocational uses, capable of controlling the wheelchair and of producing speech, graphics and music as well as printed text. To be maximally useful and therapeutically sound, it must make use of the person's strengths in such a way as to increase his independence. For example, it could recognize sounds, thereby giving purpose to the otherwise almost useless verbal efforts of a person who has some speech, but too little for satisfying communication. The recent availability of inexpensive, powerful, *small* computers makes this goal very realistic. The fact that a computer can make communication not merely possible, but *practical* means that it can expand beyond the expression of occupational needs.

2. Research in Psychology and Pedagogy:

Curricula that currently exist have what might be called cultural biases against people like Jay. They tacitly assume a very large set of experiences that Jay has not had. If we try to teach him without being sensitive to this fact, we may misinterpret his failures as indicators of his potential rather than indicators of his readiness. The fact that this kind of error can be made is illustrated well by Jay. Because Jay was *able* to type, and because he *still* showed very limited intellectual performance, some of the people working with him considered it safe to accept mental retardation as part of his diagnosis. If any part of the normal child's difficulties with math comes from his view of it as a useless exercise, how much more true that must be for someone whose major life issues have always been trying to communicate where you hurt or when to turn on the TV. Though it comes as no surprise that there are children whose intellectual potential remains largely undiscovered, each child that we work with makes it painfully clearer that the number of children whose mental life has been written off may be *vastly* greater than the most optimistic of us have ever guessed.

Research must also be directed toward understanding where some of Jay's surprising *abilities* come from. Jay's sensitivity to angle and dimension (though he has never physically manipulated these variables) needs more study. We need clearer pictures not only of what is *lacking* in the handicapped child's mental map of the world, but of what is *present*. Research in the spirit of Piaget and without our cultural biases is certainly needed.

Some specific experiments in education and habilitation must be tried. For example, while a child's progress in breath control may be a clear milestone to a trained speech therapist, it is perhaps totally undetected by the child who hopes to learn to speak. The sound recognizer mentioned above provides the child with feedback and puts each new vocal success to immediate use. The pleasure in control (and the confidence that progress *is possible*) -- not to mention the constant practice that are an expectable consequence of that pleasure and confidence -- should vastly increase the success of speech therapy, at the same time providing the kinds of educational benefits stressed elsewhere in this paper. Experiments to verify this claim are needed.

3. Experiments in vocational training:

Thought must be given both to the kinds of jobs that would be suitable for a person like Jay. Consideration must be given not only to the physical manageability of the job, but to the personal satisfaction that one can gain from it. Occupations involving the manipulation of information and which do not require much physical mobility or manipulation come to mind. For example, computer programming can be done at one's own home if mobility were a problem. Similarly, reference library work and manuscript editing can, when needed, be carried out remotely. But we cannot naively assume that by providing communication equipment alone, the remainder of educating these clients for their jobs will follow a normal course. As with other curricula, we must realize that the student's real handicap is his relatively diminished ability to adapt to our failings in teaching and that we must therefore tailor both mode of presentation and actual content in reasonable ways. This calls for a careful analysis of what knowledge and abilities are really necessary for a particular job and must of course account for the non-normal background of the student.

Formerly unemployable *pairs* of people can potentially be employed together in innovative ways. A physically handicapped person who interacts with his work via the computer may still need considerable physical aid and care which may in some cases be provided by a retarded worker whose employability and social contacts would otherwise be limited to the sheltered workshop. Thoughtful efforts must be made in developing an appropriate educational program to help a retarded person assume such a role.

4. Clinical Use and Demonstration Centers:

Experiments "*in vitro*" have the technological flexibility and staffing to generate the wildest dreams and make them come true, but do not have the variety of users (both students and teachers) that "*in vivo*" experiments do. The latter offer more feedback on the clarity of the curriculum, the generality of its usefulness or interest, and the suitability of the equipment. In addition, the different perspective of field users may furnish a wealth of ideas that are not so readily gained in a laboratory setting.

The Walden School of the Westchester, New York Board Of Cooperative Educational Services has set up an on-site computer center for the Eyeglasses-model educational habilitation of mentally, perceptually and emotionally handicapped children. The Crotched Mountain Center in New Hampshire is committed to an experiment in the use of the prosthetic computer with their severely physically handicapped children. An experiment giving language-handicapped (including autistic-like) children free access at their school to the computer system I have described is currently in the planning stage. Responsible effort must be devoted to explorations on the directions such centers might go in the future. The practical considerations of packaging an economical, maintainable, and easily exportable system for general use with severely handicapped children must be examined closely, necessitating serious on-site work and consistent use by non-technical personnel.

EFFECTIVENESS OF A DOCTOR-NURSE-ENGINEER TEAM IN THE DESIGN OF SPECIAL CARE UNITS

Edward F. Lucia, Jr., P.E.
RLC Associates, Inc.
One Merchants Plaza
Bangor, Maine 04401

Donna G. Avery, R.N., CCRN
RLC Associates, Inc.
One Merchants Plaza
Bangor, Maine 04401

Jeffrey S. Ross, E.I.T.
RLC Associates, Inc.
One Merchants Plaza
Bangor, Maine 04401

Abstract

The concept of the use of a multi-disciplinary consulting team consisting of doctors, nurses and engineers for hospital planning is presented. The team was an integral part of the Clinical Engineering Laboratory program at the University of Maine while under a grant from the W. K. Kellogg Foundation.

The effective use of the team for the design and layout of special and coronary care units is discussed with two approaches described which have been used successfully. The first is the use of the team as the hospital's consultant when dealing with the architect and builder. The second is the direct design and layout of a unit actually performed by the team. Six specific Maine hospitals are referenced where one or the other of the two methods was used.

I. Introduction

The Clinical Engineering Laboratory was partially funded under a three year grant from the W. K. Kellogg Foundation which started November 1, 1973. The concept was to establish a regional clinical engineering program for Northern New England which would assist hospitals with the use of technology. Hospital equipment is becoming more and more complex as time goes on and this is creating major problems for many hospitals. The impact of this problem is more acutely felt in sparsely populated rural areas such as those found in Maine.

The Laboratory, while under the grant, was set up with two distinct parts. The first consisted of a group of highly trained electronics technicians who were under the direct supervision of clinical engineers. Hospitals subscribing to services were visited each month by these technicians. They provided electrical safety testing, functional testing, preventative maintenance and equipment repair for much of the patient related equipment used at each facility.

The second part consisted of the unique teaming together of a technology oriented physician, a skilled CCU nurse and a group of clinical engineers.

While equipment servicing is the immediately obvious need for the hospitals, a much more fundamental need lies beneath the surface. These hospitals need help with the "management of technology." That is, planning for the phasing of new equipment into the health care system, selection of equipment which meets the long range and short term needs of the hospital, selection of equipment that can be effectively utilized by the physicians and nurses who have to use it on a daily basis, and, just as important in today's financial climate, equipment that is the most cost effective for that particular institution.

II. Discovery of a Problem

During the early stages of the program, we discovered a major problem when it came to assisting hospitals with equipment usage in special care, coronary care and critical care units. Many of the units we came into contact with were not optimally layed out for proper care of patients with minimal effort by the nursing staff. While that was to be expected in older units which were created from existing room space, it was a surprise to find the same situation in brand new construction. It became apparent to us that, while architects have great skills in designing buildings, many are structurally oriented rather than functionally oriented. They obviously never had to take care of patients in the units they designed. In all fairness one would not expect it to be otherwise.

We began to realize that our doctor-nurse-engineer team might be able to have strong impact in this area. Over a period of time we developed two techniques for dealing with these situations.

III. The Consultant Team Approach

The first method is to act as consultants to the hospital when they are undertaking new construction projects. At first, since we had just begun the program, we entered into projects well after design had taken place, and, in fact, well after construction had started in many instances. Yet we were still able to have tremendous impact in most cases. We undertook these activities at four hospitals with increasing impact as we became involved earlier in each project.

Using this method, the team starts by reviewing the architectural plans and drawings. Each component of the team is obviously interested in different aspects of the plans. The next step is to split up and meet with our associates on the hospital staff. Our physician meets with members of the medical staff to determine the level of patient care anticipated in the unit. The nurse meets with the nursing staff and discusses the nursing care contemplated. The engineers meet with administration, engineering and the clerk-of-the-works. We then tour the construction as a group.

After careful review at our offices, we begin to develop changes which we will recommend, acutely aware that in later stages of construction, change orders cost much more. Our recommendations are then reviewed with the hospital staff and final changes implemented. This technique was successfully undertaken at four hospitals.

IV. Unit Design by the Team

The second approach we utilize is to actually design the layout of special care units. This has been used for renovation projects rather than new construction. Again, we have extensive discussions with administration, nursing, engineering and the medical staff in order to formulate a unit which meets their current and future needs. We study patient loading for the types of patients which will occupy the unit and then establish a bed size for the unit.

Once the general ideas are formulated and space availability determined, unit layout is undertaken. The actual designer usually is the nurse on the team. Of course, the engineers convert this to an engineering layout with proper specifications. The physician reviews the plans and then a final configuration is settled upon. The sketches are then sent to a local draftsman who develops the final drawings. This method was very successfully utilized for one Maine hospital.

For another hospital, after four configurations were explored, we determined that creation of a unit through renovation as the hospital had desired, would result in the loss of too many general patient beds. We therefore recommended new construction be undertaken.

V. Typical Items of Concern

Below is a sample itemization of difficiencies determined in original architectural designs or items which we stress in our designs.

1. Design of adequate floor space around patient beds. With more prolific use of support equipment such as ventilators, hypothermia units, crash carts and so on, this item is essential.

2. Separate private patient rooms desired. This is not always possible in renovation projects, but even "bull pen" units can have partial partitions to help separate the patients.

3. Supplies often too far from point of use, or not enough storage space is provided.

4. Inadequate storage space for large pieces of equipment.

5. Lack of hoppers for human waste disposal. In one case, this was missing in an isolation unit.

6. Medication preparation areas missing in unit. Our design consists of a sink, light for night shift, refrigerator, double locked narcotics cabinet, storage for stock medications and work area for preparation.

7. Nourishment station within the unit. This is very important for small units so that nurses do not have to leave the area.

8. Nurses lounge near unit with STAT alarm to summon nurses back to the unit in a code situation.

9. Viewing of all patients from the central station. This should be accomplished in such a way as to minimize patients viewing and interacting with each other.

10. Lighting and noise factors must be considered. Careful planning to provide levels of lighting sufficient for proper care, yet not annoying to the patient, should be undertaken. In particular, high level lighting for code situations is essential. Proper accoustical damping is important to decrease the potential for high noise and activity in one area from upsetting other patients.

11. Design too spacious. In one case, layout of the unit was too open and resulted in many extra steps per day for the nursing staff.

12. Nurses station design considerations. Chart racks should be near the central station ECG monitors so the nurses can observe the monitors and perform necessary clerical duties. The nurse should be able to move quickly from the station toward all patient rooms in a code situation. (The horseshoe shape is a poor design.) The station should be large enough to handle nurses and paramedical personnel assisting in today's medical environment.

13. Family waiting room near unit. It is important to have such an area for the family of the critically ill. There should be a separate room for the bereaved of a patient who has just died. It is essential for families of the ill not to have to witness sorrow and crying in the general waiting room area.

14. Independent STAT alarm. This must be separate from all other alarm systems and have a tone unique to the entire hospital. There should be both an audible and visual alarm which should sound in the main area of the SCU, nurse's lounge, the main medical-surgical area in the case of a small hospital, the doctor's lounge and the switchboard. Once the alarm is activated at bedside, there should be no need for human intervention in the complete activating of the alarm. Systems which depend on response at the switchboard may have disastrous results if the operator steps away at the wrong time.

15. Insufficient oxygen, air and vacuum outlets. For today's medical procedures, there should be more than one of each at the bedside. We recommend 2 air, 2 oxygen and 3 vacuum outlets as a minimum.

16. Proper air temperature control is important. The system should be able to maintain 70°-80° F with 30%-60% relative humidity through all seasons. Outlets must be located to provide sufficient air flow, but not result in drafts.

VI. Conclusion

The use of a doctor-nurse-engineer team in the design of special care units has proven to be very effective. The medical and nursing input into the design, and the engineer's ability to integrate that function into a feasible structure which meets applicable codes and standards, is of great importance. The team very often has had the catalytic effective of encouraging similar interreaction between the medical staff, nursing staff, plant engineering and administration in the hospitals where we have worked. In many cases each group had valuable information to exchange, but communication between them was not always present.

We believe this approach to design has had, and can continue to have, tremendous impact on increasing the quality of patient care.

Acknowledgement

The authors wish to remember the late Dr. W. Page C. Clason who developed the concepts presented here. He took the time and effort to totally involve himself with the goal of improving the quality of patient care, and then passed his techniques and enthusiasm on to us. To him we are deeply indebted.

BIOMEDICAL ENGINEERING IN THE VETERANS ADMINISTRATION

Joseph E. Hoffman, Jr. Ph.D. P.E.
Veterans Administration Hospital
Providence, Rhode Island, 02908

ABSTRACT

The Veterans Administration has conducted training programs in Biomedical Engineering and Biomedical Engineering Technology for several years. The implementation of these programs is discussed from the hospital's point of view. The benefits of the program, its operation, and effectiveness are discussed with reference to several case studies.

INTRODUCTION

Medical science has made rapid progress in recent years, and the trend shows continuing progress in the future. The innovations resulting from medical research have produced a vast array of new, sophisticated medical equipment, and many new techniques for delivering better patient care. The use of this sophisticated equipment has a profound effect on the Engineering Service of the VA hospitals. Engineering personnel have the responsibility for the maintenance, repair, incoming inspection, training of personnel in the applications of special equipment and instructing them in electrical safety, and maintaining adequate records of all preventive maintenance and repairs. In order to ensure that hospital equipment is safe and is properly maintained, the Joint Commission for Accreditation of Hospitals and other agencies require that records of maintenance and repair be available for inspection. Some of the requirements can be met by placing certain items on equipment maintenance contracts with manufacturers or equipment maintenance specialists. This method is prohibitively expensive and impractical for all the equipment which must be maintained in the hospital. An alternative is to hire people with a comprehensive background in medical equipment maintenance and perform a large portion of the work with hospital personnel.

BIOMEDICAL ENGINEERING TECHNICIAN PROGRAM

The complexity and variety of modern medical equipment requires specially trained personnel to handle the maintenance and repair programs. There are not enough trained specialists to meet the demand. Several colleges and universities are attempting to fill the gap by offering programs leading to the associate degree in Biomedical Engineering Technology. Some "hands on" experience is usually included in these programs. (1),(2)

The Veterans Administration recognized the need for trained personnel and instituted a program to recruit and train Biomedical Engineering Technicians. (BMET). An experienced BMET may be recruited and hired by the hospital, but they are not readily available. The BMET program in the VA is designed to recruit and train people already employed by the hospital to undertake the duties of a BMET.

The primary source of candidates for the BMET training program is the maintenance personnel in the engineering shops. These people are usually electricians, electronics mechanics, or medical equipment repairmen, but others are considered if they show interest in the program. Candidates may also come from other areas of the hospital. The candidates usually have some experience in equipment repair, and may have repaired some of the simpler types of medical equipment. However, they generally have little or no experience in repairing complex electronic, pneumatic, mechanical or hydraulic systems. The formal education of some candidates may have ended at the high school level. Some may have technical or correspondance school training. The remainder may have attended college and may have received a degree at the associate or baccalaureate level.

Each candidate formally applies for the training program. Three letters of recommendation are required in addition to the application. Two of these letters must be from the professional medical services while the other is written by the chief engineer. The applications and supporting materials are sent to the Veterans Administration Central Office in Washington, D.C. where they are evaluated and a decision to accept the candidate or reject him is made.

A candidate becomes a trainee as soon as the application is accepted and a notification of the acceptance reaches the hospital. The trainee begins the program by taking a set of screening examinations. These are used to access the trainee's knowledge of medical terminology, amatomy and physiology, mathematics, chemistry, clinical instrumentation, troubleshooting, and electronics. The results of the examination are used to tailor an individual course of instruction for each trainee. Correspondance courses will then be given to make up for any weaknesses in the trainee's background. Those people with strong backgrounds in the above mentioned subjects will have to take no courses or olny a few of them. Others may have to take the entire series of courses if they have a weak background. The objectives of the training program are to bring each trainee to an acceptable level of competance without undue duplication of work they may have already completed, and to give them the skills they will need to perform as a BMET.

The BMET training program is directed by Dr. James Wear, and is headquartered at the Central Research Instrument Program (CRIP) facilities at Little Rock, Arkansas. The program consists of a series of correspondance courses taken at the hospital, "hands on" training in equipment maintenance, repair, and other technical subjects taken at CRIP, and on the job training in equipment maintenance at the hospital. Trainees may also be scheduled to participate in seminars and courses given by equipment manufacturers. Each trainee is alloted a portion of the day to study the correspondance material and the rest of the day consists of on the job training. However, the workload in the medical equipment repair shop may prevent the trainee from getting the study time during the day. Emergency repairs take precedence over studies. The correspondance courses supply needed theoretical information while the on the job training supplies practical experience. The training program usually takes about two or more years to complete, although some people complete it sooner. A second series of examinations is given as soon as the trainee completes his correspondance courses. They are certified by the VA as Biomedical Engineering

Technicians upon successfully completing the examinations. If they fail any part of the examination, they may retake it after one year. The BMET's training does not cease with the completion of the program as a continual effort is required to keep up with the changes in technology and the new equipment that is developed.

ADVANTAGES OF A BIOMEDICAL ENGINEERING TECHNICIAN PROGRAM

The principal advantages of a BMET program and the availability of an in house repair facility are:

(1) It is possible to respond rapidly to an emergency situation and some repairs can be made without disrupting patient care.

(2) The BMET can diagnose the problem and determine if it is possible to make repairs with available resources. If so, the repair can be made. If not, the service company can be contacted about the problem and they can bring the required parts with them.

(3) Equipment downtime can be shortened since the instrument does not have to be sent out for repairs. The time savings can be several weeks if parts are available locally. If parts are not locally available, then the downtime will be a function of the time required to obtain parts.

(4) Some emergencies can be handled on an around the clock basis by contacting the BMET. Service agencies may only be available during normal working hours. (5) The in house capability to provide a preventive maintenance program is valuable in redicing equipment downtime.

(6) Routine safety tests can be performed and documented, as required by regulations and the various accreditation commissions.

The development of a BMET program at a hospital will not solve all the maintenance problems nor reduce downtime in all areas. Many repairs will still require the services of the manufacturer or a service agency. Some equipment is so complex that the BMET never gets a chance to learn much about it, and it is critical to have the equipment available for patient care. The BMET does not have the time to learn the characteristics of the equipment and make the repairs at the same time. The manufacturer's service people specialize in one or a very few types of eqüipment and can make the repairs quickly. Even if the BMET knows a piece of equipment thoroughly and has the necessary skills, manufacturer's personnel may have to make the repairs because they have easy access to the parts required. It is impossible for any biomedical repair shop to stock the specialized parts to repair all the equipment in the hospital, but they can stock the simpler parts. Once a history of parts failure has been developed for systems the parts which fail most often can be stocked.

BIOMEDICAL ENGINEERING PROGRAM

The Graduate Clinical Engineer Training Program of the Veterans Administration is designed to recruit and train professional graduate and postgraduate level Clinical Engineers. There are two major program elements which are designed to accomodate the different training needs of these engineers.

a. The first program element is designed to indoctrinate all trainees in the basic elements of the VA Clinical Engineering program.

b. The second elent of the program is designed to provide additional training for individuals who are classified at the graduate level.

Trainees are classified into postgraduate and graduate catagories. A postgraduate trainees is one with a graduate degree in Biomedical Engineering, Clinical Engineering, Medical Engineering, Biophysics, Science, or related fields and specific work experience in clinical engineering for a minimum of two years. A graduate trainee is one with an undergraduate degree in engineering or science and little or no practical experience in clinical engineering, but who possesses the potentials to develop into a fully competent staff clinical engineer.

The first element of the training program consists of six sections, and includes much of the material which is common to the regular Graduate Engineer Training Program. Several objectives of the program are:

a. To indoctrinate the engineer to the policies and practices of the VA.

b. To provide sufficient training so the engineer can function as a staff Clinical or Biomedical Engineer.

c. Provide training and indoctrination in specific areas of clinical practice as performed in the VA hospital.

The training includes items such as Orientation, Operations, Maintenance and Repair, Security, Safety and Fire Protection, Administrative and Professional Engineering Activities, and Clinical Engineering Activities. The first element of the training program lasts from four to six months.

The second element of the program is designed to provide additional in service training for the graduate level trainees who do not have extensive experience in clinical activities. This training may be completed within one year. The material covered in this element will reflect the hospital's needs but may include the following:

a. Participation in planning long range medical programs,

b. Participation in maintenance and repair activities,

c. Participation in initial studies, design, cost estimation, and preparation of plans and specifications for new or renovated patient care facilities,

d. Supervision of new or renovated patient care facilities to see that patient care systems are properly installed,

e. Participate in medical equipment selection and application,

f. Preparation of lessions and selection of training aids for training nursing and medical personnel in clinical techniques required in medical practice such as the use of new patient care equipment. (3)

APPLICATIONS OF BIOMEDICAL ENGINEERS IN THE HOSPITAL

Every hospital will have its own problems and tasks to which a Biomedical Engineer can apply his/her talents. The following examples have been encountered at our hospital within the past year.

He assisted the nursing and medical personnel to debug a venous pressure monitoring system in the intensive care units. This involved considerable interaction between the nursing and medical staffs, maintenance personnel, and the manufacturers of the transducers and the monitoring system. The results were a standardization of transducers and techniques to connect the patient to the system. By means of appropriate training, the staff became much more confident in the operation of the system and the validity of the results.

Several staff physicians have contacted him about their research projects and may require his expertise in the near future.

The biomedical engineer was assigned the task of redesigning a nursing station which was scheduled for modernization. This task involved not only patient care systems but the demolition of the old station and construction of the new one.

Projects to convert existing space to research use are being developed. One of these involves extensive changes in a building presently used for research. The second project involves a complete functional change in the building to make it suitable for animal research.

An important aspect of the position of biomedical engineer is to keep up to date on the activities in the field. He maintains contact with BME's in other VA and private hospitals.

In order to advise medical personnel about future equipment purchases, he maintains contact with the major manufacturers of equipment within the hospital and is anxious that new equipment be the best suited to the task.

He has worked on the design for the renovation of an existing X ray room in which new equipment is being installed.

There is no doubt that the application of the Clinical, or Biomedical Engineer will be vital to the task of maintaining the standard of patient care we all wish to attain.

BIBLIOGRAPHY

(1) B. Patail, " On the Job Training (In the Department of Medical Electronics) for BMET's Affiliated with a Community College" Presested at the 10th Annual Meeting of AAMI, March 1975

(2) J.B. Finlay, and N.W. Prokos, "One Year Postgraduate BMET Program" Presented at the 10th Annual Meeting of AAMI, March 1975

(3) Veterans Administration "Graduate Clinical Engineering Training Program in VA Hospitals" TP10-20

RADIOLOGY AND THE ENGINEER: NEW CHALLENGES AND NEW OPPORTUNITIES

Edward S. Sternick, Farideh Bagne, and Bruce Curran
Dartmouth-Hitchcock Medical Center, Hanover, N.H. 03755

Introduction

A recent increase in demand by clinical Radiology for scientific and technical support has provided new and rewarding vocational opportunities for engineers having suitable educational qualifications. A majority of engineering employment openings in this branch of medicine are closely linked to the physical science subspeciality of Medical Physics which functions most effectively in the larger medical centers; and it is by means of this connection that the engineer has an opportunity to advance to professional parity with other health care workers.

Medical Physics groups commonly spend their time involved in three distinct but related areas: support of clinical activities, research, and teaching. In support of clinical tasks, the engineer contributes to instrumentation design, implements computer technology, selects and maintains equipment, and is intimately involved in the hospital's quality assurance program for radiation emitting machines and materials. The research role demands a commitment to an interdisciplinary approach to the problems of radiological science, with the engineer contributing expertise in electronic and mechanical design, and computers. Teaching responsibilities, in both basic and applied physical science, involve contact with resident physicians, medical students, technologists, and graduate students.

Because of their role as technical experts in a medical environment, radiological engineers must be prepared to acquire skills and experience beyond those normally encountered in traditional engineering disciplines. This may require supplementary didactic work in radiological physics as well as the inclusion of subjects such as anatomy and physiology into the curriculum. In addition, an internship spent with an established Medical Physics group in a busy Radiology environment is most helpful in providing the engineer with the specialized knowledge necessary for direct entry into a position of technical responsibility within the hospital.

The Need

Medical Physics can be categorized by its emphasis on the application of physical and engineering principles to health care delivery. It is thus reflective of the fact that much of basic engineering rests on a foundation of physics, while a considerable amount of engineering is visible in applied physics.

Historically, the association of Radiological Medical Physics has been primarily with Departments of Radiation Therapy. It is expected that in the United States with a population of 220 million people, 660,000 new cases of cancer will be produced each year. Of these, at least 50%, or 330,000 patients will require radiation treatments, either as the sole method of treatment or in combination with chemotherapy and surgery, which will be administered at 1100 megavoltage facilities (1). Further, it has been postulated, that the role of Radiation Therapy will be of even greater importance in the treatment of cancer in the future.

As demand for Radiation Therapy Services has increased, there has been a parallel growth in the development and installation of technologically complex equipment such as the medical linear accelerator, betatron, and computer facility. This in turn, has been reflected in an increase in demand for suitably trained physicists and engineers to participate in hospital-based Medical Physics support programs (2). It is of interest, that 60% of these positions are with University Hospitals having direct Medical School affiliation and hence imply the possibility for research opportunities and academic appointment.

In Diagnostic Radiology, demands for clinical services have been recently increasing at a compounded annual rate of 7%, and it is likely that in the next five years a number of factors will result in an expansion of these activities (3). Influences such as the implementation of a national health insurance plan which would increase demand for medical care, expanded efforts to deliver medical services to the urban poor and rural areas, and the tendency for hospitals to install larger and more complex radiological suites, is likely to be reflected in a requirement for personnel with appropriate expertise to evaluate and maintain radiological equipment.

Because of the absence of training programs specifically oriented toward radiological engineers, it is likely that the number of people with appropriate education is not now at a level which is sufficient to fill current needs, and that the anticipated expansion of Clinical Radiology will create many additional job openings in the next few years.

Duties and Responsibilities

The specific obligations of the radiological engineer may vary considerably depending on the size of the institution, the presence of an active Medical Physics group, and particular administrative arrangements of individual hospitals (4) Tables 1 and 2 summarize the variety of responsibilities delegated to Medical Physics and suggest the important niche the engineer can occupy in a busy clinical department.

TABLE 1
Radiation Therapy

1. Calibration and maintenance of radiation therapy machines and radioactive sources.
2. Calibration and maintenance of radiation dosimetry systems.
3. Provision of computerized systems for the calculation of dose distributions in patients undergoing treatment.
4. Design of monitoring devices for measuring doses delivered to the patient at selected sites.
5. Design of Radiation Therapy installations including calculation of shielding specifications.
6. Design and construction of special equipment and accessories.

TABLE 2
Diagnostic Radiology

1. Calibration, maintenance, and quality control programs for diagnostic x-ray machines.
2. Evaluation of performance characteristics of new equipment and parameters determining diagnostic imaging qualities.
3. Providing physics and engineering services in connection with newer diagnostic procedures and equipment such as computerized tomography.
4. Technical assistance in the assessment of physical and engineering aspects of various diagnostic procedures.
5. Design of Diagnostic x-ray facilities including calculation of shielding specifications.

In addition, most of the larger centers, particularly those with University affiliations, encourage research and development. Because of the multidisciplinary nature of Medical Physics, this effort usually involves cooperation with other specialists, most often in the fields of clinical radiology and radiobiology. Investigations are wide ranging and may be directed toward solutions to practical clinical problems or perhaps involve examination of the basic interactions of radiation with matter.

Education

To adequately prepare engineers for a role in Radiology, educational, programs must provide an understanding of the clinical environment as well as a sound foundation in engineering. At some point in the formal education process, additional practical experience must be gained, preferably under the close supervision of a senior Medical Physicist, so that the engineer will be able to communicate effectively with medical colleagues. Additional requirements demand a familiarity with the specialized measurement techniques and concepts which are fundamental to Medical Radiological Physics. Because the engineer may be required to provide support for research activities, formal exposure to research methodology should also be emphasized.

References

1. Manpower Goals in Radiation Oncology. Cancer 37(4), Supp. 2140-2148,1976.

2. Sternick, E.S. Report of the AAPM Placement Service July 1975-June 1976. Medical Physics 4(1): 1977.

3. Brogdon, B.G. and Moseley, R.D. The Comming Crisis in Radiological Manpower-Shortage? Radiology 114: 489-491, 1975.

4. Status and Future Manpower Needs of Physicists in Medicine in the Unites States. DHEW Publication No. (FOA) 79-8014, 1973.

REDUCING ELECTROSURGICAL HAZARDS FOR TRANSURETHRAL RESECTIONS

Glen C. Gerhard,* Ph.D., Stuart Selikowitz,† M.D.,
Richard Benedict,* and David French*

*Clinical Engineering Center and Department of Electrical Engineering, University of New Hampshire, Durham, NH 03824; †Dartmouth College - Mary Hitchcock - V.A. Medical Center, Hanover, NH 03755 and White River Junction, VT 05001

Abstract

This paper discusses electrosurgical hazards which may be present during transurethral resection (TUR) procedures. The causes of these hazards are examined and procedures are suggested for minimizing them. Additionally, two items to provide increased protection to both the urologist and the patient are discussed: a protective shroud for the resectoscope and a hazard monitor to detect excessive r.f. and d.c. potentials on the resectoscope frame.

Introduction

Electrosurgical instruments have been the subject of papers[1,2] which cite the hazards posed by the potential for radio-frequency (0.5 - 2.0 MHz) burns to both the patient and operating personnel. These include burns on the surgeon's hands due to faulty cutting or cauterizing probe insulation, burns at the dispersive electrode site due to improper or uneven skin contact, and burns produced on extremities or at EKG monitor or electrode sites due to alternate ground paths. While a new generation of solid-state equipment featuring isolated outputs for electrosurgical units and completely isolated input circuits for surgical monitors have reduced these risks, there are procedures where the risk of burns or electrical shock still have a higher probability than should be tolerated. One of the most prominent of these is the transurethral resection (TUR)[3,4] which involves one of the primary surgical tools of the urological service, the resectoscope. In this paper we will define the problem, analyze its causes, and propose several solutions.

Causes of Hazard

The resectoscope has, within its sheath, a viewing telescope with an illumination path, a fluid input channel, a fluid drain space, and a guide which holds the electrosurgical cutting loop. Because of the obvious space limitations, the cutting loop sheath diameter is rather small -- about 2.2 mm. The outer metallic sheath of this loop forms a coaxial structure with the inner tungsten wire, having a capacitance of about 35 pf. Also, during TUR procedures, the glycine solution used to flush the bladder becomes contaminated with blood and urine; this solution forms a surface film on the entire resectoscope and the surgeon's gloves during the course of the procedure which is inherently wet due to the frequent need for flushing and subsequent emptying of the bladder through the sheath which remains in place while the rest of the instrument is temporarily withdrawn. Even if this film did not provide a conductive path, it is conceivable that almost 22 mA could flow through the surgeon if a faulty glove provided a direct pathway (assuming 1000 ohms resistance of the surgeon). If this current flows through a small contact area or causes an r.f. "snap," discomfort occurs which can cause the surgeon to move involuntarily during the procedure.

Measurements made by one of the authors on the floating (loaded with 14 pf of probe capacitance) frame of the resectoscope indicate r.m.s. potentials exceeding 200 volts during cutting procedures. Utilizing the "pure cut" mode with a 0.5 MHz sine wave, isolated output electrosurgical unit has caused an approximately 150 volt d.c. component to appear on the frame with respect to ground so that the positive peak voltage on the frame was 600 volts above ground, enough to produce a "snap" to the surgeon. Additionally, cases of muscle stimulation of both the surgeon and the patient have been experienced which require a d.c. or low frequency current component. These are severe hazards during a TUR procedure since they may result in accidental perforation of the bladder by the cutting loop. Since the current for muscle stimulation would be expected to be at least 5 to 10 mA, a total loop impedance of 20,000 to 40,000 ohms is implied with a 200 volt source. However, it is strongly suspected that the d.c. arises from rectification effects at the cutting loop/tissue interface; since a d.c. component has been observed during cutting procedures, but not when the electrosurgical unit, connected to the resectoscope, was also connected to a dummy purely resistive load. If such is the case, then up to 600 volts may couple to the frame of the resectoscope; 10 mA of current would then flow through the patient and/or surgeon with a total path resistance of 60,000 ohms, a resistance which is probably obtainable with conducting liquid films covering the resectoscope and the surgeon's gloves. Added to this is the finite probability of a patient extremity making electrical contact with the grounded cystology table unit and the like probability of the urologist contacting ground (via the table and drain system), and it is apparent that the circuit can easily be completed for both d.c. and r.f. currents.

Remedies for the Hazards

The initial steps which may be taken to reduce the probability of one or more faults which increase the shock hazard to both patient and surgeon involve cystological room procedures and protocols. One resectoscope manufacturer (ACMI) has determined that certain surgeons gloves are more resistant to liquid retention and conductive film build-up than others. The patient should be isolated from accidental contact with ground even when an ungrounded (isolated) dispersive plate electrosurgical system is employed. The surgeon should likewise be isolated from both ground (including floor drains) and, whenever possible, direct electrical contact with the patient when the electrosurgical unit is activated. Of course, normal preventative maintenance checks should be made on the electrosurgical unit itself, including checks for isolation and low-frequency d.c. components. The resectoscope itself should be checked for arcing points at the hub glue-joint and within the sheath -- arcing through the sheath could, in the opinion of the writers, cause urethral strictures. An electrical test procedure can be used to check individual instruments for gross leakage prior to use; a simple lamp and resistor circuit may be obtained from a resectoscope manufacturer (ACMI). However, these procedures do not protect the surgeon from a breakdown of insulation within the resectoscope which may occur during a TUR procedure. It is difficult for the urologist to remain completely isolated from ground -- he is working in a wet environment, in close quarters with a grounded stainless steel table with an attached drain directly under his working field. Therefore, two additional methods of protection have been examined.

The first involves increasing the basic safety of the resectoscope itself -- mainly decreasing the probability of an electrical breakdown and increasing the high-frequency impedance between the active cutting loop and the frame (which cannot be grounded as it would increase the stress on the cutting loop insulation and increase the likelihood of a breakdown). Unfortunately, the only way of really accomplishing both of these goals is to completely redesign the instrument using high-mechanical strength dielectric materials. While this remains a desirable long-term objective, it became apparent that another interim approach was needed immediately.

A protective sheath was designed to cover the instrument. The sheath had to be fabricated from a dielectric of sufficient breakdown strength and thickness to withstand several thousand volts, (CSV Bovie potentials), one that could be machined and assembled relatively easily, one that could withstand several types of sterilization techniques (up to 150°C) and, preferable, one that was available and not outrageously expensive. Mechanical stiffness and strength properties were also of interest. A prototype model was made of this instrument and is currently being evaluated.

The other approach taken to increase safety was that of designing a hazard-alerting monitor -- here a monitoring device is connected electrically to the frame of the resectoscope and both the r.f. and d.c. potentials are sensed and compared with pre-set threshold values. When one of the latter is exceeded, a quiet alarm is sounded, thus alerting the surgeon to use extra caution in handling the resectoscope. The current approach is based on a 500 μA d.c. limit (assuming a resistance of 1000 ohms) and a variable r.f. voltage limit. Care has been taken so as not to deliberately provide a low impedance r.f. path to ground from the resectoscope frame for the reason previously cited. A block diagram of this monitor system is shown in Figure 1. Clinical evaluation of this system is scheduled for Spring 1977.

Summary

The reasons for electrosurgical hazards during TUR procedures have been examined. Analysis of actual incidents involving r.f. burns and/or muscle stimulation has shown that both r.f. and d.c. hazards can be present. Solutions to this problem involve modified procedures and protocols to minimize the faults which lead to such incidents. Additional clinical engineering contributions currently in process include protective sheathing for exposed resectoscope frames and a hazard monitor to warn the surgeon of impending shock potentials.

References

1. Emergency Care Research Institute: Health Devices, Vol. 2, pp. 183-227, June-July, 1973.

2. A.S. Wald, V.C.B. Mazzia, F.C. Spencer: "Accidental Burns Associated with Electrocautery," JAMA, Vol. 217, p. 916, 1971.

3. G.R. Goodman: "Electrosurgery Burns and the Urologist," Journal of Urology, Vol. 116, pp. 218-220, August, 1976.

4. R.J. Prentiss, G.W. Harvey, W.F. Bethard, et. al.: "Massive Adductor Muscle Contraction in Transurethral Surgery: Cause and Prevention: Development of New Electrical Circuitry," Journal of Urology, Vol. 93, p. 263, 1965.

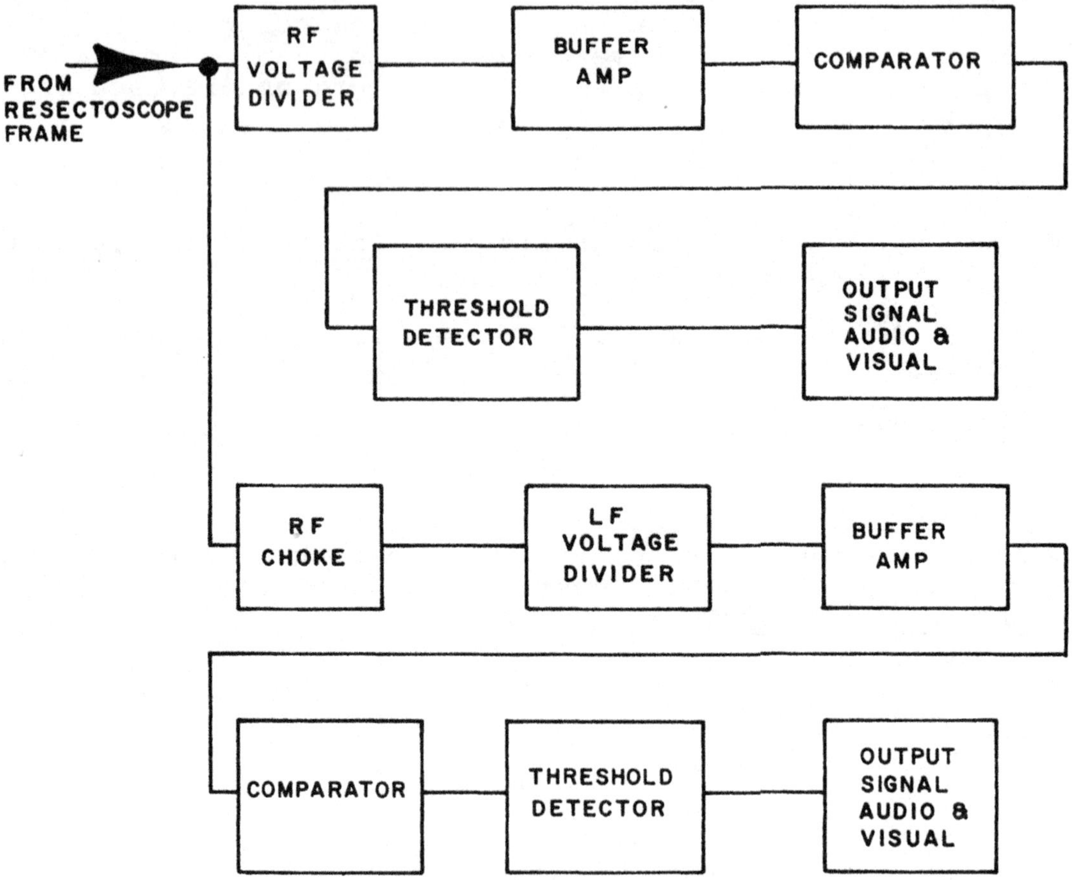

FIGURE 1: BLOCK DIAGRAM OF HAZARD MONITOR

COMPUTER-CONTROLLED PARAMETRIC STUDIES OF EVOKED RESPONSES IN HUMANS*

J. J. Brickley, Jr., S. I. Lee, A. Anné, R. N. Johnson
Division of Biomedical Engineering and Department of Neurology
University of Virginia, School of Medicine
Charlottesville, Virginia 22901

Introduction

The human cortical evoked response to sensory stimuli has been recognized as a valuable aid in clinical studies and diagnoses. The commonly used modalities are visual, auditory and somatosensory. The aim of the present study is to develop a methodology to maximize the information extracted from evoked responses, while subjecting the patient to a minimum number of stimulations.

The inherent problem in measuring the human cortical evoked response is the low level of potentials measured from scalp electrodes. These potentials, usually less than 50 μ volts, are generally below the indigenous noise level of the man-machine system. To alleviate this problem the standard procedure has been to give the same stimulus repeatedly, usually 25 or more times, and average the sampled responses. This tends to cancel the component of the signal which is not synchronized with the stimulus and leave the specific response intact. The main disadvantages of this procedure are the necessity for a large number of stimulations to determine a single response and the concomitant physical and psychological effect on the human subject. The latter includes the effect of habituation, in which the magnitude of the response diminishes with repetitive stimulation.

Despite the great effort necessary to establish the response to a single stimulus with sufficient accuracy, it is desirable to subject the patient to a broad range of stimuli, since the degree of information obtained from a system is determined by the degree of information or variety contained in the stimulus parameters.

We have conducted parametric studies of evoked responses in cats, of use in investigating drug effects and pathological mechanisms [5]. In the cat, the relatively high level of potentials measured from the surface of the cortex (i.e., 200-400 μ volts) provides single responses with an acceptable signal-to-noise ratio. Although these animal studies provide a basis for undertaking analogous studies on human subjects, such studies are hampered by the classical averaging procedures discussed previously.

The present investigation indicates that parametric studies from human scalp recordings are feasible using a modified testing and averaging procedure, combined with digital filtering. The use of a digital computer to sequence stimuli precisely, store sampled data, perform analyses and control three-dimensional plotting of results (Fig. A-D) is basic to the methodology. A key element for improvement of final resolution is Wiener filtering, discussed below.

Wiener Filtering

A conceptual construct was proposed in 1949 by Norbert Wiener [9] for the aposteriori filtering of sets of measurements of a repetitive phenomenon masked with random noise. The basic concept involves estimating the power spectra due to the noise (assumed to be a stationary random process) and to the signal

* This work supported in part by NIH Contract #NS-5-2329.

(i.e., repetitive phenomenon) in such a way as to minimize the RMS error. In 1969, D.O. Walter [8] proposed using the Wiener filtering technique for evoked responses and published an illustrative theoretical example, assuming normally distributed noise. Doyle [3] showed in 1975 that Walter's formulation of the filter, which had subsequently been used by Nogawa, et al. [6] was in error. Walter's formulation was valid for filtering single responses and not the ensemble averaged response, as proposed. Using the revised formulation, Hartwell and Erwin [4] demonstrated the effective use of the filter on visual evoked responses from human subjects.

Recently, Basar, et al. [2] and Ungan and Basar [7] have pointed out certain limitations of Wiener filtering of evoked responses and have demonstrated these with animal data. One point of contention is how well the assumption of stationarity of the background activity (i.e., noise) holds. Another difficulty arises from the consideration that the calculated signal power at a given frequency should diminish when longer time epochs are analyzed, whereas the calculated noise power should remain essentially unchanged. These factors could distort the estimates and cause erroneous filtering. Albrecht and Radil-Weiss[1] have called into question Walter's use of a factor of 1/N (N is the number of responses averaged) in his formulation, which arises from the assumption that the noise is normally distributed. They derive a mathematical representation of a time function which is a linear sum of a signal, synchronized with the stimulations, and a random background. Using this function, they demonstrate that the noise power in an ensemble averaged response is not uniformly reduced by a factor of 1/N across the spectrum, as Walter assumes. Rather, the attenuation factor is a function of frequency whose magnitude varies from zero to one. Thus, some frequencies in the noise spectrum should be unaffected by the averaging, while others are almost totally eliminated.

Methods

The detailed methodology for the computer-controlled parametric studies of thalamically evoked cortical responses in the cat has been presented elsewhere [5]. In the clinical studies, pairs of pulses were generated at 3 sec intervals from a PDP-11/10 computer. These pulses were used to trigger a Grass model PS-22 photo stimulator at one of 5 light intensities. The pulse interval (i.e., time between the two pulses in a pair) was varied from 543.9 to 22.2 msec, being stepped down under software control during the interim between pairs. Responses were sampled for 2048 msec following the second stimulus in a pair and data were asynchronously written to a disk for subsequent analysis. Each series, at a single intensity, took about 2.5 min and was followed by a 5 min rest period.

In order to minimize muscle and eye-blink artifact, the subject was seated in a comfortable position with eyes closed. Under these conditions, high levels of alpha activity (8-12 Hz) usually occur. The presence of this contaminating activity was identified from the Fourier spectrum and removed digitally, using 2048 msec of sampled data to obtain adequate resolution.

A form of "sliding-window" ensemble averaging was used to obtain the averaged evoked response (AER). The first 9 responses in a series were used to form the first AER. Responses 2 through 10 formed the second AER, and so forth. Next, the Wiener filter for each AER was calculated and applied in the frequency domain. Then the Wiener-filtered response (WAER) was reconstructed in the time domain. Since the primary response can be expected to occur within about 100 msec, the WAER was calculated over 128 msec post-stimulus. This reduces the possibility of over-filtering due to the dependence of the calculated signal power spectrum on the length of the time epoch analyzed, as mentioned above. Finally, the RMS voltage level over a given time interval was computed for each WAER and these data were plotted in three-dimensional form with pulse interval and intensity of the stimulus as independent variables and RMS response voltage as the dependent variable.

Results

A control three-dimensional plot for the cat is shown in Fig. A, with the plot in Fig. B demonstrating the effect (at a clinical therapeutic blood level) of exthosuximide, a common anticonvulsant. Note the extensive attenuation which has occurred in the post-drug profile, implying that significant changes can be expected in evoked response profiles between different clinically relevant states.

An example from initial trials in a normal human subject is shown in Fig. C-H. Fig. E and F contain raw responses after removal of alpha activity. These are 2 of the set of 9 responses used to form the AER shown in Fig. G and are indicative of the high level of noise masking the response. Note the reduced noise level of the response in Fig. H, which is the WAER corresponding to the AER in Fig. G.

Fig. C and D contain visual response planes (VRPs) demonstrating the overall results of the parametric study. In Fig. C, the VRP has been constructed from the RMS response levels over the period 5 to 30 msec post-stimulus. There are multiple peaks at each intensity. The most prominent ones are a 4.38 μvolt peak for a pulse interval of 144.3 msec occurring at intensity 1, a 4.63 μvolt peak for a pulse interval of 266.4 msec occurring at intensity 8, and a 4.61 μvolt peak for a pulse interval of 421.8 msec occurring at intensity 16. In Fig. D, the RMS response levels have been calculated for 30 to 90 msec post-stimulus. Peaking at intermediate pulse intervals occurs only at intensity 8 and there is a general upward trend in the response levels with increasing pulse interval. Note the relatively low level of the profile at intensity 2.

Discussion and Conclusions

Preliminary investigations indicate that assuming the background activity to be normally distributed and stationary is not unreasonable. For a normally distributed random process, the mean magnitude of an ensemble average of measured time epochs of the process is proportional to $1/\sqrt{N}$, where N is the number of epochs averaged. Thus, the percentage reduction diminishes with each additional epoch. In the present case, averaging 9 epochs should reduce the mean noise level in the data to about 33% of its original value. An additional 7 epochs would be necessary to reduce this level to 25%. From this perspective, 9 is a reasonable number to employ. When calculating the Wiener filter, this also implies an assumption of stationarity over 27 sec, which is only about half the time duration of many standard portocols used in evoked response determinations in humans.

Walter's factor of 1/N has been used in the filter formulation, despite the observations of Albrecht and Radil-Weiss. Although their conclusions are reasonable for continuous spectra, the discrete spectra employed here are subject to leakage, producing an average reduction across all harmonics, and the mean value of their frequency-dependent factor is 1/N.

The modified "sliding-window" form of ensemble averaging should produce some loss in signal level, both from the averaging itself and from the Wiener filtering. However, all the data are subjected to the same procedure. The basic question is how large a variation in stimulus parameters can be tolerated within a set of responses used to construct an AER. This procedure is a form of low-pass filtering and, if carried to an extreme, would produce an essentially flat response curve. The fact that this has not occurred is indicative of the reasonableness of the procedure.

The form of the final results obtained from following the proposed protocol is the ultimate measure of usefulness. It is reasonable to expect complex dynamics in the early post-stimulus period, followed by an integrated response involving peripheral cortical structures. The VRP from the preliminary trial, presented in Fig. C (5-30 msec post-stimulus), has a number of interesting features, the most outstanding of which is the progression of three prominent peaks moving diagonally back across the plane. This may be indicative of the early dynamics of the response. In Fig. D (30-90 msec post-stimulus), the VRP shows a general upward sweep in the direction of longer pulse intervals, which may be due to the development of an integrated response by the central nervous system.

The ultimate criterion for a clinical indicator is its sensitivity to changes in the state of the system. We are presently investigating the consistency and sensitivity of the VRP. One preliminary point of interest in this regard is provided by the VRPs in Fig. C and D. An independent, unbiased analysis of the EEG recorded during the test period established that the subject became drowsy during most of the series run at intensity 2. This is intimated in the VRPs by the relatively low level of the profile and absence of a prominent peak at this intensity. In conclusion, we feel that the preliminary trials of the protocol presented here provide encouraging evidence that, with further refinement, the VRP may become an effective clinical indicator and investigative tool.

References

1. Albrecht, V. and Radil-Weisss, T., Biol. Cybernetics, 1976, 24:43-46.
2. Basar, E., et al., Biol. Cybernetics, 1975, 20:127-143.
3. Doyle, D. J., Electroenceph. clin. Neurophysiol. 1975, 38:533-534.
4. Hartwell, J. W. and Erwin, C.W., Electroenceph. clin. Neurophysiol.,1976, 41:416-421.
5. Johnson, R.N. et al., Brain Research, 1976, 103:568-573.
6. Nogawa, T., et al., Electroenceph. clin. Neurophysiol., 1973, 35:375-378.
7. Ungan, P. and Basar, E., Electroenceph. clin. Neurophysiol., 1976, 40:516-520.
8. Walter, D.O., Electroenceph. clin. Neurophysiol., 1969, Suppl. 27:61-70.
9. Wiener, N., Extrapolation, interpolation and smoothing of stationary time series, J. Wiley, New York, 1949.

Legends

Fig. A: Control response profile in cat.
Fig. B: Response profile in cat following exthosuximide administration.
Fig. C: VRP from normal human subject (RMS response 5-30 msec post-stimulus).
Fig. D: VRP from normal human subject (RMS response 30-90 msec post-stimulus).
Fig. E: Raw response from normal human (521.7 msec pulse int. at intensity 16).
Fig. F: Raw response from normal human (477.3 msec pulse int. at intensity 16).
Fig. G: Ensemble average of 9 responses from normal human subjects (432.9-521.7 msec pulse intervals, mean = 477.3 msec, at intensity 16).
Fig. H: Averaged response in Fig. G after Wiener filtering.

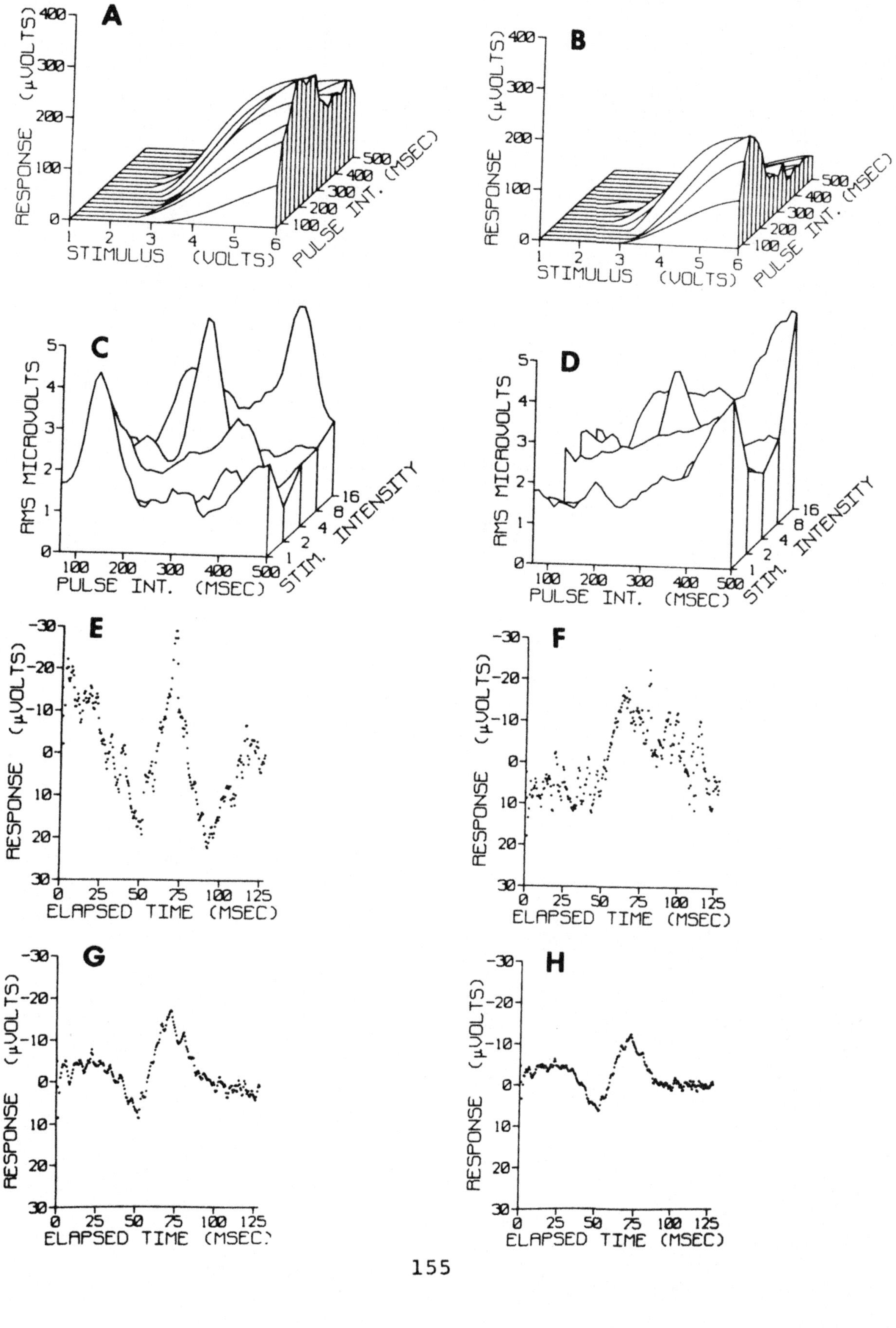
A
B
C
D
E
F
G
H
RESPONSE (µVOLTS)
STIMULUS (VOLTS)
PULSE INT. (MSEC)
RMS MICROVOLTS
PULSE INT. (MSEC)
STIM. INTENSITY
RESPONSE (µVOLTS)
ELAPSED TIME (MSEC)

COMPENSATION OF FLUID-FILLED CATHETER PRESSURE WAVEFORMS BY LINEAR PREDICTIVE ANALYSIS AND DIGITAL INVERSE FILTERING

Leland B. Jackson, Dov Jaron and Steven L. Wood
Department of Electrical Engineering
University of Rhode Island
Kingston, Rhode Island, 02881

Introduction

The non-uniform frequency response of fluid-filled catheters causes distortion of the measured pressure waveform. An effective technique for compensating the frequency response was proposed by Ciccolella et al [1]. That technique consists of filtering the data with a digital inverse filter corresponding to a second-order linear model for the catheter. The parameters of the inverse filter are determined from a "pop" test performed on the catheter before and/or after data are collected. Although quite effective, there are two drawbacks to this technique. Firstly, the properties of the fluid-filled catheter system may change between the pop test and the data collection. Secondly, the pop test requires a special apparatus and extra steps in the data collection procedure. Thus, it would be most convenient to dispense with this test, especially in the clinical situation.

We have developed a new method which allows for the calculation of the catheter parameters directly from the pressure data of interest, eliminating the necessity for the pop test. Hence, both of the above drawbacks have been eliminated. The catheter parameters are obtained from linear predictive analysis of the data using the covariance method [2]. An all-pole model of the combined catheter system and pressure signal is assumed, with two or four of the complex-conjugate poles being ascribed to the catheter. The data are then filtered by a nonrecursive digital filter corresponding to the inverse of the catheter model. The parameters of the catheter can be separated from those of the all-pole model of the pressure signal itself by implementing the linear prediction filter as a cascade of second-order filters, rather than as a single higher-order filter [3], or the system function of the higher-order filter can be rooted to separate these parameters.

Description

Linear prediction is a widely used method for parameter estimation in speech analysis, seismology, and system identification [2]. As commonly formulated, the input sequence $x(n)$ (i.e., the sampled input signal) is assumed to be predicted by a sequence $\hat{x}(n)$ given by

$$\hat{x}(n) = \sum_{k=1}^{P} a_k x(n-k). \tag{1}$$

The prediction error is then $e(n) = x(n) - \hat{x}(n)$, and the parameters a_k are determined such that some norm of the error sequence is minimized. For periodic signals such as the pressure waveforms, an appropriate norm is the total square error for an integer number of periods, i.e.

$$E = \sum_{n=1}^{N} e^2(n). \tag{2}$$

The minimization of E implies that $\partial E/\partial a_k = 0$, for $k = 1,2,\ldots,P$, which in turn yields the "normal" equations

---

Supported in part by NSF Grants ENG 76-00735 and ENG 74-21085, and by a grant from the Stride Memorial Foundation.

$$\sum_{k=1}^{P} a_k \phi(k,p) = \phi(0,p), \qquad p = 1,2,\ldots,P \tag{3}$$

where

$$\phi(k,p) = \sum_{n=1}^{N} x(n-k)x(n-p). \tag{4}$$

The solution of the simultaneous linear equations in (3) yields the estimated parameters a_k, and this is known as the "covariance method" of linear prediction.

The error e(n) is effectively the output of a nonrecursive inverse digital filter with system function

$$H(z) = \sum_{k=1}^{P} a_k z^{-k} \tag{5}$$

where the term "inverse filter" implies that the generating system is modeled as an all-pole filter with system function 1/H(z). Hence, the P zeros of H(z) correspond to poles of the system model. Some of these zeros correspond to poles of the pressure waveform itself and some to poles of the catheter. If we can identify M zeros σ_m, m = 1,2,...,M, as being associated with the catheter model, we can remove the distortion on x(n) due to the corresponding catheter poles by filtering x(n) with the inverse filter

$$H_I(z) = \prod_{m=1}^{M} (1 - \sigma_m z^{-1}). \tag{6}$$

The zeros σ_m are obtained either by rooting H(z) as determined from (3) and (5), or by determining H(z) directly in factored form as described in [3]. We have used the latter technique. This involves the expression and evaluation of H(z) in "cascade form" as

$$H(z) = \prod_{j=1}^{P/2} (1 + \alpha_{1j} z^{-1} + \alpha_{2j} z^{-2}) \tag{7}$$

where the coefficients α_{ij} are real. The nonlinear normal equations associated with (7) are solved by an iterative, gradient-based algorithm [3]. Letting the first M/2 factors in (7) be identified with the catheter, the inverse filter used for compensating x(n) is then

$$H_I(z) = \prod_{m=1}^{M/2} (1 + \alpha_{1j} z^{-1} + \alpha_{2j} z^{-2}). \tag{8}$$

Examples

We have applied linear prediction of order P = 4 and 6 to four sample pressure waveforms obtained from animal experiments with inverse filtering of order M = 2 and 4. An additional zero of H(z) was fixed at $\sigma_{P+1} = 1$ to remove any d.c. bias and to give a 6-db/oct preemphasis in x(n). An additional zero of $H_I(z)$ was fixed at $\sigma_{M+1} = -1$ to attenuate high-frequency noise components [1]. Fig. 1 is a typical example of the four zero locations in the z plane for P = 4, where the estimated zeros are denoted by squares and the large circle corresponds to $|z| = 1$. The paths of the zeros from their initial to final locations using the iterative solution technique in [3] are also shown. Note that two zeros are real and two are complex conjugates, and because of the typically underdamped nature of the catheter distortion, we ascribe the complex zeros to the catheter and the real zeros to the pressure signal itself. Fig. 2 shows the corresponding pressure waveforms obtained

(A) from the fluid-filled catheter, (B) from a catheter-tip manometer, (C) from our compensation technique with P = 4 and M = 2, and (D) by using the pop test and a second-order filter.

Fig. 3 shows the zero locations for the same data with P = 6. Note that all six zeros are complex. The two zeros with the lowest natural frequencies (i.e., nearest the positive real axis) are associated with the pressure signal itself. The next-lowest-frequency zeros are the most underdamped (i.e., closest to the unit circle) and correspond to the principal poles of the catheter model. Hence, these are selected for the compensation filter when M = 2; while the highest-frequency zeros are added to $H_I(z)$ when M = 4. Figures 4 and 5 present the corresponding waveforms as described above for M = 2 and M = 4, respectively. Note the noticeable improvements in our compensation results as P and M are increased. For the data samples studied to date, no improvement has been gained by using P > 6.

A second data example is presented in Figures 6 thru 10, where the sequence of figures is the same as for Figures 1 thru 5. Note from Fig. 8 that for P = 6 there are two real zeros, which we associate with the pressure signal itself. Note also from Figures 7, 9 and 10, that no significant improvement is obtained by increasing P from 4 to 6 for this data unless M is also increased from 2 to 4. Interestingly, the compensated waveform obtained from the pop test and M = 2 is comparable to our result for M = 4.

Conclusions

We have shown that linear prediction can be used to determine the poles of a fluid-filled catheter model, without the necessity of a separate test of the catheter. Digital inverse filtering will then remove most of the corresponding distortion. Two or four complex-conjugate zeros are employed in the inverse filter. Sixth-order linear prediction appears to be most appropriate.

References

1. S. Ciccolella, L. Jackson, A. Most and D. Jaron, "Compensation of Fluid-Filled Catheter Response using Digital Filter Techniques", Proc. 4th New England Bioengineering Conf., pp. 265-268, May 1976.
2. J. Makhoul, "Linear Prediction - A Tutorial Review", Proc. IEEE, Vol. 63, pp. 361-380, April 1976.
3. L. Jackson, R. Rao and S. Wood, "Parameter Estimation by Linear Prediction in Cascade Form", Proc. 1977 IEEE Conf. on Acoustics, Speech, and Signal Processing, Hartford, Conn., May 1977.

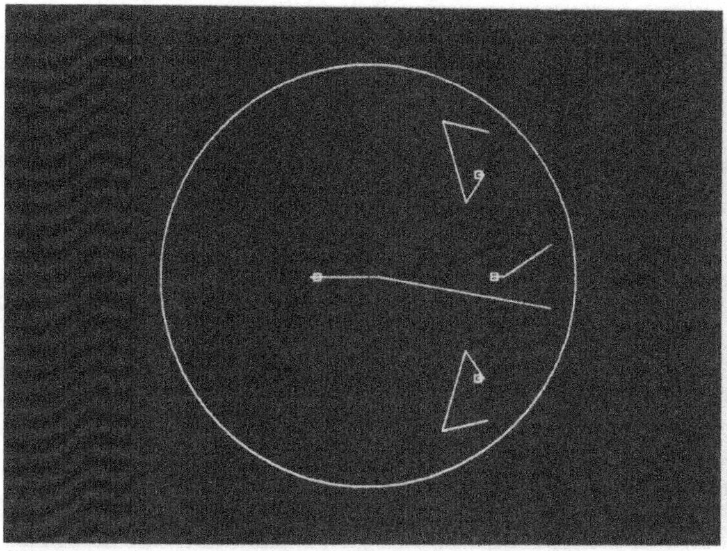

Fig. 1 - z plane for P = 4

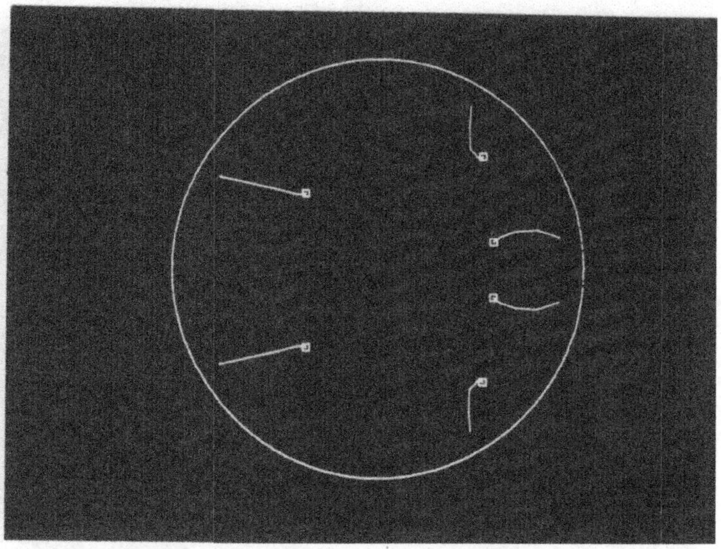

Fig. 3 - z plane for P = 6

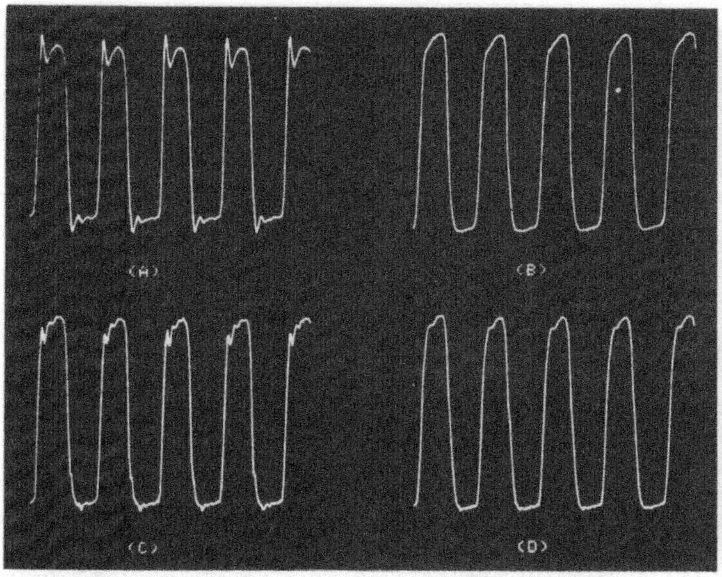

Fig. 2 - Waveforms for P = 4, M = 2

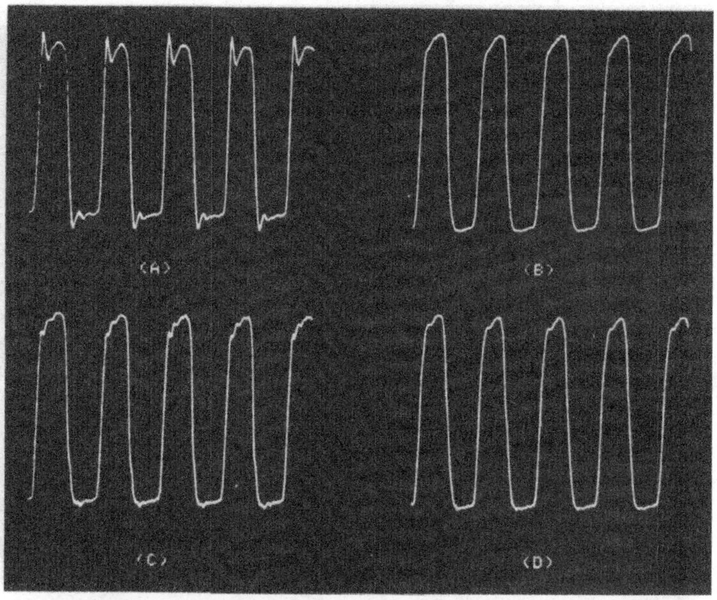

Fig. 4 - Waveforms for P = 6, M = 2

Waveform Descriptions:

(A) - Fluid-Filled Catheter

(B) - Catheter-Tip Manometer

(C) - Linear Prediction and Inverse Filter Compensation

(D) - Pop Test and Inverse Filter Compensation

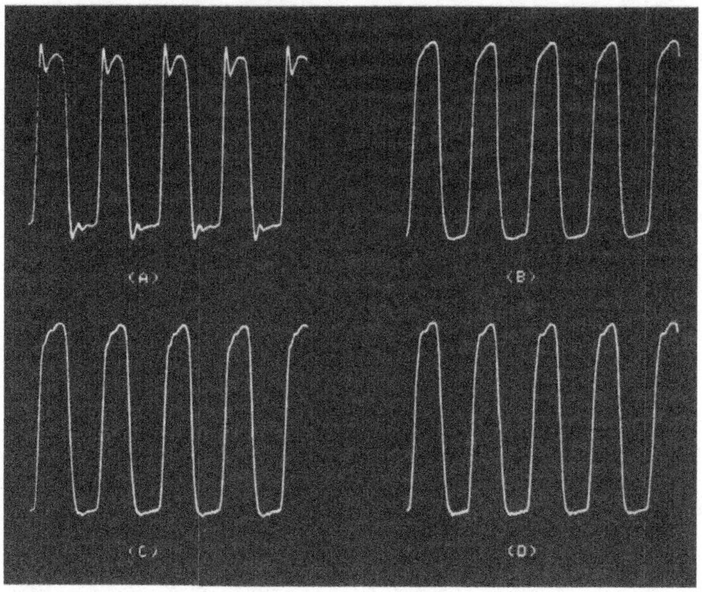

Fig. 5 - Waveforms for P = 6, M = 4

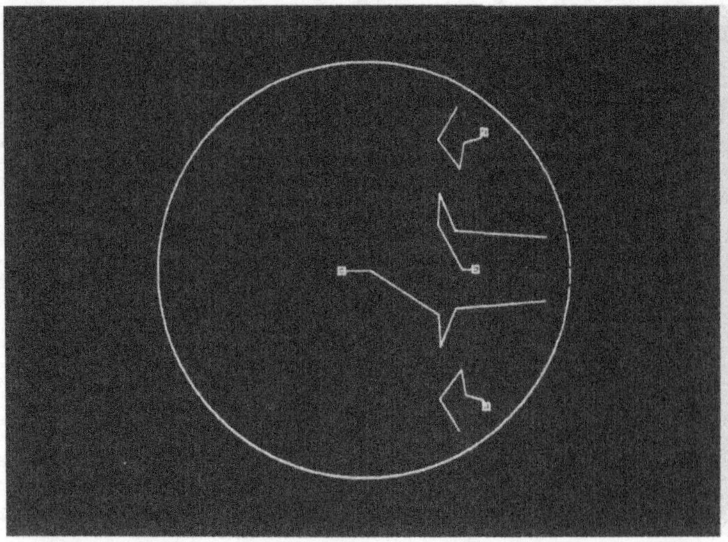

Fig. 6 - z plane for P = 4

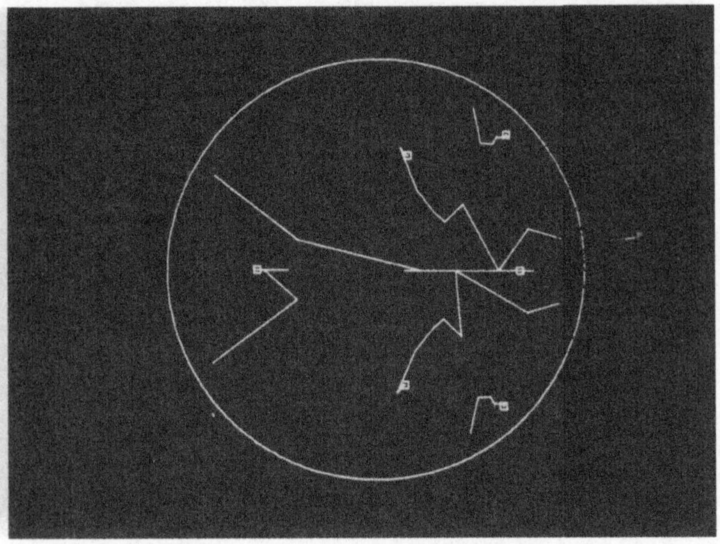

Fig. 8 - z plane for P = 6

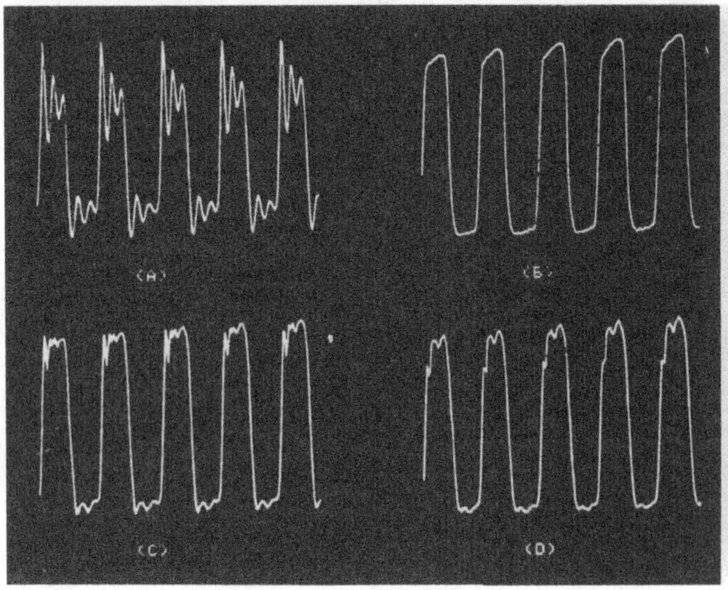

Fig. 7 - Waveforms for P = 4, M = 2

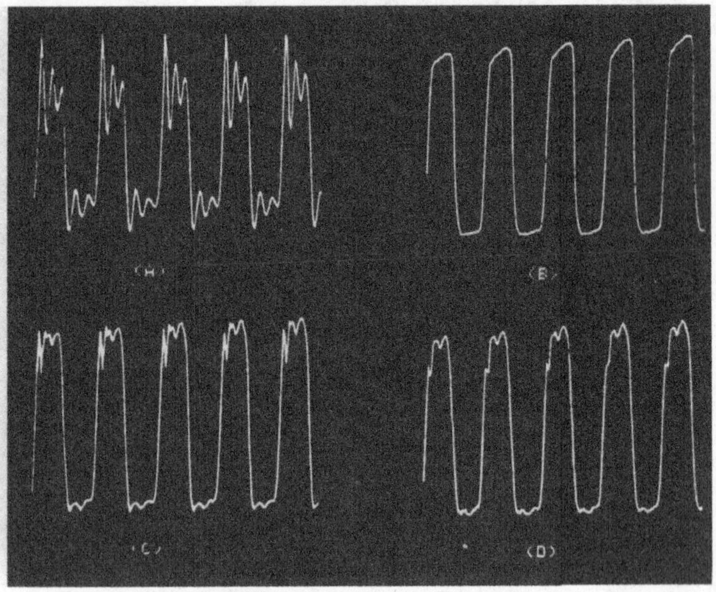

Fig. 9 - Waveforms for P = 6, M = 2

Waveform Descriptions:

(A) - Fluid-Filled Catheter

(B) - Catheter-Tip Manometer

(C) - Linear Prediction and
Inverse Filter Compensation

(D) - Pop Test and
Inverse Filter Compensation

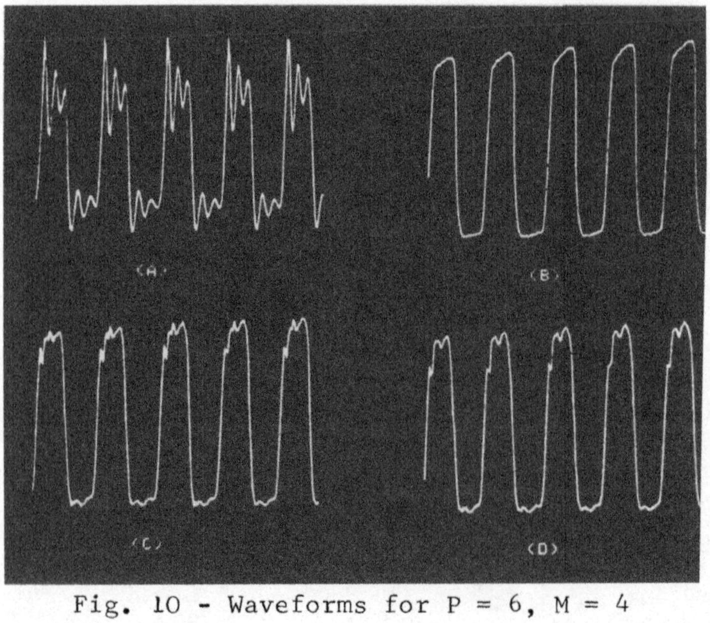

Fig. 10 - Waveforms for P = 6, M = 4

IMAGE ENHANCEMENT OF CONVENTIONAL TRANSVERSE-AXIAL TOMOGRAMS

Carter H. Yates, John W. Strohbehn, Chris Walker
Thayer School of Engineering, Dartmouth College 03755

and Bruce H. Curran and Edward S. Sternick
Norris Cotton Cancer Center 03755

Summary

This paper describes the authors' attempts at improving the diagnostic utility of conventional transverse-axial tomograms using digital image processing techniques. Conventional tomography attempts to image a single cross-sectional plane of the patient by focusing on this plane and blurring all adjacent planes to varying degrees. The end result is an image badly degraded with unwanted blur. Prominent features of interest such as bone-tissue boundaries are often not clear and are difficult to interpret. Therefore, digital processing techniques were used to determine 1) if the desired information has been irretrievably lost and 2) if not, can it be enhanced. Representative tomograms were digitized and placed on magnetic tape. These digitized images were then processed using several spatial filtering algorithms and the results written back onto a magnetic tape. The contents of this tape were then displayed on a video monitor. It was found that linear filtering techniques improved the perceived information within the conventional tomograms used in this study. The limits to linear enhancement filtering arise from the noise level in the image.

Introduction

In the field of radiation cancer therapy, there is an important need for high-quality imaging techniques. At the Norris Cotton Cancer Center, use is made of several imaging devices with the interior body information coming primarily from x-ray tomography techniques (Sternick et al., 1974). Using these techniques, density variations within a single anatomical plane or cross-section are recorded as a true 2-dimensional image. However, as can be seen in figure #4-a, it is difficult to distinguish the important features using conventional tomography techniques. These important features are primarily the bone-tissue boundaries. Although increasing use has been made of computed tomography due to the better image quality, cost considerations provide an incentive to improve the usefulness of conventional tomographic images.

The principle behind linear tomography is shown in figure #1-a. As the x-ray source moves from S_1 to S_2, the film moves from F_1G_1 to F_2G_2 such that the image of AB (in the tomographic plane) remains stationary on the film. However, the image of $A'B'$ (outside the tomographic plane) moves relative to the film's motion thereby producing a blurred image of itself. Hence, the conventional tomographic image contains an "in-focus" image of the density variations within the tomographic plane and blurred images of the density variations in adjacent planes. Qualitatively, the resulting poor quality can be attributed to the limited exposure range of the film and the degree of blurring from nearby planes covering up the desired detail. Transverse-axial tomography is similar in concept to linear tomography except that it is optimized to get lateral cross-sections as opposed to longitudinal planes. The geometrical configuration of this unit is shown in figure #1-b.

The imaging properties of the transverse-axial unit can be described quantitatively using Fourier decomposition techniques (Orphanoudakis & Strohbehn, 1975). The modulation transfer function for each plane parallel to the tomographic plane can be represented as in figure #2. Each point in this graph represents the amplitude response to a density variation with spatial frequency components k_x and k_y in a plane located a distance "z" from the tomographic plane. These values are normalized by the response to the same density variation within the tomographic plane. We see that the imaging process acts as a low-pass filter to all planes but the tomographic plane, with a decreasing cut-off frequency as the distance from the tomographic plane is increased.

This study suggests one possible enhancement scheme. The high spatial frequency information within the film has been shown to be characteristic of only the planes near the tomographic plane. Low-frequency information is generated from all planes. Therefore, since the desired information is primarily contained in edges and since high-spatial frequencies contain much edge information, a strategy of emphasising the high spatial frequencies within the image would enhance the edge information in the tomographic plane.

The Image Processor Design

It was decided to use digital processing techniques (as opposed to photographic or optical techniques) because of the potential power and flexibility available. The films used were first digitized and stored on magnetic tape by Optronics Intl. using their micro-densitometer. Since Dartmouth does not presently have a processing system that is easily adaptable to image processing or display, the computing was done using the Dartmouth Time-Sharing System (DTSS). The output from DTSS (a magnetic tape) was then carried to the Ohio-Nuclear computed tomography unit that resides in the diagnostic radiology department of Mary Hitchcock Memorial Hospital (MHMH). The display facilities connected with this unit were used to provide a video display of the processed output.

The primary image processing algorithms that were implemented were:

1) Spatial Frequency Filtering

2) Histogram Modification
3) Spectrum Calculation
4) Addition of Images
5) Image Resampling and Windowing

Several supplementary routines are also available such as filter design programs and a half-tone output program to be used with a high-speed printer. The most complicated (and most used) of the algorithms mentioned above is the spatial frequency filtering program. The number of calculations is large as is the data-handling requirements so that great care was taken to optimize the operation of this algorithm. A frequency domain approach was used as opposed to a convolution for the resulting savings in computations. Usually, each image filtered was 200 X 200 points and was broken up into 4 sections when being processed to reduce the data storage requirements. At the present time it takes about 150 seconds and 4000 Input/Output units to filter one image. This costs $7.50 using the non-commercial rates at DTSS.

Results of Tomographic Image Enhancement

The tomogram shown in figure #4-a was one of those used in the enhancement process. Figure #4-b shows the same image after being digitized and displayed on a video screen. Note the loss of image quality in the display.

The first step taken in enhancing the tomographic images was to analyze the original images using quantitative measurements such as the image spectrum and histogram. The picture spectrum revealed that most of the image energy resides in the lower frequencies (<.1 cycles/mm.) with a sampling frequency of about .5 cycles/mm. used. Other research has shown that this is a general pattern followed by most images; however, this effect seems to be stronger in this family of images. A study of the histograms of the digitized tomograms shows that most of the points within the image lie at the extremes of the film exposure curve. This means that the contrast in the areas populated by these points is lessened since the film density does not change as much for a given increment of exposure. These two image measurements indicate that the amount of high frequency information (edge information) is relatively small and that there is a problem of low contrast due to the large intensity range of the transmitted x-rays.

With this problem in mind, a series of high-pass filters with a finite impulse response and linear phase were designed. The sampling process that provided a 200 X 200 point image acted as a low-pass filter so that the net filter being used was actually a band-pass filter. Two parameters were varied; the degree of attenuation in the low-frequency "stopband" and the value of the cutoff frequency. Figure #3 provides a graphic look at the various filters tried, when plotted using the two parameters just mentioned. Qualitative judgements as to the results are noted on the graph. Representative results are shown in figure #4-c,d.

Conclusions

Several conclusions can be drawn from the research just described. Although the tomographic image is badly masked with blur and the contrast within the image is poor, the linear filtering process does enhance the edges that are being looked for. However, it is hard to tell whether any information that was previously unrecognizeable to the trained radiologist has been brought out. Finally, it is concluded that linear spatially invariant filtering procedures are limited in their usefulness by the noise level in the image. To get further improvements, a non-linear spatially-variant procedure should be tried.

Acknowledgements

This work was supported by NIH Grant GM 22081. We want to thank Dr. Peter Spiegal and Dr. Robert Jeffery of the Diagnostic Radiology Department at MHMH for permission to use the Ohio-Nuclear Delta Scanner.

References

Orphanoudakis, Stylianos C. and Strohbehn, John W., "Mathematical Model of Conventional Tomography", Medical Physics, Vol. 3, No. 4, 1976, pp. 224-232.

Sternick, Edward S., Hughes, M., Hogan, R., and Goodhue, L.R., "External Beam Treatment Planning on a Time-Sharing Graphics Terminal System", Radiology, Vol. 110, No. 2, 1973, pp. 467-472.

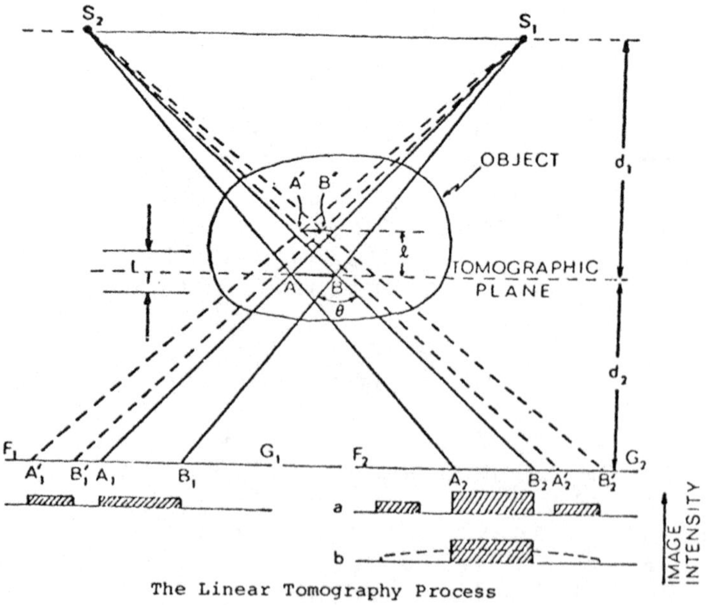

The Linear Tomography Process

(a)

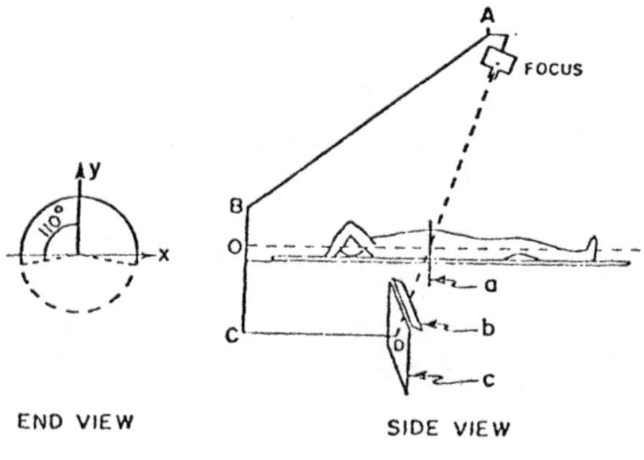

The Transverse-Axial Tomographic Process
a= tomographic plane, b= wedge grid used to block scattered radiation, and c= film

(b)

FIGURE #1

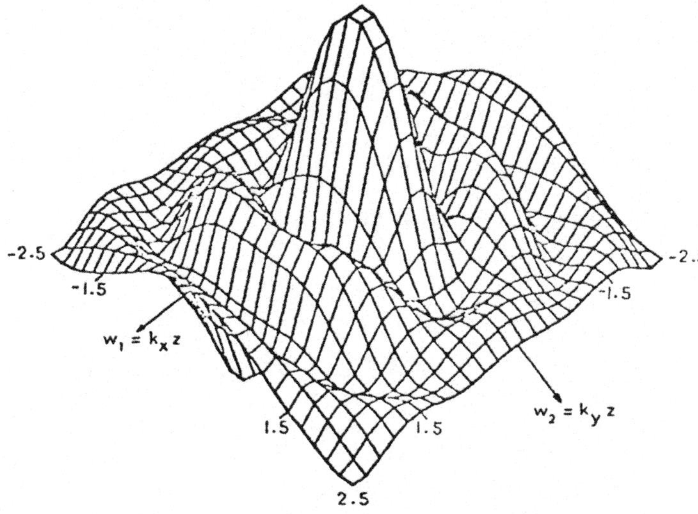

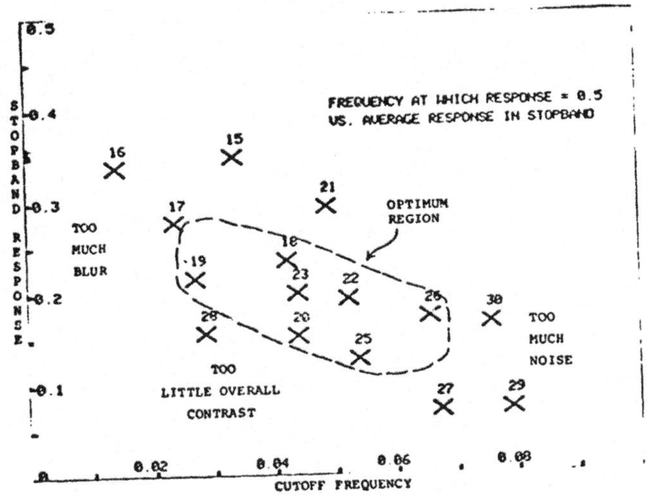

Normalized Amplitude Response of Transverse-Axial Tomography with 220° Rotation

FIGURE #2

FIGURE #3

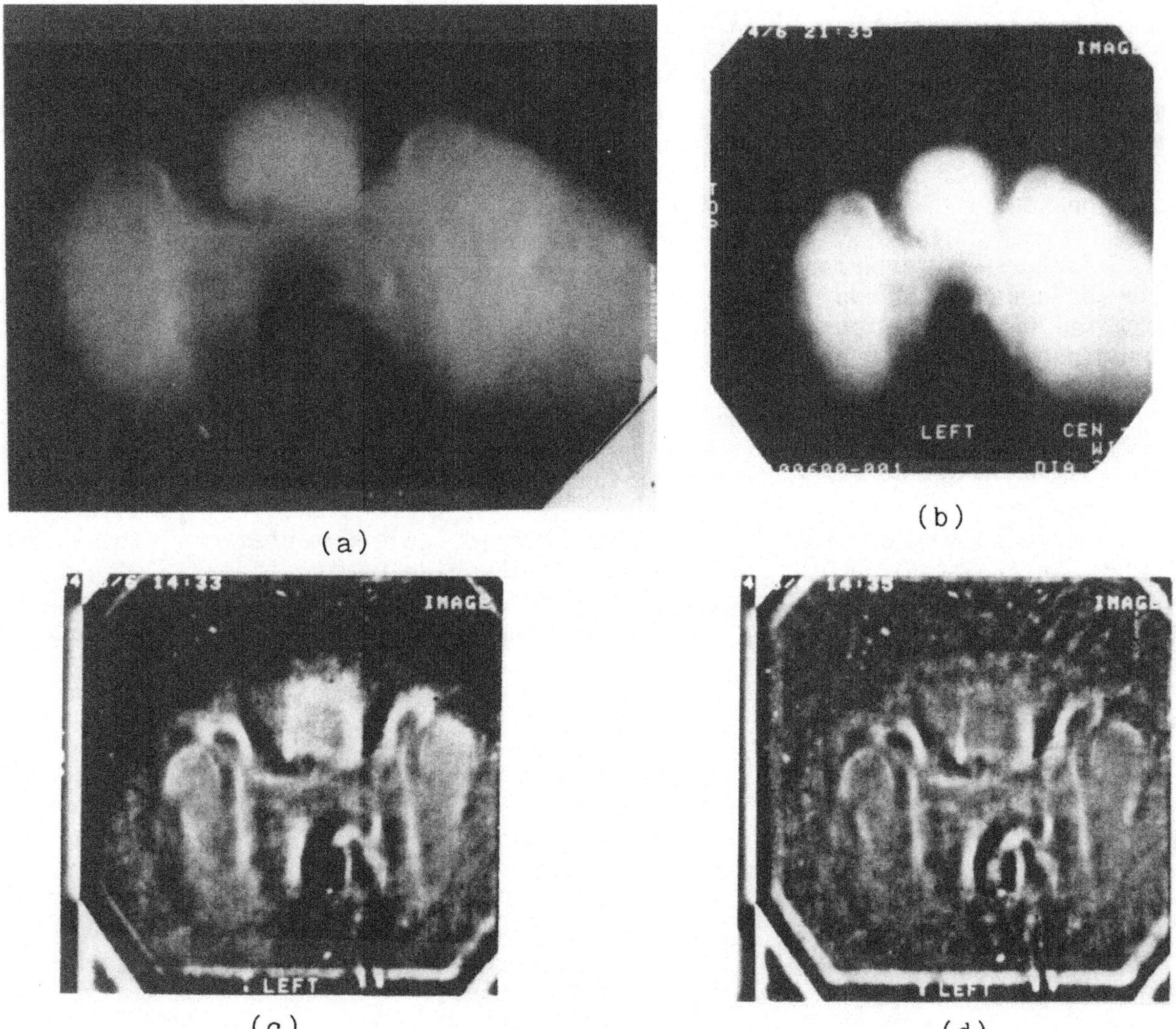

(a)

(b)

(c)

(d)

FIGURE #4

REAL-TIME COMPUTATION OF EMG POWER DENSITY SPECTRA BY FAST WALSH-HADAMARD TRANSFORMATION

Calman Gold
Thayer School of Engineering, Dartmouth College
Hanover, New Hampshire 03755

Introduction

The use of EMG frequency power spectra as corroborative measures in clinical electrodiagnosis has been hampered by wideband spectral estimates, statistical inconsistency, and the necessity of employing off-line data handling by medium to large scale computers. The latter requirement precludes an instantaneous analysis of several sites in each muscle probed during an electrodiagnostic examination. Statistically unsmoothed spectra are notoriously inconsistent estimators. Thus, some form of averaging is required to reduce the variance of power spectra to acceptable levels. Because very little is known a priori about the coherence of the EMG signal, a single raw EMG spectrum is essentially meaningless. The requirements for providing Fourier transformation, spectrum computation, and averaging in real-time are severe both because of the computational complexity and as a result of the high sampling rates dictated by the EMG signal bandwidth.

This paper describes a method employing Walsh-Hadamard transformation for computing narrowband statistically consistent EMG power spectra in real-time using minicomputers of simple architecture or inexpensive microprocessors. Further details of the method and the results of clinical studies suggesting the diagnostic utility of the method are reported elsewhere(1).

The use of the Walsh Transform for signal processing is based upon the theory of generalized Fourier transformation employing the complete set of orthogonal Walsh functions as basis vectors, and upon the concept of a generalized frequency or "sequency" domain and the sequency power spectrum(2).

The most appealing feature of the discrete Walsh transform is the simplicity of its implementation. Multiplication of an input data array by a Walsh ordered Hadamard matrix requires only additions and subtractions. When contrasted with the multiplication by exponential factors required in discrete FFT operations, the Walsh transform offers considerable computational advantages. Fast Walsh Transform algorithms(3,4), provide computationally efficient, though not necessarily optimal, means for rapid digital signal processing in the sequency domain when applied to finite sequences of quantized data.

Sequency theory suggests that the information content of sequency domain EMG power spectra should be comparable to that of frequency domain representations; but with added computational advantages. In this work, these advantages were realized in EMG signal processing by combining the simplicity of Walsh transformation with optimal memory utilization, the elimination of floating point software, and other simplifications of existing algorithms.

Method

The prototype signal processing program consists in three major sections: (i) an interpretive parameter entry routine; (ii) a data-acquisition and processing section; and (iii) an output section providing numerical tabulation and graphics. Computer software was written in PAL-III symbolic assembly language and was implemented on a Digital Equipment Corp. PDP-8/e central processor equipped with 4K memory, teleprinter, programmable real-time clock, 2 channel A/D converter (10-bit bipolar, ± 5 volt range), and extended arithmetic element.

The program was designed to be a useful laboratory tool permitting EMG power spectra to be computed on-line from the amplified EMG signal. Operating at data acquisition rates of up to 4,096 samples per sec., the program provides simultaneous data-acquisition and Walsh transformation in addition to spectral computation and ensemble averaging of sequency power spectra. Capabilities are a maximum sequency resolution of 8 zps over the range of 0 to 1,024 zps, and a maximum resolution of 16 zps over the range of 0 to 2,048 zps.

The operation of the program is as follows. The sequency range, the resolving bandwidth, and the statistical criteria governing the number of spectra to be averaged are specified by the user of the program via keyboard entry prior to the EMG test. Provisions are also included for entering clinical comments, data, and patient identification on the printout. From the input values the program computes and initializes the program parameters, the signal sampling rate, the required EMG record length, and the number of spectra to be averaged. Data acquisition may be triggered by either a start pulse from an electrical switch closure or by a keyboard command. The EMG record is processed serially in epochs, the length of which are a function of the desired resolving bandwidth. The hardware interrupt system is used to acquire new data samples during each epoch while simultaneously transforming and processing the values acquired during the previous epoch. The processing continues until a total EMG record length sufficient to meet the statistical criteria has been acquired. Printed output results (Fig.1) include both a numerical tabulation of relative power density versus sequency and a normalized decibel plot of the spectrum in 0.5 db increments. The range of the db scale is adjustable by altering one constant in the graphics routine. The linear sequency axis is automatically adjusted as a function of the range and the resolving bandwidth employed.

Transformation and processing is performed in double-precision integer format. Decimal printout and decibel plotting computations are performed by simple algorithms requiring neither floating-point software nor logarithm computation. Normalization of the spectrum and all output computations are performed during printing, where the time required for these calculations is not critical and is negligible with respect to the character output time.

The total memory requirement of the program is 3,458 (12-bit) storage locations. This requirement includes parameter input routines, data acquisition and processing algorithms, data buffering and transformation arrays, and output routines. Walsh transformation and spectral averaging buffer areas occupy 1,666 memory locations allowing for a maximum matrix size of length N=256. A free space of 638 locations remains in the 4,096 location memory field.

Results

The results of measurements to determine the duration of the computational operations in each epoch are presented in Table 1.

The performance of the prototype program is remarkable when one considers the computational requirement and data acquisition rate imposed upon the limited single address architecture of the PDP-8/e processor. Other architectures are eminently more suited for real-time spectral computations. However, the simplicity of Walsh transformation and the design of efficient software permit such signal processing to be performed rapidly on less powerful computers with an additional economy of memory usage.

The results of Table 1, indicate that true real-time processing is not achieved at all combinations of sequency resolution and signal sampling rates, the computational performance at higher sampling rates is apparently degraded. The time allotted for computation (the epoch length) is equal to the reciprocal of the resolving bandwidth. At fixed resolving bandwidth, an increase in sampling rate requires an increase in matrix size. Because computational time increases rapidly with the array size, an increase of sampling rate at fixed resolution results in a disproportionate increase of processing time required with respect to a fixed time allowed for processing. For example, with a matrix length of N=64, a spectrum can be computed at a resolution of 32 zps at a sampling rate of 2,048 sec^{-1} in 34 msec; an excess over real-time of only 8.8 percent. When the sampling rate is doubled to 4,096 sec^{-1}, the same matrix must be computed in an allowed epoch duration of only 15.625 msec; an excess over real-time of 117.6 percent.

The overall effect of these factors at high sampling rates is the presence of a gap between each processed epoch of the EMG record. Because the ENG interference pattern is a random variable, and the spectra are statistical estimates based on random samples of the EMG signal, the presence of these gaps are not detrimental to the ensemble averaged spectrum. However, when several hundred spectral estimates are being computed, the total time attributable to these gaps builds rapidly.

Conclusions

The simplicity of the computational routines and the economy of memory allocation in the prototype program suggest that the same method can be implemented on commercially available micro-processors and that special purpose dedicated systems for use in the electrodiagnostic laboratory could be developed. THe software developed could, with minor modifications, be implemented on an Intersil IM6100 micro-processor, a 12-bit processor exhibiting software compatibility with the PDP-8/e minicomputer. This microprocessor would have to be used at a maximum clock rate and supply voltage to achieve the processing speed demonstrated in this study. A commercially available peripheral hardware multiplier would also be required. Other microprocessors (e.g. bit-slice architectures), may provide greater advantages because they exhibit more flexible instruction sets and higher speed. A dedicated implementation would preferably employ non-volatile read-only memory for the storage of programmed instructions and constants. The random access memory requirements could then be reduced to accommodate only transformation arrays and other alterable memory locations. Typed keyboard commands could be eliminated by

```
SUBJ: 301N
TEST: 2
ZH: 2048
DELZ: 64
N=0064
DF: 512
SPECTRA AVGD =0256
??GO

TOTAL POWER= 0218294

SEQUENCY   REL. POWER*100

0000       0092.529
0064       0047.656
0128       0100.000
0192       0086.401
0256       0069.799
0320       0046.875
0384       0032.250
0448       0027.031
0512       0015.649
0576       0015.356
0640       0014.111
0704       0014.941
0768       0014.941
0832       0010.815
0896       0007.543
0960       0005.517
1024       0002.148
1088       0001.586
1152       0002.001
1216       0002.050
1280       0002.416
1344       0002.661
1408       0002.465
1472       0002.465
1536       0002.612
1600       0002.978
1664       0002.807
1728       0002.978
1792       0002.978
1856       0002.246
1920       0001.538
1984       0001.025
2048       0000.219
```

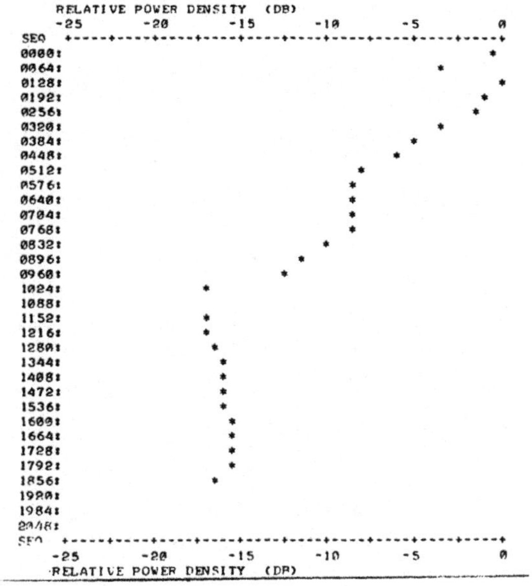

FIGURE 1 - Sample printout

Table 1. Computational Time Measurements

Sampling Rate $f_s(sec^{-1})$	Sequency Resolving Bandwidth Z (zps)	Matrix Size N	Epoch Length T_i (msec)	Measured Computational Time T_c (msec)	% Excess Over Real-Time $100x(T_c-T_i)/T_i$
	8	256	125.0	164	31.2
	16	128	62.5	76	21.6
2,048	32	64	31.25	34	8.8
	64	32	15.625	16	2.4
	128	16	7.813	8	2.4
	16	256	62.5	164	162.4
	32	128	31.25	76	143.2
4,096	64	64	15.625	34	117.6
	128	32	7.813	16	104.8
	256	16	3.906	8	104.8

TABLE 1- Computational performance

substituting keypads for the command operations. As the spectral range, resolving bandwidth, and number of spectra to be averaged are all integer powers of two, these parameters could be set by switches which could be interrogated by the microprocessor. Several desirable alternatives exist which would increase the output rate. A high-speed miniature line printer and/or a CRT display would be useful features. Using a display, an investigator could obtain an essentially instantaneous view of the computed spectrum. Furthermore, as the computation is performed on-line and virtually in real-time, a spectrum from the EMG recorded at each of several sites in a muscle could be produced without any significant delay in diagnostic procedures. Preliminary clinical results of a study (1), involving both normal and diseased individuals and undertaken in cooperation with the Jerry Lewis Neuromuscular Disease Center of the Institute of Rehabilitation Medicine (N.Y.U. Medical Center), suggest the clinical utility of the technique in providing statistically consistent spectral estimates corroborating electrodiagnostic findings in both health and disease.

References

1. Gold, C.; "Design and Experimental Evaluation of a Method Employing Fast Walsh-Hadamard Transformation for Real-Time Computation of EMG Power Spectra by Digital Minicomputers" Doctoral Dissertation. New York University, Feb. 1977. (Also University Microfilms, Ann Arbor, Michigan.)

2. Harmuth, H.; "A Generalized Concept of Frequency and Some Applications", IEEE Transactions on Info. Theory, Vol. I 11-14, 1969, pp. 375-382.

3. Shanks, J. L.; "Computation of the Fast Walsh-Fourier Transform", IEEE Transactions on Computers, Vol. C-18, 1969, pp. 457-458.

4. Ulman, L. J.; "Computation of the Hadamard Transform and R-Transforms in Ordered Form", IEEE Transactions on Computers (correspondence), Vol. C-19, 1970, pp. 359-360.

Acknowledgements

The author gratefully acknowledges the encouragement and assistance of Dr. Neil I. Spielholz, Dr. E. R. Tichauer, and Dr. June N. Barker, of the Institute of Rehabilitation Medicine, N.Y.U. Medical Center for making this work possible.

TELEVISION, COMPUTERS AND THE BEHAVIOR OF SICK BUGS

John O. B. Greaves
Assoc. Prof. of Electrical Engineering
Southeastern Massachusetts University
North Dartmouth, MA 02747

ABSTRACT

This paper will describe the goals, instrumentation and users view of a second generation television - computer system to observe and quantify the behavior of healthy and environmentally stressed organisms. This behavior studying machine, labeled the Bugsystem, is still undergoing development for the analysis of motion in three dimensions of fish in a 4-foot diameter tank and is being used by research biologists in two dimensions to observe and attempt to classify normal and abnormal behavioral patterns for a variety of micro-organisms. The system relies heavily upon the use of an interactive minicomputer system with graphics capabilities, a specially built piece of hardware dubbed the Bugwatcher that preprocesses the video image and greatly reduces it information content, a FORTRAN based monitor for interpreting user commands from a keyboard and a sophisticated virtual file system for accessing and creating data sets.

DESIGN GOALS

The prototype Bugsystem software was written in assembler language to run on an IBM 1800 computer and is described in the literature (1,2). A similar system using much of the same hardware and software was implemented at the Jet Propulsion Lab in Pasadena to study the behavior of micro-organisms. This second generation system was developed under a research grant (R803244) from the Environmental Protection Agency's Environmental Research Lab in Narragansett, Rhode Island. The work began in 1974 and is continuing until May of 1977 with the following design goals:

1. Maintainable hardware and software. The video to digital preprocessor was redesigned with more modern circuitry and documentation. The software was to be implemented in a high level language for documentation, ease of implementation and software maintenance purposes.

2. Software portability. While the first computer to implement the system was known to be a PDP 11/45 under the DO9 operating system at Southeastern Massachusetts University (we later switched to the RT-11 operating system) the computer that was to be used at the EPA lab was unspecified at the onset of the project. Hence the decision to use the FORTRAN IV language as the implementation language. The EPA lab obtained a Data General Corporation ECLIPSE S/200 computer and within 3-4 months the Bugwatcher hardware interface was built, the new machine was learned and the software was operational. New software can be designed and easily added due to our isolation of implementation details from the applications program.

3. Capability of adding new keyboard functions (applications programs) in a high level language with minimal knowledge of the innards of the software. The applications programmer uses standardized software interfaces to deal with special NON-FORTRAN System calls such as character string comparisons and programs that talk directly with the specialized hardware such as the Bugwatcher.

4. A disk-based virtual file system to allow the user to create and access very large data files and to free the applications programmer from the constraint of working in a small (mini-) computer's memory. Again through the use of carefully conceived and standardized software interfaces, the application programmer can CREATE, OPEN, READ, WRITE, CLOSE and DELETE files that can be several times the size of the computer's memory. Through the use of a Table of Contents at the beginning of each file, each vector (either a video page or a function of an organism's path trajectory) can be directly accessed for runtime speed considerations. The details as to when to swap parts of a file into and out of the computer's memory onto disk is transparent to the applications programmer and of course to the user of the system as well.

5. The availability of HELP messages to the user of the system to allow him or her to learn or refresh himself on how to use the system. For example, typing HELP on the keyboard will cause the names of all the keyboard functions to be printed on the terminal (PLOT, LIST, INPUT, CENTROIDS, PATHS, STATISTICS, etc.) while a separate HELP message can be printed for each of the functions such as the PLOT program by specifying PLOT/HELP.

THE BEHAVIORAL RESPONSE LANGUAGE - THE USER'S VIEW

The user of the system is typically a research biologist who has designed a set of experiments with one of two goals: (1) To observe and quantify the effects of an organism cultured in or subjected to a polluted environment (such as different concentrations of one or more heavy metals), or (2) To observe and quantify the immediate effects of an external stimulus such as polorized light on the behavior of a group of organisms. In either case, several sets of experiments may be run through the analysis programs, the results merged and the desired averages, standard deviations and other statistical parameters may be displayed in a graphic or tabular form. The user directs the input and analysis of his data through a set of over 40 commands, or keyboard functions which are underlined in the text. Typical of the commands, or keyboard functions, are: BWIN BUGS.VI 50 (CR) which direct the system to the Bugwatcher Input Program to input a video file (.VI) of 50 blocks in length (roughly 50 video pages) according to the frame rate and image seen on the TV monitor as the recorded experiment is played back on the video tape deck. The name of the file is BUGS.VI. When this is complete as signified by the * prompting character, he may elect to plot the entire file by specifying PLOT (CR) or optionally PLOT BUGS.VI (CR). The first 10 pages may be viewed by typing PLOT 1, 10 (CR). The next step is typically one to reduce the outline of each organism, represented as a series of X-Y coordinate pairs as presented by the Bugwatcher (video to digital processor) to their respective centroidal points via the CENTROID function, which may also be plotted. Then the path finding algorithm is invoked which restructures the data into a sequence of time ordered path trajectories. Both the CENTROID and PATH keyboard functions have the "intelligence" to filter out noise from the original recording, but the user is the end judge and he may elect to CHOOZE the paths or path segments to save and he can JOIN together paths that may have been separated by a collision on two or more organisms. In any one experiment there may be between 5 and 25 such decent trajectories, thus to collect a statistically significant number of observations for a given experimental condition, these operations are repeated perhaps 5 to 10 times and the results can be MERGED into a single file for futher analysis. Some of the more obvious mathematical functions that can be derived from these path trajectories include: LVEL and AVEL to calculate linear and angular velocities for each in two space: NGDR which computes the Net to Gross Displacement Ratios for each path and RCDI for the Rate of Change of Direction. Some of the statistical keyboard functions include: STAT to accumulate the statistics for each path or the entire data file onsay the linear velocities; RUST to calculate the Running Statistis (such as linear velocity, etc.); ANOVA to compute the Analysis of Variance and CHI2 for a Chi-Squared Test. In addition to these mathematical functions, the user may plot these data at any time or print the results on his terminal (LIST) or on the line printer (LIST/LP).

CONCLUSIONS

In retrospect, this hardware-software behavior studying machine that was constructed out of lower level elements has fairly well lined up to its design goals. The system is fairly maintainable in terms of hardware and software, although a resident programmer who understands and can use the software interfaces is necessary for further expansion. It is being used by the biologists who are getting results much quicker and simpler than was possible with older manual or semi-automatic methods for collecting and analysizing their behavioral data. The 3-D collection an analysis of data with two TV cameras positioned above a 4-foot diameter fish tank is currently under development and indications are that with 3-D versions of many of the above programs and a mapping function to account for the air-water interface, we will be able to account for motion in the fish tank in three dimensions.

REFERENCES

1. Davenport, Culler, Greaves, Forward, Hand, "The investigation of the behavior of microorganisms by computerized television", IEEE Trans. Biomed Eng., Vol. BME-17, pp. 230-237, July 1970.

2. Greaves, "The Bugsystem: The software structure for the reduction of quantized video data of moving organisms", Proc. of IEEE, Vol. 63, No. 10, pp. 1415-1425, October 1975.

COMPUTERIZED PROGRAM FOR THE INVESTIGATION OF THE ELECTROPHYSIOLOGICAL CHARACTERISTICS OF THE HEART

R. Erickson, B. Avitall Ph.D.
Department of Electrical Engineering, Tufts University, Medford, Massachusetts 02155 and
New England Medical Center, Boston, Massachusetts 02111

The excitability of the heart muscle during the phase of repolarization is an important parameter measurement in the evaluation of the heart vulnerability to ventricular fibrillation during myocardial ischemia. The amount of current required to reactivate the muscle at different times during the phase of repolarization is called the strength interval (SI) curve. It is desirable that this curve be generated rapidly in order to describe the time course of excitability at the time of ischemia, hypoxia or during reoxygenation.

The computer program devised to generate the SI curve is dependent upon the detection of 2 external timing sources called S1 and R1. S1 is the pacemaker pulse while R1 is the muscle response to the S1 stimulus. The program generates one secondary stimulus per beat designated S2. Upon detection of a secondary response the program allows for recovery. These include suppressing the normal stimulus S1 for one beat and omitting the S2 stimulation for several more.

Specifications include the acceptance of up to 50 different R1-S2 delays and 50 ascending current levels for S2 to be entered by the user. During experimentation for each of the R1-S2 delays the program outputs the ascending current levels until a response is detected or the last current level is reached. A voltage level stored during calibration is associated with each current level. The associated voltage level is output through the Digital to Analog converter which drives a constant current unit. The constant current unit then feeds the current desired to the load. The stimulation output is a pulse of 3.5 milliseconds duration.

In order to save time and to generate valid SI curves a current level tracking feature is implemented so that the program need not start each time from the lowest current level. If a response is detected on the first current level at a R1-S2 delay the current level is decreased by a number of levels specified by the user. The stimulation is attempted again at the same R1-S2 delay, and the program will track down if another response is detected. If no response is detected the program starts to increment the current levels until a response is detected. If the lowest current level evokes a response the program continues with the next R1-S2 delay.

Response detection occurs by comparison of control sampling where there is no S2 stimulus with that of sampling where the S2 stimulus occurs. The control sample is taken at the beginning of each R1-S2 delay. During the sampling period, the sampling occurs in groups of 8 samples at the rate entered. Each group is averaged and if the difference between the average of the control sample group and the average of the experimentation sample group exceeds the tolerance specified a response is found. Sampling takes place between S2 plus a delay to avoid S2 artifact until either a response is detected or S1 occurs. The individual sample is obtained from an Analog to Digital converter channel to which a detector electrode has been attached.

Upon the detection of a response the program stores the R1-S2 delay, the calibrated current level, the "measured" current level and the conduction time associated with that response. The "measured" current level is obtained by sampling the voltage from an A to D channel across a resistor in the current path during stimulation. This value should correspond closely to the calibrated current level and is ideally identical. The conduction time is defined as the time between the S2 stimulus and the detected response. For this program, the conduction time is determined between S2 and the half way time point of the sample group during which the response was detected.

The program is designed to operate in two general modes. In the first mode called the operator mode the user is able to change the parameters of experimentation such as sampling rate, print sets of response data and enter variables such as the current levels and R1-S2 delays. The user may change the number of current levels or R1-S2 delays to be used and may enter single values or a series of values at any time. Each current level is automatically calibrated without user intervention. The current unit output is incremented until it matches the specified current.

In the second mode various run time conditions may be set both before and during experimentation. These features include rapid experimentation where data printout is suppressed and a series of SI curves are generated without interruption. The user may also decide manually via the keyboard that a response occured which the program did not detect. A return to operator mode and the end of rapid experimentation may also be specified.

The program is implemented in assembly language on a standard PDP-8/e lab computer which includes a real time clock, an A to D converter, a D to A converter and buffered input-output channels.

Program Parameters and Variables for User Entry

up to 50 different R1-S2 delays
up to 50 different current levels
delay after S2 but before sampling
sampling rate
number of current levels to track
resistance for measuring current levels
number of recovery beats after response detection
response detection tolerance

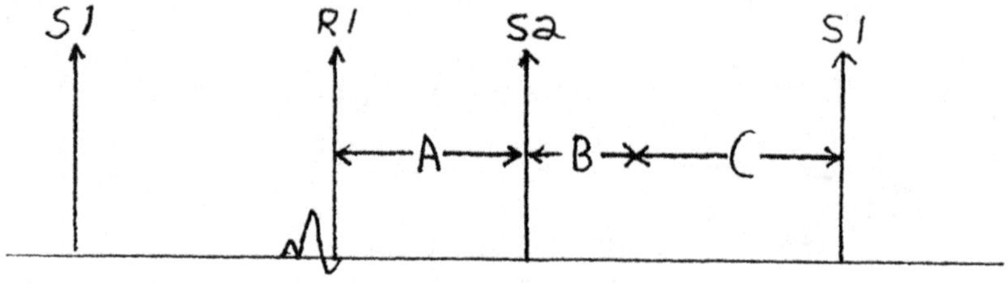

Fig. 1

A Period: R1-S2 delay
B Period: Stabil zation delay
C Period: Sampling for response occurs
S1,R1 are timing pulses recieved at the computer

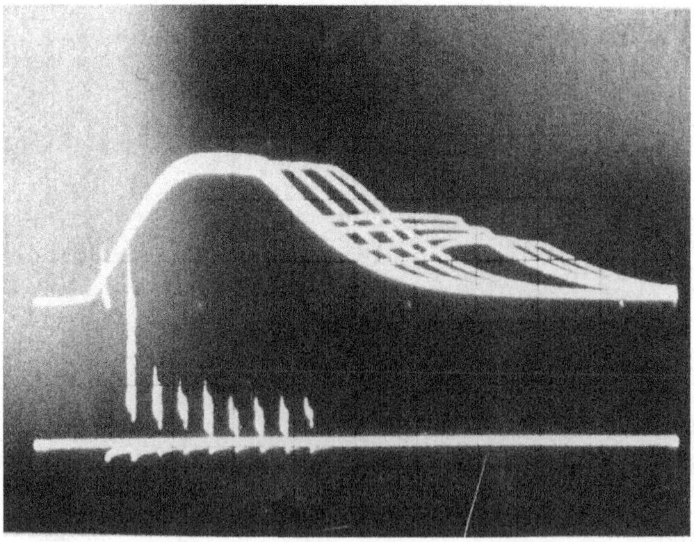

Fig. 2

The SI curve(bottom) and the tension response(top) of an isolated preparation of rat papillary muscle.

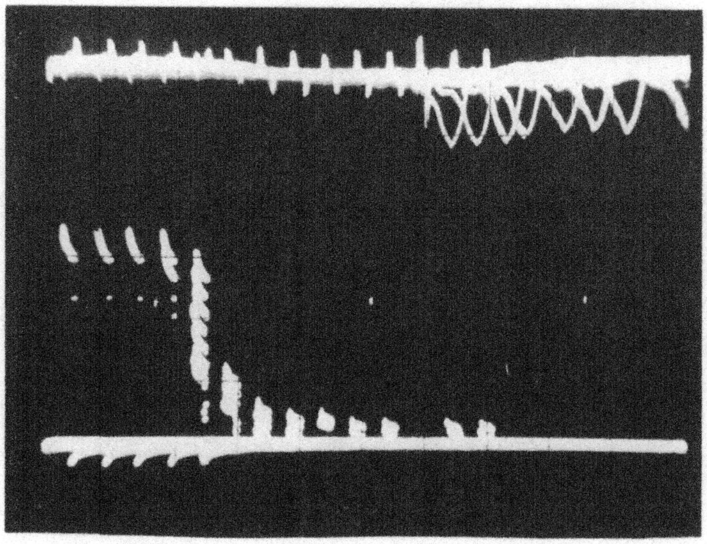

Fig. 3

The SI curve(bottom) and the electrical response(top) of an intact dog heart.

Figures 2 and 3 represent multiple tracings displayed on a storage oscilloscope. The lower curves represent the current pulses applied to the tissue, the envelope of which represents the SI curve. The top traces are the responses to the primary and secondary stimulations.

References

1. Avitall, B. Measuring the electrical properties of the heart during and following regional myocardial ischemia: An interactive instrumentation system. Proceedings of the Third New England Bioengineering Conference. Tufts University, Boston, Mass., May 1975, Pg. 171-183.

2. Avitall, B., S. Naimi and H.J. Levine. Conduction times, refractory periods and thresholds during and following release of coronary ligation. Clin Res 23:379A, 1975.

3. Avitall, B., S. Naimi, J. Mieszala and H.J. Levine. Time course of strength-interval curve and conduction times during myocardial ischemia and reperfusion: Possible mechanism for changing electrical stability. Circ 52:71, 1975.

4. Avitall, B. Computerized investigation of the electrophysiological characteristics of the heart during acute myocardial ischemia and abrupt reperfusion. Ph.D. Dissertation, Tufts University, Boston, Mass., 1975.

LEARNING PAINS IN BEGINNING A COMPUTER AIDED CARDIOVASCULAR LABORATORY

James A. Cox and Schuyler E. Hardin
Shands Teaching Hospital
University of Florida
Gainesville, Florida 32610

Abstract - The rapid onset of minicomputers in the medical field has given great impetus to the quality of medical care. Coincident with this growth, cardiovascular laboratories have greatly increased their numbers in the preceding decade. The symbiotic relationship that has culminated between these two areas is lending itself to accomplishment of even the most difficult tasks.

This paper discusses such an implementation of the minicomputer in medicine. The Shands Teaching Hospital in Gainesville, Florida has recently installed a Honeywell Computer Catheterization Laboratory System (CCLS) in its cardiovascular laboratory. Acceptance of this new diagnostic tool by the attending physicians is surveyed. Ease of use, reliability of data calculations, optimum use of software, and training time required are all emphasized in this discussion as it affects the physician. A comparison of the physicians' manually calculated data versus that of the computer's is shown. The need for interaction between the physician and programmer is demonstrated to insure that updated algorithms and software corrections are implemented.

In addition to the diagnostic potential of the minicomputer, the capabilities as concern didactic training of medical students and ancillary help are explored. Finally, the data storage and retrieval ability of the CCLS system is demonstrated as it regards the updating and logging of patient files.

Primarily, this paper is intended to point out the interactions needed between the physician who is generally inexperienced in the computer area and the engineer who generally does not have the expertise in medicine. A joint effort of both fields will provide the desired end product: better health care delivery.

I. CAPABILITIES OF A DIGITAL COMPUTER

Of the many uses found by the digital computer, its basic function as applies to medicine can be seen in the following areas (1):

1. Data acquisition
2. Storage and retrieval
3. Assimilation and organization of information
4. Data reduction and transformation
5. Mathematical operations
6. Pattern recognition
7. Limit detection
8. Statistical analysis of data
9. Data presentation
10. Control functions

The Honeywell 316 Computer System combines these capabilities to provide an effective clinical diagnostic instrument, the CCLS.

II. THE COMPUTERIZED CATH LAB

The Shands Teaching Hospital purchased the CCLS system supplemented with the Patient Oriented File System (POFS) and the Angiocardiography Analyzer peripherals. The core of the system, the CCLS unit, provides the data acquisition, data analysis, organizational, and formatting functions. The POFS subsystem is a convenience peripheral which provides data storage to and retrieval from the CCLS core via mag tape. The Angiocardiography Analyzer consists of a sonic two dimensional "GRAF-PEN" and cine film viewer unit which provides heart volume data which when fed to the CCLS unit in conjunction with stored pressure data provides additional hemodynamic function results. Control of the entire system is implemented at one of three independent terminals located in the different procedure rooms. The cardiovascular technician at the terminal is in constant audiovisual communication with the attending physician during the cath. The patient data is sampled and recorded upon request of the physician or based on established protocol. This sampled data may be analyzed at that moment or stored for later analysis. As the data is logged into the CCLS memory, a chronological record is maintained in the patient file at the time of data acquisition. This characteristic is helpful in later analysis to determine the time of the onset of certain events. Up to ten patients may be held in the core memory of the CCLS unit. The "active" patient file status is given to the patient upon which data sampling and analysis is being performed. Patients in core memory must be transferred to active status before data analysis can be implemented. Dumping of any patient files into bulk storage is achieved using POFS. The POFS system will be discussed in the later section of this paper. A block diagram of the CCLS system is shown in figure 1.

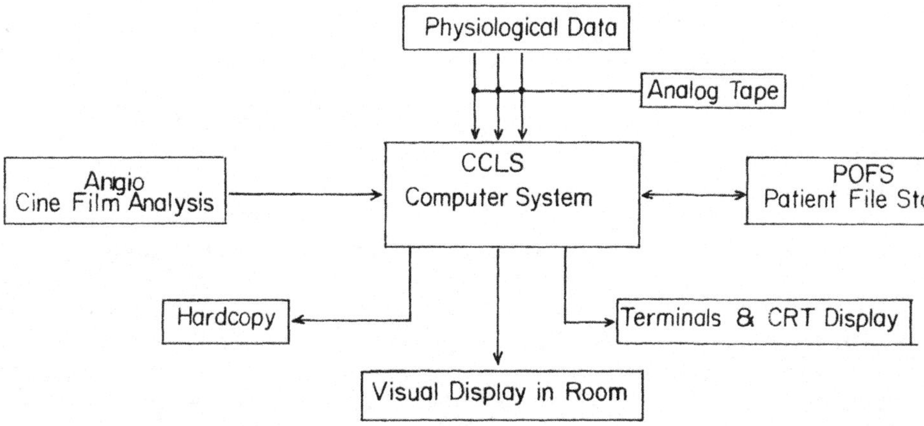

Fig. I Block diagram of the Honeywell Computer Catheterization Laboratory System (CCLS).

III. PHYSICIAN ACCEPTANCE

A system designed to evaluate even the most elaborate data can only prove its effectiveness when it fulfills the designed function. Of major importance in achieving this end, the user must be shown the advantages of such a system. Teaching an old dog new tricks can be quite a chore. The physician wants to know the time saved, accuracy gained, and additional diagnostics which can be handled by this "black box." Down-time of the system becomes a major drawback for initiating interest by the physician. Several inconveniences during a single procedure as a result of equipment failure will rapidly negate any positive factors gained over a period of weeks. The new kid on the block has to prove himself. The physician demands that new medical instrumentation serve a superior function over its predecessors in addition to creating fewer interruptions in procuring more accurate physiological data. The sudden rush of gadgets into the medical field can only leave the physician wary of that equipments' superiority over reliable, older, less sophisticated equipment and methods.

Because the physician is dedicated to the care of his or her patient, any factor which contributes to patient care and the reduction of patient anxieties would certainly be scrutinized. The computer offers a means of providing instantaneous data to the physician. With the data on hand, the physician has more parameters available to indicate the need for concurrent testing. He or she, however, must be ready to implement alternate routes should real-time data values indicate unscheduled procedures. Presently, procedures are decided on before the actual study and the collected data analyzed afterward. Conceivably, rapid implementation of unscheduled studies may lead to a reduction in recatheterizations of patients. Manual analysis of recorded data can prove to be tedious and time-consuming. Uneventful realization of the need for additional studies after the procedure only leads to increased patient anxieties. Hospital layover, increased bills, physician time and productivity loss, facility use, and increased odds of procedure complications are heavily weighted factors in favor of computerized cath labs. The increase in productivity from the health care team narrows the increasing imbalance of patient/physician/ancillary ratios. As a result of the above reasoning, the Cardiovascular Lab chose to engage the services of a CCLS system.

IV. ANCILLARY CONSIDERATIONS

Physician/technician team interaction with the computer is essential for proper operation of the system. The physician's familiarity with the computer's diagnostic algorithms coupled with the technician's adeptness in implementing these selected studies yield maximal results. Cardiovascular technicians are the brains behind the CCLS. It is the technician who must sample data for the computer, enter comments into the patient's file, provide the physician with quasi-real-time results in the procedure room via a slave CRT monitor, choose the correct signal processing sensitivities, alert the physician to nondiagnostic recorded data and system malfunction. Consequently, data acquisition must be catered to the technician while data analysis must be catered to the physician. The technician keyboard console is therefore the most important consideration needed to provide a viable data acquisition pathway. The operation of the CCLS has been greatly simplified by using a pneumonic keyboard listing for primary function headings. When selected, these headings provide a CRT display of further listings and commands for implementing calculations, sampling data, etc. While simple in concept, this flow mechanism may become somewhat involved for even simple operations. Technician response to various command listings and indirect routings should be held to a minimum. Confusion is directly proportional to the number of manual steps required to perform a desired function regardless of the simplicity of each step. The time required for selecting the proper routines needs to be considered as there may be a minimal time window for collecting the data. For this very reason, the user should have the flexibility of obtaining customized listings for their terminals. Further reductions in command listings still need to be optimized.

Training time necessary for a cardiovascular technician who is for the most part unfamiliar with the computer world becomes a major concern. All laboratory technicians should be able to use the system. The importance of this statement can be seen from the aspect of radiation safety. Personnel rotation through all job duties within the laboratory minimizes total time in the procedure room and, therefore, reduces radiation exposure to all individuals. Total training time in this situation supports to a greater degree the need for minimal training requirements. Training of a separate computer technician should be avoided as that individual is usually unable to provide the necessary physician/technician interaction so fundamental in data collection.

Recognition of the above factors by the user will aide the computer designer in meeting the user's needs; likewise, the designer must recognize the plight of the user.

V. COMPUTATIONS: MAN VERSUS MACHINE

For the most part, the computer uses the same algorithms for computation of valve areas, cardiac output, etc., as does the physician. Values for variables in these algorithms may vary though, depending on interpretation of the pressure waveform. The computer overrules any variations in its interpretive mechanism when arriving at its physiological data values. Each waveform analysis is performed in the same manner at each testing. Human interpretation leaves room for variation among different individuals in procuring data values from the same waveform. Computer sampling and analysis provides uniformity in data collection. When computing mean values of pressures, the computer calculates the exact area under the curve and in turn gives an accurate mean. The physician is able to manually measure an electronic mean value of a pulsating pressure. To arrive at this mean, the height (calibrated to mmHg) of the mean line from the base line is measured with a ruler. As shown in table 1, computer and physician mean values generally correlate well since little interpretation is needed for manual analysis. Systolic, diastolic, and varying positions on pressure waveforms, however, leave room for interpretation and, consequently, the physician's values vary to a greater degree from the computer's assessment (see table 1). Likewise, cardiac outputs derived from dye dilution studies involve extrapolations of data and consequent interpretation. Table 2 reveals comparisons of cardiac outputs.

At this point in the CCLS's use, actual diagnosis is still made by the physician using the collected data. Future efforts will undoubtly engage the services of pattern recognition schemes to provide rapid diagnostic results. As mentioned earlier, manual analysis consumes valuable time of the cardiologist in addition to prolonging patient anxieties. The computer can provide the necessary physiological values and calculations accurately, repeatedly and quickly. This then allows the cardiologist to proceed to the patient diagnosis stage bypassing the time normally devoted to data acquisition and, thereby, increasing direct patient contact hours. No information is lost by using the computer since any and all waveforms from which the data is derived is available in hardcopy form or an analog tape. Should inconsistencies be present, the physician can then perform the necessary manual analysis.

Uniformity in the hardcopy format of calculations also will reduce wasted time in the diagnostic reports. Standard formats can be readily perused by staff members when evaluating a patient's case. Again, the computer has proven its usefulness.

Table I

Sample pressure values in mmHg comparing automated versus manual interpretion

Aorta	Systolic Peak		Diastolic		Mean	
	Comp	Phys	Comp	Phys	Comp	Phys
a	69	77	49	49	58	58
b	113	110	69	65	88	90
c	116	120	68	65	91	90
d	63	70	35	40	49	50
e	74	80	43	48	57	60
LV	Systolic Peak		Begin Diastolic		End Diastolic	
	Comp	Phys	Comp	Phys	Comp	Phys
a	76	73	14	9 to 17	21	0 to 18
b	80	73	15	9 to 17	23	8 to 19
c	80	80	13	12 to 24	27	8 to 23
d	77	77	16	0 to 21	24	9 to 19
e	80	85	0	-5 to 6	4	-6 to 5
f	104	103	-4	-3	9	1 to 12
g	99	100	-3	0	10	7
PA	A Wave		V Wave		Mean	
	Comp	Phys	Comp	Phys	Comp	Phys
a	2	4	1	0	0	1
b	34	36	7	11	18	19
c	45	47	8	14	23	23
d	23	28	1	10	14	13
e	18	20	9	12	14	14
f	17	22	1	5	9	15

Table 2

Cardiac output (L/min) as derived from manual and automated analysis of dye dilution curves

CARDIAC OUTPUT (L/MIN)					
	Computer	Physician		Computer	Physician
a	7.69	5.35	f	3.01	3.2
b	8.37	7.5	g	9.76	7.7
c	4.43	3.3	h	5.79	4.5
d	3.0	3.48	i	5.31	5.0
e	6.68	7.1	j	5.26	5.11

VI. DATA STORAGE AND RECALL

The Patient Oriented File System (POFS) is an efficient storage and recall peripheral. POFS enables the operator to store patient files, analog samples, Angiocardiography Analyzer data, and progress notes. Stored data may be recalled using the patient's name, hospital number, or disease classification. The entire inventory of recorded files which normally holds 100 patient files can be displayed and hardcopied for cataloging a library listing of the POFS' mag tape contents. In addition, each recorded file unit is given a file number which can also be used to recall specific files. Conventional storage of recorder paper and records is very cumbersome and occupies valuable space. The mag tape filing system reduces the amount of storage space needed. Even more important, any portion of any file may be recalled and analyzed with minimal efforts if the need should arise. Deterioration of data also diminishes when mag tapes are used instead of conventional light-developed recorder paper.

VII. DIDACTIC CAPABILITIES

The CCLS system can be placed in a "Teaching" mode to reduce the amount of instructor contact time required for residents, interns, and technicians. Certain patient files can be designated as teaching files. When the terminal is in the teaching mode, the permanent files cannot be altered. This feature can be of value to the student in comparing manually calculated data with the more accurate results supplied by the computer. The teaching mode also provides a training stage for the student prior to initiation as a member of the physician/technician team.

VIII. OPTIMUM USE OF SOFTWARE/MODIFICATIONS

Even the best designed systems need modifications once they are installed. Since most medical system designs are generated to serve a general user, specific facilities may find a system provides some form of the necessary functions but must be altered to fit specific needs. The technician keyboard console as mentioned previously is the bottle neck in data acquisition and manipulation. Optimal use of software is most easily accomplished if the most used algorithms, sampling schemes, and file commentaries require a minimal number of steps to be implemented at the terminal.

Modifications should be implemented with minimal complications to the present system. For instance, computer listings for drugs, diagnostic categories, catheter types, etc. should be designed for easy deletion and/or insertion of old and new materials into said lists. Similarily, the canned algorithms should have the capability of being altered should new techniques evolve.

IX. USER GROUPS

At present there are six (6) institutions in the United States which use the CCLS system. In an effort to strengthen the capabilities of this sytem for each user, the coauthor has distributed a proposal to the five (5) other institutions. In this User Proposal, the joint employment of a programmer-engineer by the aforementioned institutions is suggested. Benefits which are expected to evolve from the said proposal include:

1. State of the art library of cath programs
2. Updating and customized programs for individual labs
3. Routine maintenance of equipment at reduced cost
4. Communication between users: common problems - their prevention and solution
5. Increased productivity and flexibility of the system

A priority on programming time is suggested with highest weighting given to common group benefits. Customized programming for individual labs is in turn determined using a priority formula. Cooperation is needed between all the users if indeed a viable user network is to evolve.

X. CONCLUSIONS

Entry of the computer into the medical field is now a reality. Utilization of its capabilities to serve the medical community can only become a reality if the health related personnel are educated to its operation and potential benefits.

Once installed, the system must be reliabile (minimal down-time) and excel in its designed use over older equipment, or the new system will eventually end up as surplus, unused stock. A joint effort between designer and user should be made to generate a versatile system which meets not only the present day needs, but also future needs if obsolescence is to be prevented. The interaction of the physician or user group and engineer must bridge any gaps formulated in these areas if the two fields are going to complement each other.

REFERENCES

(1) Leslie Cromwell, et al., Medical Instrumentation for Health Care, New Jersey: Prentice-Hall, 1976, pp. 353-356.

(2) CCLS Maintenance Training Course Manual, Honeywell, March, 1976.

COMPUTER APPLICATIONS IN BREAST CANCER TREATMENT*

Roger L. Brotman, M.D.
Peter Grossman, M.S.
Herbert W. Mower, Sc. D.

Department of Therapeutic Radiology
Tufts-New England Medical Center Hospital
Boston, MA

INTRODUCTION

New computer applications in medical science are being reported daily which make the practice of medicine less costly, tedious, and time-consuming and more accurate and effective. The system described herein has been operational for 8 months at the Department of Therapeutic Radiology, Tufts-New England Medical Center Hospital to aid in the treatment of patients with breast cancer. This system has three parts: decision making, dosimetry, and verification. A great deal of attention has been given to the human engineering problems of ease of use and clarity of output.

THE TREATMENT PLANNING SYSTEM

An automated treatment planner, capable of selecting the appropriate plans for a course of radiotherapeutic treatment for a breast cancer patient, has been written based upon the breast cancer treatment protocol of the Department of Therapeutic Radiology. Before a patient begins treatment the patient data necessary for prescribing a treatment plan is supplied by the radiotherapist by completing a form containing initial examination information. A transcriber then codes the data contained on the form and enters it in the patient's computer file.

In one mode of operation, the treatment planning program accesses a list of all Initial Examination forms entered during a given day and selects all those patients for whom it can produce treatment plans. These plans are automatically printed and returned to the radiotherapist for use as an aid in prescribing treatment.

In cases where the radiotherapist wishes to run the treatment planner on a patient lacking a computer file or to rerun a patient using modified information (i.e., if a patient's tumor was slightly larger than previously stated and thus might result in a different treatment plan), the treatment planner can be run in a conversational mode with the user supplying all patient data.

In the event that a patient's file contains a missing or insufficient datum the treatment planner will substitute all possible replacements and produce the corresponding treatment plans. A sample output is shown in Figure 1.

The Treatment Planner

The planner has been written in the MUMPS programming language and is implemented

*This work was supported in part by grant #CA 12178, awarded by the National Cancer Institute, DHEW.

on a PDP 11/55 minicomputer. It consists of eight modules occupying about 3K core each (not including the MUMPS interpreter), only one of which needs to be in core at any particular time, and a data structure stored on disk. The data structure is called a "MUMPS GLOBAL" and in effect can be simulated by a sparse three dimensional array. It is composed of approximately 70 records or "frames" which contain pointers into a formatted set of data items stored in a patient's data file as well as treatment dosages and notes, and branching information to other frames.

In deriving a treatment plan, the system programs "walk" through the data structure collecting treatment information and travel to other frames under the direction of branching data contained in each frame. In this way a given treatment plan is the result of a unique path through the global. The addresses of the frames through which the planner has travelled are stored in a stack in order to provide the physician with a readable tracing of how the final treatment was derived. In the case where there is insufficient data for a particular frame (question), this stack is used to follow all traversible paths.

The design goals for producing the treatment planner were: 1) to provide a reliable and accurate treatment plan, 2) to detect situations where insufficient information prohibited the derivation of reliable plans and 3) to allow for easy modification and updating of the data structure as the departmental protocol was updated.

Benefits of the System

Extensive testing of the treatment planner has revealed that it is an accurate representation of the protocol in use. In addition, the departmental protocol (a 20 page document) representing a highly intricate flow chart with steps numbered to four decimal places, has been reduced to an organized data structure consisting of 70 frames and about a dozen possible treatment plans. It is hoped that this data structure will be of use when the departmental protocol is updated or rewritten.

It has also been suggested that the treatment planner could be used as a teaching device for residents so they may be shown more graphically how the treatment prescription relates to particular types of tumors.

PHYSICAL ASPECTS OF TREATMENT PLANNING

In most clinical settings, a key factor in the therapy of cancer by radiation is the adequate treatment of the lymph node drainage areas. In breast cancer, this is of prime importance, and the computerized dosimetry system described herein is designed to insure adequate doses. The nodal areas of interest in treating patients with breast cancer are the axillary, supraclavicular and internal mammary nodes. They are best treated with a single anterior field, in order to avoid the problems of hot and cold spots with adjoining fields. In order to do this, one must use the large, irregular "L" shaped or "hockey stick" field shown in Diagram 1. When such a field is used, there are major variations in dose delivery due to three main factors: 1) the patient's irregular contour leads to varying source to skin distances (SSD's), 2) there is considerable beam fall-off from side to side (especially with Cobalt 60), and 3) the lymph node areas are at varying depths. A member of the clinical physics staff has made exhaustive dosimetric measurements on a suitable phantom and has derived equations which express the depth dose to several points of interest, given such data as overall field, size, patient thickness,

PROTOCOL RX PLAN FOR MATILDA MASTECTOMY (AGE 40) JANUARY 25, 1977
123456

INITIAL EXAM ON 8/23/1974

SITE	DOSE		FRACT.	RAD. TYPE
PT. A (L FIELD)	5000	G. D.	25	CO
PT. B (SUPRACL)	5000	G. D.	25	CO
PT. C (AXILLA)	4500	T. D.	20-25	CO
		SEE NOTE "A" BELOW		
PT. D (INT MAM)	5000	T. D.	25	CO
BR & CW (TANG)	4500-5000	T. D.	20-25	CO

ADDITIONAL NOTES :
IN SELECTED CASES, ELECTRONS SHOULD BE USED INSTEAD OF COBALT.
SEE DEPARTMENTAL PROTOCOL (SEC. 2.04, 2.05).
"A" IF TUMOR EXTENDED INTO AXILLARY FAT THEN A TOTAL DOSE OF 5500-6000 RADS SHOULD BE GIVEN. THIS ADDITIONAL DOSAGE SHOULD BE DELIVERED THROUGH THE POSTERIOR AXILLARY PORTAL IN 5 TREATMENTS, AFTER COMPLETION OF THE INITIAL 20-25 TREATMENTS.
IF SURGICAL MARGINS WERE INADEQUATE THEN ELECTRON BEAM BOOSTS OF 500-1000 RADS AT 5 OR 10 MEV SHOULD BE APPLIED TO THE SCAR OR TUMOR BED. FIRST 4 WEEKS; THEY MAY BE DISCONTINUED EARLIER IF A BRISK REACTION OCCURS

****

DATA USED IN REACHING THE ABOVE TREATMENT

1 DISTANT METASTASES PRESENT? (M STAGING)	NO
2 IS DISEASE RECURRENT? (YES OR NO)	NO
3 AIM OF RADIOTHERAPY	POST-OPER
4 TYPE OF SURGERY	MASTECTOMY, RADICAL
5 ARE ANY OF THE FOLLOWING PRESENT	
A. MULTIPLE TUMORS	
B. SKIN INVASION OR PEAU D'ORANGE BEYOND THE PALPABLE LIMITS OF THE TUMOR	
C. ULCERATION OR SKIN INFLAMMATION	
D. TUMOR FIXATION TO PECTORALIS MAJOR	
E. HIGHLY ANAPLASTIC TUMOR	
F. PERIPHERAL TUMORS (PARASTERNAL, TAIL, OR INFRAMAMMARY FOLD).	
G. INADEQUATE SURGICAL MARGINS.	
H. POORLY DIFFERENTIATED TUMOR	NO
6 SIZE OF TUMOR	<= 2 CM.
7 TUMOR LOCATION	INNER
8 HOW MANY NODES WERE EXAMINED?	>14
9 WERE THERE ANY POSITIVE NODES?	YES
10WERE THERE LESS THAN 5 POSITIVE NODES?	YES

FROM TERMINAL FRAME 70.60
RESPONSIBLE STAFF M.D. IS MUNZENRIDER, JOHN E.

FIG. 1 PROTOCOL TREATMENT PLANNER

HOCKEY STICK FIELD DOSIMETRY

PLEASE ENTER HOSPITAL # FOR MATILDA MASTECTOMY: 123456

WIDTH OF RECTANGLE: 16 CM
LENGTH OF RECTANGLE: 26 CM
TABLE HEIGHT AT A (REFERENCE POINT): 20.5 CM
TABLE HEIGHT AT B (SUPRACLAVICULAR): 16 CM
TABLE HEIGHT AT C (AXILLA): 16 CM AP DIAM AT C (AXILLA): 13.5 CM
TABLE HEIGHT AT D (INTERNAL MAMMARY): 21 CM
CO I (1) OR CO II (2) OR WCH (4) ? 1
SSD: 80 CM
TOTAL GIVEN DOSE TO A: 5000 RADS (50 GRAYS)
TOTAL # OF FRACTIONS: 25
PLANNED DOSE TO B (TYPE 'NB' IF NO BOOST DESIRED): NB
PLANNED DOSE TO C (TYPE 'NB' IF NO BOOST DESIRED): 6000 RADS (60 GRAYS)
PLANNED DOSE TO D (TYPE 'NB' IF NO BOOST DESIRED): 5000 RADS (50 GRAYS)

*** FIELD # 1.0 ***

TOTAL TUMOR DOSE TO POINT A = 5000 RADS (50 GY) 100.00 % AT 0.5 CM
TOTAL TUMOR DOSE TO POINT B = 4450 RADS (44.50 GY) 89 % AT 0.5 CM
TOTAL TUMOR DOSE TO POINT C = 3099 RADS (30.99 GY) 61.98 % AT 6.75 CM
TOTAL TUMOR DOSE TO POINT D = 4436 RADS (44.36 GY) 88.72 % AT 3.0 CM

EQUIVALENT SQUARE = 19.20 CM
OUTPUT FACTOR = 0.8082
GIVEN DOSE TO POINT A = 200 RADS (2 GRAYS)
NEMCH COBALT #1 TREATMENT TIME = 1.69 MIN

*** FIELD # 1.1 (AXILLARY BOOST) ***

BOOST DOSE REQUIRED = 2901 RADS (29.01 GRAYS)

IN HOW MANY FRACTIONS ? 15
FIELD SIZE: 7 CM X 8.5 CM
SSD: 80 CM
DOUBLE LUCITE ? YES

EQUIVALENT SQUARE = 7.70 CM
OUTPUT FACTOR = 0.8854
GIVEN DOSE = 283 RADS (2.83 GRAYS)
TUMOR DOSE = 193 RADS (1.93 GRAYS)
DEPTH = 6.75 CM % DEPTH DOSE = 68.42
NEMCH COBALT #1 TREATMENT TIME = 2.51 MIN

FIG 2 DOSIMETRY PROGRAM

*** NO ADJUSTMENTS NECESSARY TO POINT B (SUPRACLAVICULAR) ***

*** FIELD # 1.3 (INTERNAL MAMMARY BOOST) ***

BOOST DOSE REQUIRED = 564 RADS (5.64 GRAYS)
IN HOW MANY FRACTIONS ? 2
FIELD SIZE: 10 CM X 6 CM
SSD: 80 CM
DOUBLE LUCITE ? YES

EQUIVALENT SQUARE = 7.50 CM
OUTPUT FACTOR = 0.8874
GIVEN DOSE = 323 RADS (3.23 GRAYS)
TUMOR DOSE = 282 RADS (2.82 GRAYS)
DEPTH = 3.00 CM % DEPTH DOSE = 87.35
NEMCH COBALT #1 TREATMENT TIME = 2.87 MIN

DO YOU WISH TO DO CHEST WALL CALCULATIONS ? NO

MORE DOSIMETRY FOR THIS PATIENT ? NO

FIG 2 (CONTINUED)

WEEKLY PHYSICS CHECK FOR HOK PATIENTS

PATIENT'S NUMBER: 123456
MATILDA MASTECTOMY......OK ? YES

HAVE ALL TREATMENTS DELIVERED THE PRESCRIBED GIVEN DOSE ? YES

TOTAL # OF RX'S TO FIELD 1.0: 20
TOTAL # OF RX'S TO FIELD 1.1: 5
TOTAL # OF RX'S TO FIELD 1.2: 0
TOTAL # OF RX'S TO FIELD 1.3: 3

MATILDA MASTECTOMY JANUARY 5, 1977
DATE OF INITIAL CALCULATION: DECEMBER 29, 1976

*** HOCKEY STICK DOSES TO DATE ***

FIELD 1.0: HOCKEY STICK

POINT	% GIVEN DOSE OF POINT A	DOSE
A (REFERENCE POINT)	100.00	4000
B (SUPRACLAVICULAR)	89.00	3560
C (AXILLARY)	61.98	2479
D (INTERNAL MAMMARY)	88.72	3548

*** CUMULATIVE DOSES AT DEPTH -- INCLUDING BOOSTS ***

BOOST FIELD	RX'S TO DATE	DESIGNATION	DEPTH	% GIVEN DOSE OF BOOST	BOOST GD	TUMOR DOSE
1.1	5	AXILLARY	6.75	68.42	1413	3446
1.2	0	SUPRACLAVICULAR	0.50	*** NO BOOST SPECIFIED ***		
1.3	3	INTERNAL MAMMARY	3.00	87.35	968	4394

FIG 3 PHYSICS VERIFICATION

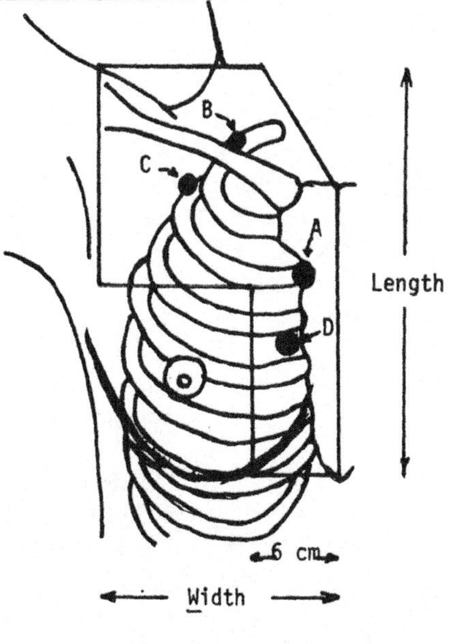

Diagram 1 Hockey Stick Field

and lymph node depth. These equations were tested on actual patients with thermo luminescent dosimeters (TLD) and found to be accurate to within ± 5% for a wide range of patient dimensions. A computer program has been written to allow the physician or physicist to access these equations for a particular patient, and receive a neatly tabulated sheet containing all the necessary dosimetric information.

The "L" shaped field is drawn on the patient using the proper anatomical limits. Four points of interest are marked: point A is the overall reference point for machine set up and given dose and is defined at the junction of the second rib with the sternum at a depth of 0.5cm; point B lies at 0.5cm depth, just above the middle 1/3 of the clavicle (representing the supraclavicular nodes); point C lies at the apex of the axilla, at the patient's midplane; point D lies 3.0cm deep, 3.0cm lateral to the midline at the 3rd costal interspace (representing the internal mammary nodes). The physician then takes measurements of the field size and of the patient size at the four points, and using these, along with the dose prescriptions supplied by the protocol planning system, runs the dosimetry program. The program devises any necessary boost fields which may be required to complete the treatment as prescribed. An example is shown in Figure 2. The output provides: 1) a record of the initial prescribed dose to all points of interest, 2) a tabulation of the fractional depth doses and cumulative contributions to all points of interest, 3) a summary of all required boost fields, and 4) all daily dose and treatment time data necessary for the technician to treat the patient.

WEEKLY VERIFICATION OF TREATMENT DELIVERY

When the above calculations are performed, the patient's name is stored in a file of those "hockey stick" patients currently under treatment. Each week, when all patients undergoing therapy have their treatment sheets reviewed and checked by the clinical physicists in preparation for chart rounds, this list is used to identify this group of patients.

The computer uses the information stored during the initial calculations and the patient's daily treatment record to determine the dose contributed from the "hockey stick" field to points A,B,C, and D. If any boost fields have been treated, their contribution to the tumor dose is also tabulated as well as the cumulative tumor dose to date for each point.

A sample calculation is given in Figure 3. The patient's hospital number is entered and the patient's name is echoed for verification. If, for some reason, there has been a variation from the prescribed treatment at any time, the physics check must be done manually. Allowance for variations from the prescribed treatment dose will be incorporated in the near future. The only input now required is the number of treatments to each field: #1.0 (the hockey stick), #1.1 (axillary boost), #1.2 (supraclavicular boost), and #1.3 (internal mammary boost). The output is a well tabulated summary, as seen. In addition to the total tumor dose to each site, the contributions from each field are also shown. The previously stored fractional depth doses are printed for manual comparison if desired. The cumulative tumor doses for each site are entered on the daily treatment sheet for the physician's use.

When the patient completes treatment, another program is called which removes the patient from the list of those under treatment and stores the patient's name, number, and prescribed doses for future follow-up and evaluation studies. This group of programs saves time for the physicist, reduces the chance of error, and prepares a neat and orderly summary of the patient's treatments to date.

FUTURE EXPANSION OF THIS SYSTEM

A concurrent project in our department is computer verification of treatment set-up and delivery. When this system is fully operational, it will no longer be necessary to enter the number of treatments per field, departures from planned treatments, or completion date. This will all be recorded automatically and printed for cross-referencing with the technician's daily treatment sheet.

CONCLUSIONS

Close cooperation among clinicians, physicists, and programmers have resulted in a clinical application of computers which meets the following criteria:

1) provides the clinician with useful data
2) encourages use by clinicians and other non-technical personnel
3) aids the physicist in routine calculations and verifications.
4) reduces chances for human error
5) increases efficiency of the department resulting in better patient care at reduced cost to the patient

REFERENCES

1. Conklin, J., Munzenrider, J.E., Neurath, P.W., and Ross, W.M.: Computer Aided Medical Decision Making in Radiotherapy. (Submitted for publication to Radiology).
2. Pauker, S.G., Gorry, G.A., Kassirer, J.P. and Schwartz, W.B.: Towards the Simulation of Clinical Cognition - Taking a Present Illness by Computer. The American Journal of Medicine, 60:981-996 (1976).

ACKNOWLEDGEMENTS

The authors acknowledge the contributions to this system by Carl Atkins, Mabini M. Castro, Ph. D., Jeff Conklin, John E. Munzenrider, M.D., Peter W. Neurath, D. Sc. and Winifred M. Ross, M.D.

EFFECT OF RIB CAGE PARAMETERS ON LOW ORDER MODEL APPROXIMATION OF RESPIRATORY MECHANICS: A LIAPUNOV MODELING PROCESS

Charles J. Maffeo and Antharvedi Anné
Division of Biomedical Engineering
University of Virginia
Charlottesville, Virginia 22901

Introduction

In order to provide adequate ventilation to critically ill patients, reliable models of the respiratory system are required. A number of optimal ventilation techniques have been proposed which are based on the assumption that the respiratory mechanics can be completely described by a first order model relating lung volume to the applied pressure. Furthermore, the time constant of the model is assumed to represent the lung mechanics. In various pathological conditions leading to deformation of the rib cage, the above assumption may not be valid and rib cage damping and compliance may strongly influence the relationship between tidal volume and ventilator pressure.

The present theoretical analysis was undertaken in order to provide insight in developing low order models of the respiratory system for clinical use and to test the sensitivity of the models to variations in rib cage parameters. The basic analysis philosophy is to develop a system description of respiratory mechanics and then analytically determine a lower order model that satisfies the same performance criteria as the system. This approach shows exactly how the parameters of the system are incorporated into the lower order model and provides an indication of how the model will change with variations in the system parameters.

Mathematical Description of Respiratory Mechanics

The mechanical analog used to describe the respiratory mechanics is shown in fig. A. The elements M, R and C represent mass, resistance and compliance respectively. The subscripts distinguish the major structures, lung (l), rib cage (rc), abdomen (ab) and pleural cavity (pl). During spontaneous inspiration the forces produced by the rib cage and diaphragm muscles increase the volume of the thoracic cage and reduce the interpleural pressure. The lungs are therefore inflated as a result of the positive pressure gradient existing between the trachea which is at atmospheric pressure and the pleural cavity which is below atmospheric pressure. A slightly different situation exists when a patient is breathing in synchronous with a ventilator. In this case the forces produced by the rib cage and diaphragm muscles are negligible and the lungs are inflated as a result of the positive pressure produced by the ventilator.

In order to simplify the system description the following assumptions were made:

1. The acceleration terms are negligible compared with the viscous and elastic terms.

2. The pleural cavity provides a stiff coupling between the thoraco-abdominal and lung structures. It therefore has a very low compliance compared with the other compliances in the system.

3. The parameters in the system are linear and time invariant. Thus, the resistance of the lung does not include the effects of turbulence in the airways.

The equations describing the system dynamics when a patient is on a ventilator are:

$$R_l \dot{V}_l + V_l/C_l = P - P_{pl} \qquad (1)$$

$$R_{rc} \dot{V}_{rc} + V_{rc}/C_{rc} = P_{pl} \qquad (2)$$

$$R_{ab} \dot{V}_{ab} + V_{ab}/C_{ab} = P_{pl} \qquad (3)$$

$$V_{pl} = V_{ab} + V_{ab} - V_l \qquad (4)$$

$$P_{pl} = -V_{pl}/C_{pl} \qquad (5)$$

where V and P represent volume and pressure repsectively. These equations describe the pressure changes about an atmospheric pressure operating point. The symbols used follow the convention established in the literature. It is necessary to determine a mathematical description relating the applied pressure, P, and lung volume, V_l. One method of obtaining this relationship is to represent the system equations by a flow diagram in which P and V_l are the input and output signals respectively. The application of Mason's gain formula provides the required transfer function.

$$\frac{V_l(S)}{P(S)} = \frac{K\,(S^2 + G_2 S + G_1)}{S^3 + a_3 S^2 + a_2 S + a_1} \qquad (6)$$

where the a_i's are functions of the parameters in all three structures, lung, abdomen and rib cage, and the G_i's are only functions of the rib cage and abdominal parameters. Expressing eq. (6) in phase variables yields

$$\begin{bmatrix} \dot{X}_1 \\ \dot{X}_2 \\ \dot{X}_3 \end{bmatrix} = \begin{bmatrix} 0 & 1 & 0 \\ 0 & 0 & 1 \\ -a_1 & -a_2 & -a_3 \end{bmatrix} \begin{bmatrix} X_1 \\ X_2 \\ X_3 \end{bmatrix} + \begin{bmatrix} 0 \\ 0 \\ K \end{bmatrix} P \qquad (7)$$

$$V_l = [G_1 \quad G_2 \quad 1] \begin{bmatrix} X_1 \\ X_2 \\ X_3 \end{bmatrix} \qquad (8)$$

where X's are the phase variables. It is this system description given by eq. (7) and (8) that is to be approximated by a low order model. The system equations can be represented in matrix notation as

$$[\dot{X}] = [A][X] + [B]P \tag{9}$$

$$V_1 = [G][X] \tag{10}$$

Low Order Model of the System

The model technique is divided into three parts. First, modifying the output vector such that V_1 is independent of the higher order derivative. Second, determining the proper input vector for the nonautonomous model. Third, modeling the third order autonomous system by a second order model.

The first two parts may be accomplished simultaneously. Equation (8) indicates that V_1 is a linear combination of the three phase variables. Therefore, in developing a second order model approximation of the system it is necessary to modify eq. (7) and (8) such that V_1 is a linear combination of only X_1 and X_2 [1]. The next step, involving the selection of an input vector that only controls x_1 and x_2, is also obtained by changing eq. (9). The above changes are accomplished simultaneously by first modifying the matrix B in eq. (9) from $[B]^T = [0 \quad 0 \quad K]$ to

$$[B]^T = [b_1 \quad b_2 \quad 0] \tag{11}$$

where b_1 and b_2 are functions of the system parameters, Secondly, matrix G in eq. (10) is modified from $[G] = [G_1 \quad G_2 \quad 1]$ to

$$[G] = [kG_1 \quad G_2 \quad 0] \tag{12}$$

where k is a function of the system parameters.

The second step involves matching Liapunov V functions of the autonomous portion of the third order phase variable system (eq. 7) and the second order model. This approach forces the system and model to have the same performance index. A geometric representation of the Liapunov modeling technique is shown in fig. B. If the V surface of the third order system is an ellipsoid then the "best" second order model is the model whose V function, which is a curve in this particular case, matches the ellipse formed by the intersection of the ellipsoid and the $X_1 - X_2$ plane. The V surface of the model in 3 space is an elliptical cylinder. In order to match the shapes of two ellipses it is necessary to match the major and minor axis since these two parameters determine the geometry of an ellipse. Several methods are available to perform this type of matching [2]. Since the second order model has two adjustable parameters, it is possible to match the shape of the ellipses although their orientations may not be the same. The tip angle shown in fig. B provides a good indication of the accuracy of the model [3].

The resulting model for the autonomous system is

$$\begin{bmatrix} \dot{X}_1 \\ \dot{X}_2 \end{bmatrix} = \begin{bmatrix} 0 & 1 \\ -m_1 & -m_2 \end{bmatrix} \begin{bmatrix} X_1 \\ X_2 \end{bmatrix} \tag{13}$$

The low order model is obtained by combining eq. (11), (12) and (13).

$$\begin{bmatrix} \dot{X}_1 \\ \dot{X}_2 \end{bmatrix} = \begin{bmatrix} 0 & 1 \\ -m_1 & -m_2 \end{bmatrix} \begin{bmatrix} X_1 \\ X_2 \end{bmatrix} + \begin{bmatrix} b_1 \\ b_2 \end{bmatrix} P \qquad (14)$$

$$V_1 = [kC_1 \quad C_2] \begin{bmatrix} X_1 \\ X_2 \end{bmatrix} \qquad (15)$$

Sensitivity Analysis

A sensitivity analysis was conducted to determine the model perturbations due to variations in rib cage damping, R_{rc}, and compliance, C_{rc}. The second order model is characterized by two parameters: the resonant frequency and damping ratio. The resonant frequency and damping ratio of the model are plotted in fig. C and D respectively as a function of the compliance ratio C_{rc}/C_1. The analysis was repeated for different resistance ratios, R_{rc}/R_1. It is evident that a decrease in either the compliance ratio or resistance ratio results in an increase in resonant frequency and damping ratio. However, the damping ratio remains relatively constant over a wide range of compliance ratios and then increases extremely rapidly for low values of C_{rc}/C_1.

Discussion

The results of the model development and sensitivity analysis demonstrate that the model is insensitive to small perturbations in C_{rc}/C_1 and R_{rc}/R_1. However, the model parameters are significantly changed due to large perturbations in rib cage parameters. These results agree with the clinical observations made on patients with diseases that decrease thoracic cage compliance, such as kyphoscoliosis, pectus excavatum and arthritic spondylitis. Agreement was also found in obese patients with the Pickwickian syndrome [4].

When a patient is connected to a ventilator the work of breathing is reduced. However, the use of pressure controlled ventilators in conjunction with optimal ventilation techniques that do not include the effects of rib cage parameters in their low order models, will not adequately ventilate critically ill patients. Thus, extreme caution should be used in selecting a low order model description of respiratory mechanics as well as an optimal ventilation technique.

References

[1] D. G. Schultz and J. L. Melsa, State Functions and Linear Control Systems. New York: McGraw-Hill, 1970

[2] R. E. Kalman and J. E. Bertman, "Control system analysis and design via the "Second Method" of Liapunov," Journal of Basic Engineering, 371-393, 1960.

[3] D. M. Nordahl and James L. Melsa, "Modeling with Liapunov functions," J.A.C.C. preprints, 1967.

[4] J. H. Comroe, R. E. Forster and A. B. Dubois, The Lung. Chicago: Yearbook Medical, 1962.

List of Figures

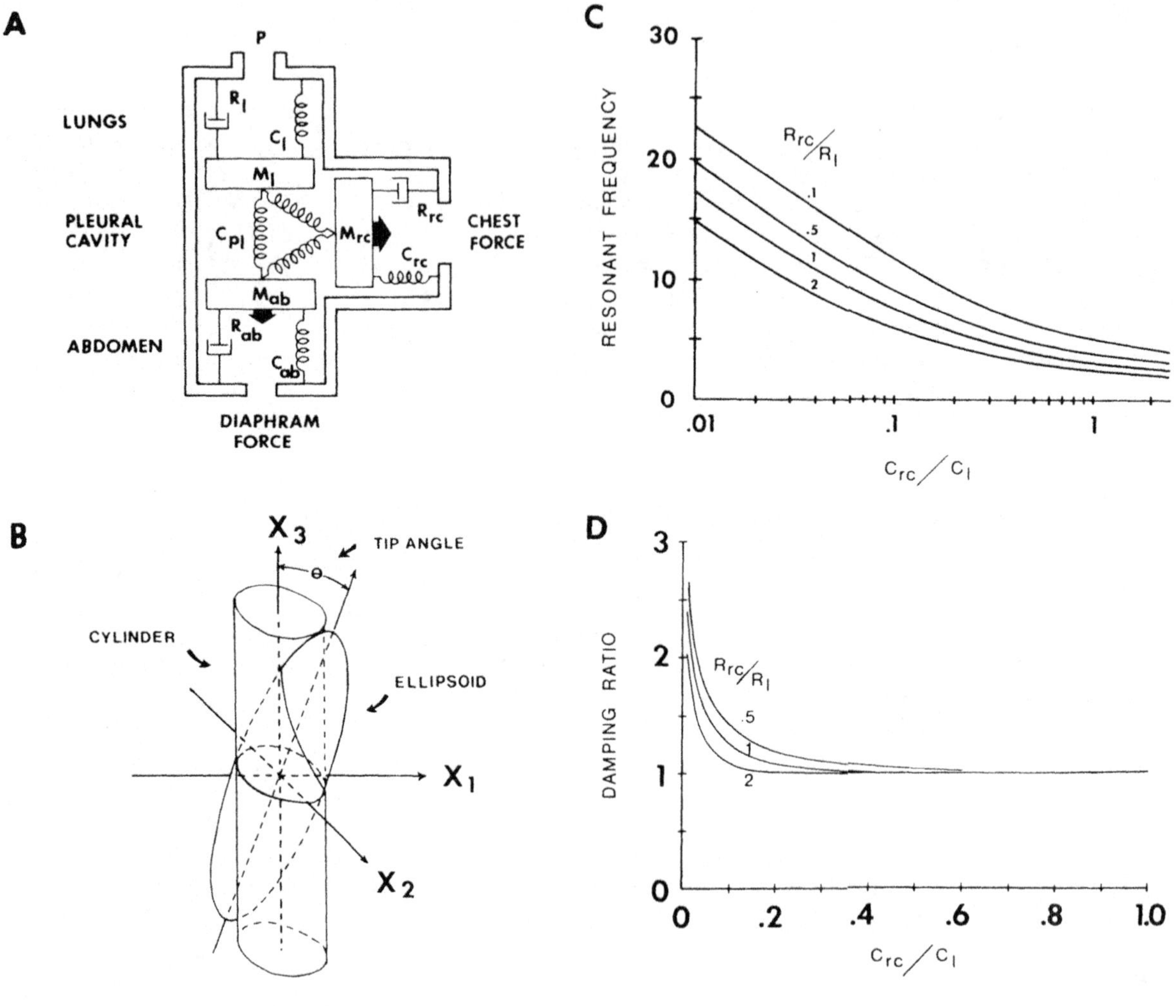
A
P
LUNGS
R_l
C_l
M_l
PLEURAL CAVITY
C_{Pl}
M_{rc}
R_{rc}
C_{rc}
CHEST FORCE
M_{ab}
R_{ab}
ABDOMEN
C_{ab}
DIAPHRAM FORCE
B
X_3
TIP ANGLE
θ
CYLINDER
ELLIPSOID
X_1
X_2
C
RESONANT FREQUENCY
30
20
10
0
R_{rc}/R_l
.1
.5
1
2
.01
.1
1
C_{rc}/C_l
D
DAMPING RATIO
3
2
1
0
R_{rc}/R_l
.5
1
2
0
.2
.4
.6
.8
1.0
C_{rc}/C_l

FEEDBACK MECHANISMS AND EPILEPSY: POSSIBLE ROLE OF THE CEREBELLUM

J. D. Charlton, G. C. Theodoridis, G. R. Hanna, and R. N. Johnson

Division of Biomedical Engineering and Department of Neurology
University of Virginia, School of Medicine
Charlottesville, Virginia 22901

Introduction

Many investigators have considered the presence of regulatory mechanisms which act to prevent or limit epileptic activity. In this regard, Julien [5] has recorded bursts of cerebellar purkinje cell activity following cerebral cortical epileptiform discharge. Additionally, he has demonstrated that the average firing rate of purkinje cells rises with the intravenous administration of phenytoin, a common anticonvulsant. Hence, at a cellular level, cerebellum appears strongly coupled to epileptiform activity. Indeed, Dow [2] has suggested such a regulator function for the cerebellum as a controller of convulsive discharges. Fernández-Guardiola [3] and associates have postulated the involvement of the rubro-olivo-cerebellar loop in the control of epileptiform discharges. Recently Cooper [1] et al, have demonstrated the effectiveness of chronic cerebellar stimulation in man for the control of epileptic seizures unresponsive to standard drug therapy. Although many conflicting results have been obtained in experimental animal studies on the effects of cerebellar stimulation, it appears likely that cerebellum has an inhibitory or regulatory effect on epileptiform activity.

In previous studies by our group [4], we have shown that cerebellar stimulation can dramatically reduce the excitability level of the thalamocortical motor system and raise threshold level for an evoked response on cortex. We have also found [6] that during the development of an acute penicillin focus a phenomenon of regulation occurs; where, following periods of epileptiform activity, the response profile is attenuated and an extensive increase in threshold level occurs resembling the effects of cerebellar stimulation. In some cases [6], we have also observed a stage of dynamic regulation during penicillin induced epileptiform activity where the system seems to abort spike discharges by reducing excitability temporarily, so that the response goes down and back up again resulting in a very irregular looking profile (see Figure E).

In this present study, we have sectioned the feedback pathways to cerebellum following penicillin induced epileptiform activity in order to test the degree of cerebellar regulation. In addition, the effects of removing cerebellum entirely on the excitability of the thalamocortical motor system has been evaluated.

Methods

Adult female cats between 2.5 and 3 kg. were anesthetised with a chloralose-urethane mixture (40 mg/kg α-chloralose i.v.) and a bipolar stimulating electrode was stereotaxically placed in the ventrolateral nucleus of thalamus (VL). Paired stimuli under computer control were delivered to VL and the resulting responses on sensory motor cortex (MC) to the second or test stimuli were sampled, averaged and displayed for online analysis, and stored for future use. The stimulus voltage of the second pulse was varied between 70 and 500 msec. With stimulus intensity and pulse interval as the independent variables and the average evoked response as the dependent variable, a computer plot of a three-dimensional response surface was obtained (Figures A-D). If the stimulus voltage is held constant for both pulses and the time between the two is the only variable, response amplitude will be a two-dimensional excitability curve (Figures E and F). Cerebellar stimulation was accomplished through a pair of epidural electrodes placed on the contralateral paramedian lobule and stimuli were delivered through a constant current stimulus isolation unit at 500 μa, 1 msec pulse duration and at 100/sec. The penicillin focus was achieved through subpial injection of 10 to 20 μl of 250,000 Units/ml sodium penicillin-G in the homologous contralateral sensory motor cortex.

Results

In previous work by our group [4] it has been shown that cerebellar stimulation dramatically reduces thalamocortical excitability levels and raises the threshold of activation in this system. Figure A gives a control three-dimensional response profile and Figure B shows the effects of stimulating the contralateral paramedian lobule of the cerebellum, where the stimulus position was obtained to give the maximum inhibitory effect on the VL-MC system. Note the reduced response level and the increased threshold (Figure B). We have also found that intravenous administration of phenytoin reduces the response level but has a minimal effect on threshold level [4]. This inhibitory effect of phenytoin could in part be mediated through cerebellum, as cerebellar purkinje cell firing rate has been shown by Julien [5] to increase after phenytoin administration. In studies employing a penicillin focus in contralateral MC, we have found a definite increase in threshold level and a slight reduction in response amplitude following periods of epileptiform discharge [6] (Figure C). This is comparable to the effect of cerebellar stimulation and suggests that cerebellum is involved in the tonic reduction of thalamocortical excitability for the regulation of epileptiform activity.

In the presence of a penicillin focus, a phenomenon was observed where the response level shifted suddenly from a relatively high level to almost zero or would alternately be high and then low resulting in an irregular looking response profile [6]. This effect was seen in the present study and is shown in Figure E, taken after penicillin, as compared to its control (Figure D). Quint and Johnson [6] have considered this phenomenon to be a form of dynamic regulation to maintain the system in a stable state, where, when the excitability becomes too high, the regulator comes in to reduce the level of excitability and thus maintains the system in a stable mode. Under continued destabilizing influences this regulator may also act in a more continuous manner to raise the threshold and reduce the response level as is the case of Figure C where the system has regained stability after a long period of epileptic discharge.

In order to test the supposition that cerebellum functions as a dynamic regulator, the following approach was employed. First, the dynamic regulation of penicillin induced epileptiform activity was demonstrated. This can be seen by comparing the initial control plot of Figure D with Figure E, which was taken after induction of penicillin epileptiform activity. Second, the contralateral inferior and middle cerebellar peduncles were sectioned, to remove feedback input to the cerebellum. As can be seen in Figure F, this resulted in a cessation of the dynamic regulation in the VL-MC system and a relatively smooth response profile in comparison to Figure E taken before the feedback pathways were sectioned. Third, the superior peduncle was also sectioned, thus removing cerebellar influences entirely. Figure G shows the initial control (smooth curve) and the associated response (dashed curve) after penicillin introduction and before sectioning of the cerebellar peduncles. As can be seen by comparing the dashed curve of Figure H (same as in Figure G) taken before section of the peduncles with the smooth curve of Figure H, taken after section of the inferior middle and superior peduncles, a larger response amplitude resulted, suggesting that cerebellum can be a source of tonic inhibition to the VL-MC system.

Conclusions

These studies suggest that dynamic as well as tonic regulation occurs in response to penicillin induced epileptiform activity in the mammalian brain. The similarity of such regulation with cerebellar stimulation, as well as the loss of regulation with sectioning of the cerebellar peduncles, suggests that the cerebellum can operate as a regulator of epileptiform activity.

References

1. Cooper, I.S., Amin, I., Riklan, M., Waltz, J.M., Poon, T. P.: Chronic cerebellar stimulation in epilepsy, Arch. Neurol. 33:559-570, 1976.

2. Dow, R. S.: Some novel concepts of cerebellar physiology. Mt. Sinai J. Med. 41:103-119, 1974.

3. Fernández-Guardiola, A., Contreras, C.M., Calvo, J.M., Ayala, F., Brailowsky, S., Solis, H., and Salgado, A.: Changes in spontaneous neuronal firing in cerebellum red nucleus and raphe nuclear complex during convulsive activity. In Cooper, I.S., Riklan,M., Snider, R.S., Eds. The Cerebellum, Epilepsy, and Behavior, Plenum, N.Y. 19-36, 1975.

4. Johnson, R.N., Englander, R.N., Quint, S.R. and Hanna, G.R.: Control of excitability and threshold in the thalamocortical motor system of the cat by cerebellar stimulation. Brain Res. 103:568-573, 1976 .

5. Julien, R.M.: Experimental epilepsy: cerebrocerebellar interactions and antiepileptic drugs. In Cooper, I.S., Riklan, M., Snider, R.S., Eds., The Cerebellum, Epilepsy, and Behavior, Plenum, N.Y., 97-117, 1975.

6. Quint, S.R., and Johnson, R.N.: Regulation of experimentally induced epilepsy. The Physiologist, 19:331, 1976.

Figures

Figure A: Control profile for comparison with Figures B and C.
Figure B: Cerebellar stimulation, elevation of threshold, reduction of amplitude.
Figure C: Threshold elevation after cessation of penicillin epileptiform discharge.
Figure D: Control profile for Figures E and F.
Figure E: Dynamic regulation.
Figure F: Loss of dynamic regulation with tonic amplitude reduction after sectioning inferior and middle cerebellar peduncles.
Figure G: Two-dimensional plots at 6V stimulus level: solid curve is a control, and dashed curve is after penicillin focus development.
Figure H: Two-dimensional plots at 6V stimulus level: dashed curve same as in G, solid curve is after section of superior cerebellar peduncle.

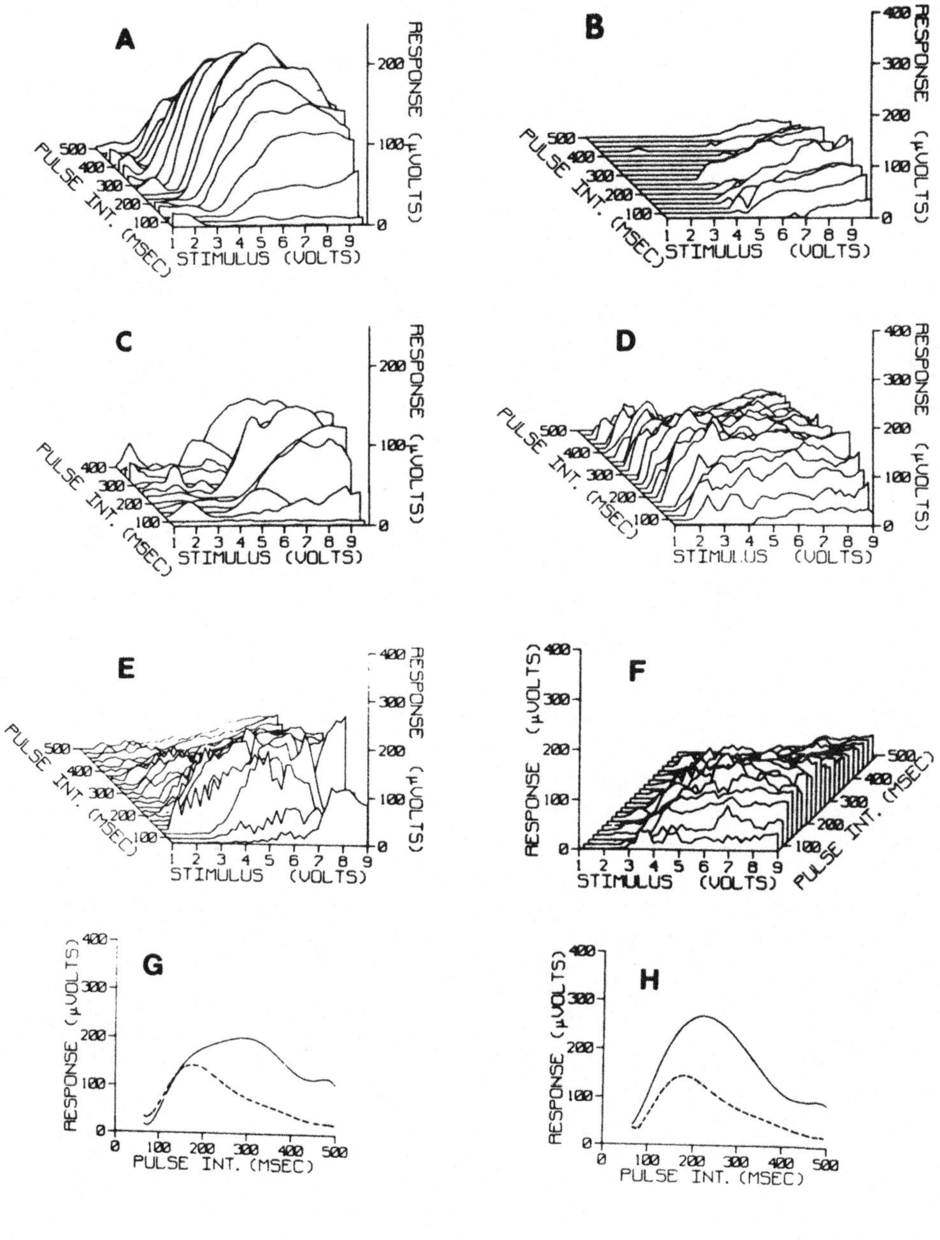

A
B
C
D
E
F
G
H
RESPONSE (μVOLTS)
PULSE INT. (MSEC)
STIMULUS (VOLTS)

A NON LINEAR MODEL OF INTRAAORTIC BALLOON PUMPING*

William J. Ohley, Dov Jaron and Karl E. Karlson

Biomedical Engineering Program
Department of Electrical Engineering
University of Rhode Island
Kingston, R. I. 02881

and

Department of Surgery
Rhode Island Hospital
Providence, R. I. 02903

Introduction

Intraaortic balloon pumping has been used clinically for a number of years as temporary in-series cardiac assistance. In general, in-series assistance is capable of decreasing left ventricular systolic pressure, increasing cardiac output and enhancing coronary perfusion. However, the optimum effects and the methods to achieve them are still in question.

Most of the information concerning the interaction of balloon pumping with the vascular system is solely the result of in-vivo studies. Results of such studies by different investigators have varied widely. In part, this is caused by the inability to precisely control vascular conditions and the operation of the assist device.

In an effort to improve our understanding of the dynamics of intraaortic balloon pumping, several analytical studies have been conducted. Previous approaches have considered the system by the use of a lumped linear model (1,2). In this study we have utilized a multi-element non-linear model of the arterial system to analyze the effects of intraaortic balloon pumping. We have used this model previously to study another form of in-series assistance: external counterpulsation (3). This paper presents preliminary results produced by a model of the intraaortic balloon pump which was incorporated into the non-linear multi-element vascular model.

Model Development

The mathematical model consists of 3 functional elements: the arterial system, the heart, and the balloon.

The model for the arterial system has been previously described in detail elsewhere (3) and can be briefly summarized: It is based on the Navier-Stokes equation for fluid flow in distensible tubes. The major assumptions used here are that blood is a Newtonian fluid; the velocity profile is rectangular; the

*Supported in part by NSF grant No. ENG 74-21085 and by a grant from the Stride Memorial Foundation.

stress-strain relationship in the vessel wall is linear; the radial acceleration forces are negligible; and that radial motion of the wall is finite. The aorta and its major branches were considered a series of finite sections. The finite difference method was used for solving the model equations.

The resulting set of equations is unique by the manner in which radial wall motion is considered. This is done by assuming that during a particular time interval the blood volume entering a given section produces an expansion of this section in the radial direction only. The new value for the vessel radius is then used to compute new hydraulic parameters during the next time interval. Thus, for the Nth section:

$$P_N = (1/C_N) \int (F_N - F_{N+1})\, dt \tag{1}$$

$$F_N = (1/L_N) \int [(P_{N-1} - P_N) - R_N F_N]\, dt \tag{2}$$

$$r_N = [(1/\pi\, \Delta Z) \int (F_{N+1} - F_N)\, dt]^{\frac{1}{2}} \tag{3}$$

$$R_N = 10.1\ \mu\ \Delta Z/\pi r^4 \tag{4}$$

$$L_N = 2.25\ \rho\ \Delta Z/\pi r^2 \tag{5}$$

$$C_N = 1.5\ \pi r^3\ \Delta Z/Yh \tag{6}$$

where P_N = pressure, F_N = flow, r_N = radius, ΔZ = segment length, ρ = blood density, μ = blood viscosity, Y = Young's modulus, and h = wall thickness.

These equations were implemented for digital solution by dividing the cardiac cycle into a number of finite time intervals. During each interval new values for R_N, L_N and C_N are calculated by applying Equations (3-6). These values are used in calculating P_N, F_N and r_N for the ensuing time interval. The heart was considered as a simple time varying pressure source.

Balloon action in the aorta is a complex phenomenon in which inflation within the descending aorta causes displacement of blood both retrograde and antegrade. Retrograde flow during diastole causes a pressure rise in the aorta between the balloon and the closed aortic valve. To simplify the analysis, we postulated that the balloon acts as an independent pressure source in an arterial segment. We also assumed that its activation adds directly to the pressure already present as a result of heart pumping. Equation (1) may then be modified to:

$$P_N = (1/C_N) \int (F_N - F_{N+1})\, dt + P_B(t) \tag{7}$$

where $P_B(t)$ is the equivalent pressure in the vessel generated by the balloon.

In general, the maximum flow generated by the balloon is limited by the total displacement volume of the balloon. This constraint is applied to the equation set by specifying that

$$V_N \geq V_{BN} \tag{8}$$

where V_N is the blood volume and V_{BN} is maximum balloon volume within the Nth segment.

Equations (1-6) were applied to each element of the model which does not contain the balloon. Equations (2-8) were applied to three sections which were assumed to contain the intraaortic balloon.

Results

The model was implemented on a Digital Equipment Corporation PDP 11/40 computer system. Using the model, the effects of varying the equivalent balloon pressure waveform were investigated by keeping all vascular parameters constant and maintaining the ventricular pressure waveform. The waveshape and timing of the balloon pressure were varied during the cardiac cycle. Thus far, we have investigated only the effects of the rate of balloon deflation on aortic root flow.

Figures 1 and 2 show typical pressure and flow waveforms produced by the model. In each figure, the equivalent pressure generated by the balloon, aortic root pressure and root flow are plotted as functions of time for several cardiac cycles. Pressures are given in mmHg and flow in ml/sec.

Figure 1 demonstrates the response to a pressure with relatively long rise and fall times. Figure 2 shows the result of long rise time followed by short fall time.

Discussion

When the rise and fall times of the equivalent pressure generated by the balloon are similar to those generated by existing pump systems, aortic pressure and flows generated by the model are similar to those observed clinically and in animal experiments (Fig. 1). When the equivalent pressure rises and falls abruptly, aortic pressures and flows in the model exhibit large amplitude transients not seen in-vivo. Fig. 2 is a sample of such a phenomenon, showing a sharp transient in aortic end diastolic pressure as a result of abrupt balloon deflation. Furthermore, the results suggest that as time of balloon deflation is shortened, aortic root flow is increased. This is particularly noticeable if deflation occurs close to the beginning of systole (Fig. 2).

In this study, the heart was specified as a time varying pressure source. Hence, effects of assistance on cardiac energetics in the model will appear only through changes in flow. Ventricular systolic pressure remains unchanged. In the model, sharp deflation increases cardiac output by suddenly forming a low pressure area in the aorta. In effect, the pressure in the segments containing the balloon and the radius of these segments decrease. This enables the heart to eject an increased quantity of blood. By contrast, if the equivalent balloon pressure during deflation decreases slowly, aortic end diastolic pressure does not diminish as much as with abrupt deflation. With slow deflation, the effective blood volume in the aortic segments containing the balloon increases before aortic valve opening. This happens since blood from adjacent segments begins to fill the low pressure area prior to the beginning of the next systolic interval. Thus, the results suggest that a sudden pressure drop in the aorta during deflation may be desirable if we wish to increase cardiac output.

This model should be useful in studying a number of other effects of the intraaortic balloon. Among these are the energy relationship between the device and the heart, balloon position and size, relationship of various vascular conditions to pumping effectiveness, and testing of on-line automatic control algorithms. In addition, since the balloon model has been incorporated into the same arterial model as that previously used to examine external counterpulsation, a direct comparison of the two methods of circulatory support will be possible.

Conclusion

This paper presents preliminary results of a model for intraaortic balloon pumping. The model produces results similar to those obtained from in-vivo measurements. In cases where the model tends to deviate from observed experimental results, interpretation of these differences yields valuable insight into the dynamics of in-series cardiac assistance. One such observation suggests that increasing the rate of balloon deflation can improve cardiac output.

References

1. Jelinek, J., "Hemodynamics of Counterpulsation: The Study of a Lumped-Parameter Computer Model", J. Biomechanics 5:511-515, 1972.

2. Jones, R. T., Petschek, H. E., Kantrowitz, A. R., "Elementary Theory of Synchronous Arterio-Arterial Blood Pumps", Med. Biol. Engin. 6:303-308, 1968.

3. Ohley, W. J., Birtwell, W. C., Braun, L., Bicker, A., Soroff, H. S., "Computer Analysis of External Counterpulsation by use of a Non-linear Mathematical Model of the Cardiovascular System", J. Assoc. Adv. Med. Inst. 10:228-231, 1976.

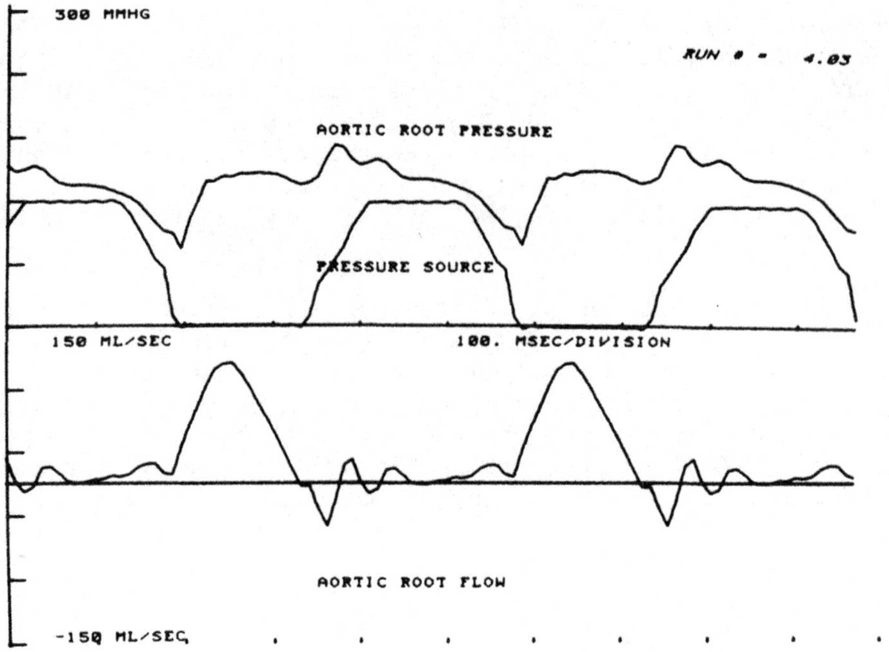

Figure 1. Aortic pressure (top tracing), balloon pressure (middle tracing), and aortic flow (bottom tracing), generated by the model.

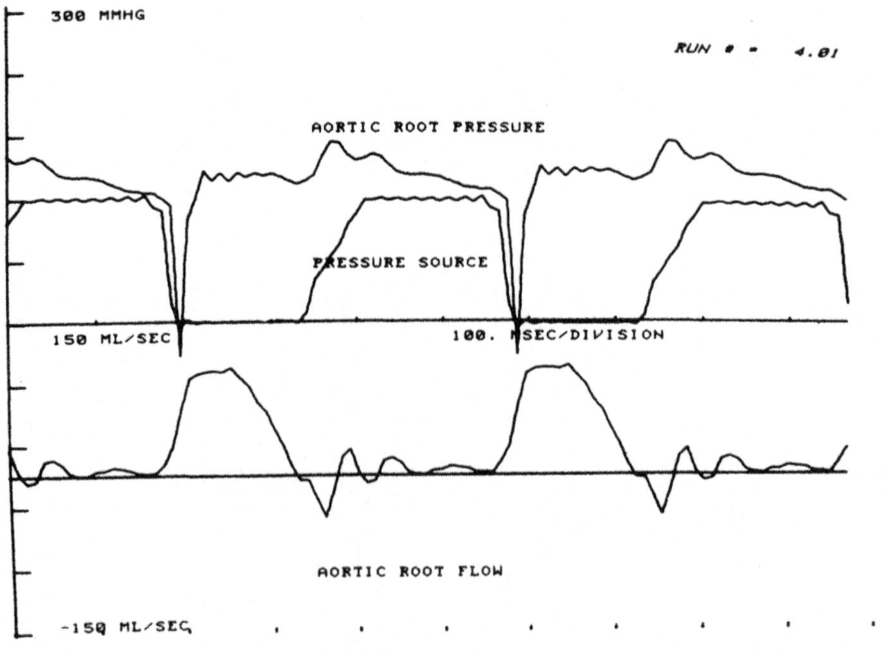

Figure 2. Aortic pressure (top tracing), balloon pressure (middle tracing), and aortic flow (bottom tracing), generated by the model.

SIMULATION ANALYSIS OF A NEW CONTROL UNIT FOR IN-SERIES CARDIAC ASSIST DEVICES*

Peter J. Martin, William J. Ohley, Karl E. Karlson and Dov Jaron

Biomedical Engineering Program
Department of Electrical Engineering
University of Rhode Island
Kingston, Rhode Island, 02881

and

Department of Surgery
Rhode Island Hospital
Providence, Rhode Island, 02903

Introduction

The development of an automatic control system for in-series cardiac assist devices is desirable in order to increase the effectiveness of cardiac assistance [1]. However, control parameters which can be measured and which are clinically useful must be found before a control system can be developed. One parameter which has been suggested for use is the afterload phase angle [2]. This phase angle is defined as the phase difference between the fundamental components of aortic pressure and flow. A means of controlling the phase angle is needed before its relationship to the hemodynamic effects of assistance can be studied.

The objectives of this work were twofold. The first was to design a control unit which allows accurate and convenient control of the afterload phase angle. The second was to determine the effects of the control parameters associated with the new control unit on the afterload phase angle.

Design of the Control Unit

The purpose of the control unit is to provide a driving waveform for the assist device which is in synchrony and out of phase with the heart cycle. Phasing of this driving waveform is controlled with a single adjustment.

Conventional units employ two timing adjustments for controlling an assist device. The first is the delay from the ECG R-wave to pump activation. The second is the duration of pump systole. A change in either the delay or the duration will cause a change in the phase of the driving waveform which results in a change in the afterload phase angle.

The phase of the driving waveform can be controlled with a single adjustment by replacing the conventional delay with the time interval from the ECG R-wave to mid pump systole (TMPS). Consequently, the duration, which is symmetric about the new TMPS adjustment, can be altered without disturbing the phase of the driving waveform. A comparison of the timing adjustments between the new and conventional control units is shown in Figure 1.

---

*Supported by NSF Grant ENG 74-21085

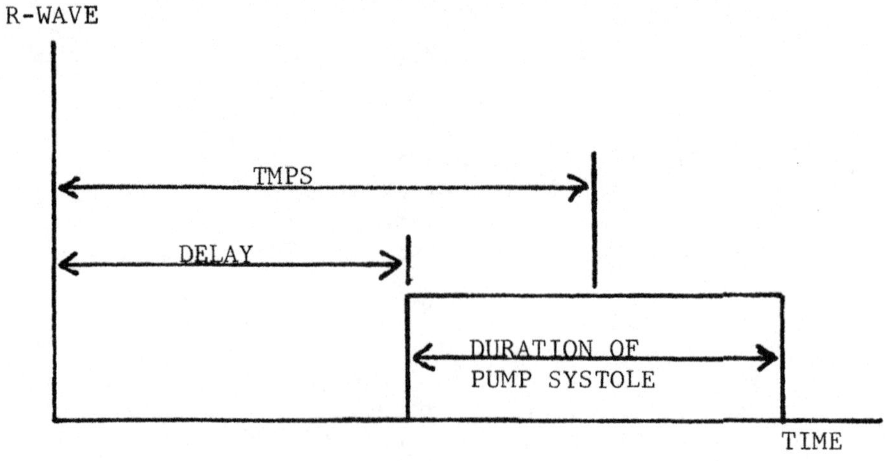

FIGURE 1.

The two adjustments which will control the assist device are TMPS and one half the duration of pump systole (1/2 DUR). The time of activation (T_{act}) and the time of deactivation (T_{deact}) are given by:

$$T_{act} = TMPS - 1/2\ DUR \tag{1}$$

$$T_{deact} = TMPS + 1/2\ DUR \tag{2}$$

Four other major criteria were included in the design of the control unit. The first was the accuracy of the phasing. Since the control unit is to be used in experiments to determine the effects of the afterload phase angle on hemodynamics it was decided that the phase angle should be accurate to one degree. Timing adjustments accurate to one msec satisfy this criterion for heart rates up to 166 beats per minute. The second criterion was that a safety feature should be included to deactivate the assist device if a premature ventricular contraction occurs. This insures that the heart will never eject against an active assist device. The third consideration was to design the unit so that it could be operated either manually or by a computer as part of an automatic control system. Finally, the unit was designed so that under computer control all past hemodynamic data, including those collected during the previous heart beat, could be used to calculate new timing adjustments.

The control unit was designed and built in a modular fashion using digital circuitry. This allows flexibility for alteration or expansion of logic functions, and for quick trouble shooting and repair. The control unit has been described in detail elsewhere [3].

Computer Simulation

A simulation of cardiac assistance based upon the operation of the new control unit was performed. The purpose was to determine the effects of the new control parameters on the afterload phase angle. The simulation was used to find the afterload phase angle as a function of TMPS, duration, diastolic pressure, and the slope of the diastolic pressure pulse.

A digital computer was used to generate aortic pressure waveforms which approximate the shape of waveforms found during assistance. Unassisted aortic pressure and flow data from a previous dog experiment were used as a basis for this. The test pressure waveforms were generated by modifying the diastolic portion of the unassisted waveform to include the pressure pulse associated with in-series assistance. This pressure pulse was given a duration equal to twice 1/2 DUR and was centered around TMPS. The maximum value of diastolic pressure and the slope of the rising and falling edges of the pressure pulse were also controlled. The aortic flow waveform was assumed to remain unchanged throughout the simulation.

A series of test runs were performed by varying the four control parameters independently of each other. The results of the simulation are summarized in Figures 2 through 5. In each of these figures TMPS and duration are given in percent of the cardiac cycle, diastolic pressure is given in mmHg, and the slope of the pressure pulse is given in mmHg/percent of the cardiac cycle. Normalization of control parameters to percent of the cardiac cycle allows comparison of in-vivo measurements by maintaining the phase angle as the heart rate changes.

The afterload phase angle is shown as a function of TMPS for three values of duration in Figure 2. Peak diastolic pressure and the slope of the pressure pulse were held constant. As anticipated, the afterload phase angle increases as TMPS increases. It can also be seen that the sensitivity of the afterload phase angle to TMPS decreases as the duration is increased.

Figure 3 shows the afterload phase angle as a function of duration for four values of TMPS. Peak diastolic pressure and the slope of the pressure pulse were held constant. Note that once a threshold value in duration is reached the afterload phase angle remains nearly constant as the duration is increased further. One can also see that as TMPS approaches approximately 80% of the cardiac cycle both this threshold and the sensitivity of the afterload phase angle to duration decrease.

The afterload phase angle is plotted as a function of duration for three values of diastolic pressure in Figure 4. In this case TMPS and the slope of the pressure pulse were held constant. This figure shows that as diastolic pressure is increased both the threshold for duration and the sensitivity of the afterload phase angle to duration decrease.

In Figure 5, the afterload phase angle is shown as a function of diastolic pressure for three values of duration. TMPS and the slope of the pressure pulse were held constant. This figure shows that there is a threshold value for diastolic pressure which must be exceeded before the afterload phase angle can be controlled within a useful range. The effects of the duration threshold on the afterload phase, as was shown in Figure 3, can also be seen in this figure. When the duration is above the threshold value the afterload phase angle can be maintained approximately at a fixed value. When the duration is below the threshold value the afterload phase angle remains much lower.

A series of tests were also performed to determine the effects of the slope of the rising and falling edges of the diastolic pressure pulse on the afterload phase angle. The results of these tests showed that the afterload phase angle increases as the slope decreases.

Conclusion

The simulation has shown that there are threshold values for diastolic pressure and duration which must be exceeded before accurate control of the afterload phase angle can be achieved. Diastolic pressure must be above a given threshold in order to control the afterload phase angle within a useful range. The afterload phase angle will then remain constant as the duration is changed as long as the duration is above its threshold value. In addition to the threshold values, increasing diastolic pressure causes both the sensitivity of the afterload phase angle to duration and the threshold in duration to decrease.

These results indicate that under the proper conditions, this control unit will accurately control the afterload phase angle with a single adjustment. In-vivo testing of the control unit is needed to substantiate these results and to establish the threshold values for duration and diastolic pressure which are necessary to obtain accurate control.

References

1. Clark, J. W., et al: On the Feasibility of Closed-Loop Control of Intra-aortic Balloon Pumping. IEEE Trans., BME, 20:404, Nov. 1973.

2. Jaron, D., et al: Measurement of Ventricular Load Phase Angle as an Operating Criterion for In-Series Assist Devices: Hemodynamic Studies Utilizing Intraaortic Balloon Pumping. Trans. ASAIO, 16:466, 1970.

3. Martin, P.: A Digital Afterload Phase Angle Controller for In-Series Cardiac Assist Devices. M.S. Thesis, University of Rhode Island, December 1976.

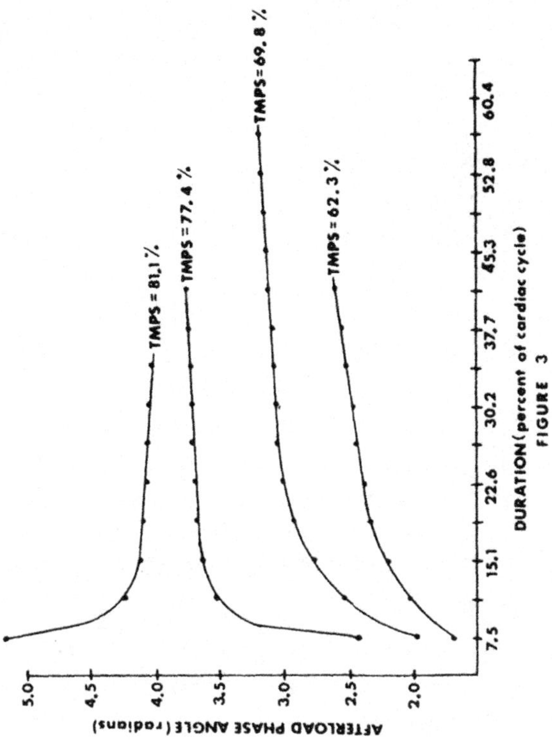

FIGURE 3

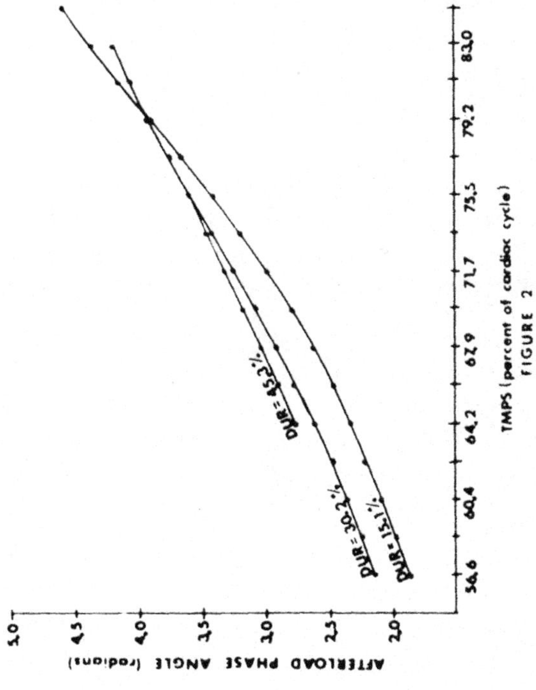

FIGURE 2

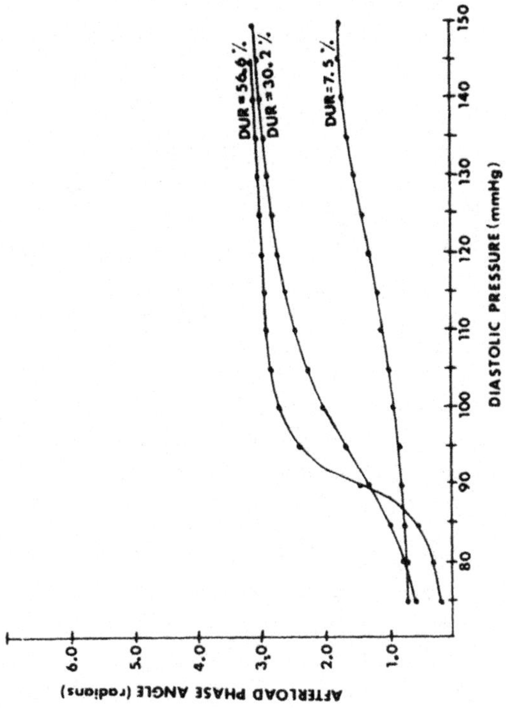

FIGURE 5

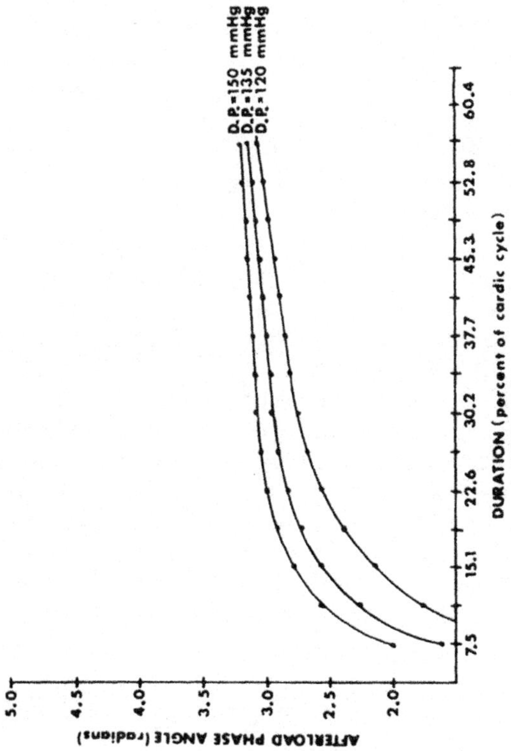

FIGURE 4

A DECISION-THEORETIC ANALYSIS OF THE INDICATIONS OF MEDICAL EXAMINATIONS

Stelios C. Orphanoudakis, Ph.D.
Department of Diagnostic Radiology, Yale University School of Medicine
New Haven, Connecticut 06510

Andrew J. Dwyer, M.D.
Department of Radiology, Hospital of St. Raphael
New Haven, Connecticut 06510

INTRODUCTION

In recent years, the medical diagnostic process has been studied by a number of investigators both as a decision and as a classification problem. Concepts and measures extracted from decision and information theory have been injected into the clinical context and attempts have been made to construct a diagnostic algebra.[1] The potential value of a decision (information) theoretic approach to the analysis of clinical judgements and the medical diagnostic process has been recognized and discussed by previous workers.[2-7] In this paper, the medical diagnostic process is viewed as a decision problem and is formulated in terms of the reduction in expected clinical regret achieved by the accrued clinical information. Implicit is a proposed measure of the appropriateness of a specific test as a function of its cost and the reduction in expected clinical regret.

Medical diagnosis is the process of reduction in the uncertainty of the state of the patient in order to administer appropriate therapy and to generate predictions of the pertinent future (complications, morbidity, mortality, prognosis). Each step in the medical diagnostic process may be characterized by a set of possible disease entities and their associated prior probabilities, a set of possible therapeutic actions (treatments), and a set of possible additional tests (test is used in its broad sense to indicate any modality of patient data acquisition).

Clinical medicine is replete with the question "Should I choose a course of therapy now, or should I perform additional tests". More formally phrased, "Given the set of possible disease entities with their associated prior probabilities, and given the set of possible therapies and the set of possible additional tests, does optimal strategy dictate a choice of a course of therapy now or after additional testing?" Central to this decision is a measure of the expected cost and benefits of each course of action. Once such measures have been defined, the choice of optimal strategy follows trivially from a comparison of their numerical values. The appropriateness of the decision is dependent on the validity of these measures. We propose the reduction in EXPECTED CLINICAL REGRET as a measure of the clinical efficacy associated with reduction in the undertainty of the patient's disease state. The question of whether one should test or treat may be analyzed by comparing the expected cost of testing against the expected reduction in clinical regret associated with the test.

NO-DATA DECISIONS UNDER UNCERTAINTY

As alluded to in the previous section, the elemental decision of the diagnostic process is a choice between electing treatment without further testing, or choosing a test and then electing to treat based on the test outcome. For simplicity, the number of possible tests will be restricted to one. Then, the problem of interest in this paper can be restated as follows:

Given a set of possible disease entities $D = \{D_1, D_2 \ldots, D_n\}$, the set of their associated prior probabilities $P_o = \{P_o(D_1), P_o(D_2), \ldots, P_o(D_n)\}$, the set of possible therapies $\Theta = \{\Theta_1, \Theta_2, \ldots, \Theta_m\}$, and the test t, is it the better course of action to choose the best therapy out of Θ initially, or rather to perform test t, and, based on its outcome, choose the best therapy out of Θ?

Let us initially examine the choice of an optimal course of therapy given the sets D, P_o, and Θ. Such a problem is discussed in decision theory as a no-data decision under uncertainty.[8] Its solution demands a measure of the efficacy of each choice of therapy given the possible disease entities and their prior probabilities.

Associated with each disease-therapy combination is a certain value of utility (i.e., the sum total morbidity, mortality, cost, etc. associated with treating a specific disease with a certain therapy). We shall denote this as u_{ij}, the utility attributed to treating disease D_i with therapy Θ_j. u_{ij} is assumed to be positive for all values of i and j. Considering any specific course of therapy Θ_j, the utility u_{ij} varies with the disease entity D_i (e.g., the utility associated with the treatment of regional enteritis and accute appendicitis with appendectomy.) The expected utility of therapy Θj, denoted by U_j, is given by

$$U_j = \sum_{i=1}^{n} P_o(D_i)u_{ij} \tag{1}$$

and it is a measure of the utility assigned to a particular course of action(therapy) given the prior probabilities of all possible disease entities. Under Bayesian criteria, the expected utility may be used to obtain an ordering of the possible therapeutic actions that permits the selection of an optimum therapy. In particular, the optimum therapy is that for which expected utility is maximum. Since there is a certain cost or regret associated with treating a specific disease state with a therapy which is not optimal, one can also select an optimum therapy based on a minimization of the expected regret associated with each therapy. In this paper, the expected regret, R_j, of therapy Θ_j is defined as

$$R_j = \sum_{i=1}^{n} P_o(D_i)r_{ij}, \tag{2}$$

where r_{ij} is obtained from

$$r_{ij} = \max_j (u_{ij}) - u_{ij}. \tag{3}$$

It can be shown that a Bayes choice of therapy based on minimizing the expected regret will be the same as one based on maximizing the expected utility.

An intuitive attribute of the minimum expected regret of a set of therapies is that it is zero when there is no uncertainty concerning the disease state present, since the therapy of choice for the known disease state can then be elected. Therefore, the minimum expected regret at any point in the diagnostic process is a measure of the value associated with pursuing the workup to the point of no residual uncertainty (the point when the patient's disease state is known with certainty.) Moreover, the generalized cost of a specific test may be compared to the reduction in expected clinical regret achieved by the results of the test and this comparison may be utilized as a criterion for determining the indication of the test.

As Eq. (2) implies, the expected regret associated with a particular choice of therapeutic action is a function of the prior probabilities of disease, $P_o(D_i)$. For convenience, let us examine the simple case of two possible disease entities such as the presence and absence of a particular disease and the choices of giving or withholding a specific therapy Θ_D. Then, the expected regret of each course of action is a linear function of the prior probability of disease and it is graphically represented by a straight line. In Figure 1, the expected regrets of the particular choices of therapeutic action "to treat (Θ)" or "not to treat ($\sim\Theta$)" disease D with its therapy of choice Θ_D are plotted as functions of the prior probability of disease $P_o(D)$. There is no expected clinical regret if the choice is to treat with Θ_D when the disease is present with probability $P_o(D)$ equal to 1. The regret is maximum for the choice to treat when the probability of disease is zero. Similarly, there is no expected clinical regret associated with the choice not to treat when the probability of disease is 0 and the clinical regret is maximum for that choice if the probability of disease is equal to 1. C and B, the two values of clinical regret associated with the decisions to treat when the disease is absent and not to treat when the disease is present have been called the "cost" and "benefit" of therapy respectively and point I in Figure 1 has been referred to as the "indifference point".[3] At that particular value of the prior probability of disease, $P_I(D)$, both decisions "to treat" and "not to treat" have the same expected regret. If the probability of disease is greater than $P_I(D)$, the decision "to treat" yields the minimum expected regret (maximum expected utility) and is preferable under Bayesian decision criteria. Similarly, if the probability of disease is less than $P_I(D)$, the decision "not to treat" yields the minimum expected regret and is the optimal action.

TEST EFFICACY

The performance of a test may be looked upon as the accrual of knowledge concerning additional patient attributes (clinical, historic, roentgenologic, anatomic, physiologic, etc.). Knowledge of a test outcome modifies the probabilities of the possible disease states. The probability of a disease state given a specific test outcome, also known as the posterior or conditional probability, may be calculated from the prior probability and the probability of the test outcome occuring in the presence of that disease state by invoking Baye's theorem. The posterior probability of a particular disease state is greater than its prior probability given a positive test outcome and lower than its prior probability given a negative test outcome. In fact, this is often used as a definition of the positiveness or negativeness of test outcomes.

The optimal choice of therapy following knowledge of a specific test outcome is the therapy with the least expected regret, based on the posterior probabilities of disease given that test outcome. The decision processes both prior and posterior to testing are analogous. The test outcome simply modifies the expected regrets of the different therapeutic options by modifying the probabilities of the disease states. The expected posterior regret may be obtained as the product of the regret associated with the optimal therapeutic choice given a test outcome multiplied by the probability of occurrence of that particular test outcome and summed over all possible test outcomes.

The reduction in clinical regret, afforded by knowledge of the test outcome, is a measure of test efficacy. It is defined as the difference between the minimum expected regret prior to testing and the posterior regret. If the reduction in clinical regret exceeds the cost of testing, then the testing procedure and therapy assignments based on its outcome are indicated. If the cost of testing exceeds the reduction in clinical regret, then the testing procedure is not indicated. The choices to treat or test and then treat are considered equivalent when the cost of testing and the reduction in expected clinical regret are equal.

A mathematical analysis of the regret criterion produces results which allow determination of the indication of a test based on the conditional probabilities of its outcomes given the possible disease states, its cost, and the set of possible therapies. Moreover, certain results demonstrate interesting relationships between the expected reduction in clinical regret and the indifference probability. For example, consider again the case of two possible disease states, D and $\sim D$, and two possible courses of action, Θ and $\sim\Theta$. The indifference probability partitions the range of values of the probability of disease, P(D), into two intervals (see Figure 1). For P(D) less than $P_I(D)$, the indifference probability, the optimum choice is not to treat. For P(D) greater than $P_I(D)$, the optimum choice is to treat. Furthermore, it is easily shown that if both the prior probability of disease and the posterior probabilities of disease for all outcomes of a given test lie in the same probability interval relative to $P_I(D)$, performing the test results in no change in clinical regret. Since there is some cost associated with any test, such a test is not indicated. This is in keeping with the clinical practice of anticipating possible modification of therapy following a test as a necessary criterion for its indication. However, it is important to note that if two or more tests are performed, each single test may not be indicated by the above criterion, while collectively the tests may be indicated.

The expected reduction in clinical regret accompanying a specific test may be obtained by averaging the reductions produced by individual test outcomes over all possible test outcomes. In mathematical notation, when the prior probability of disease is less than $P_I(D)$, it can be shown that the expected reduction in clinical regret, which we denote by $\overline{\Delta r}$, is given by

$$\overline{\Delta r} = (B+C)\sum_i P(t_i)[P(D/t_i) - P_I(D)], \tag{4}$$

where B and C are the benefit and cost of therapy respectively and $P(D/t_i)$ denotes the posterior probability of disease given test outcome t_i. The summation Σ is only over those test outcomes for which $P(D/t_i)$ is greater than $P_I(D)$. Thus, the expected reduction in clinical regret is calculated based only on those posterior probabilities which lie on the opposite side of the indifference probability relative to the prior probability of disease. Similarly, when the prior probability of disease is greater than $P_I(D)$, $\overline{\Delta r}$ is given by

$$\overline{\Delta r} = (B+C)\sum_i P(t_i)[P_I(D)-P(D/t_i)], \tag{5}$$

where the summation is now over all test outcomes for which $P(D/t_i)$ is less than $P_I(D)$.

In both Eqs. (4) and (5), $\overline{\Delta r}$ can be expressed as a function of $P_o(D)$. When this is done, one can show that:

1. The expected reduction in clinical regret for a given test is maximum when the prior probability of disease is equal to the indifference probability.
2. The expected reduction in clinical regret is a piecewise linear function of the prior probability of disease. Breakpoints of this function occur at values of $P_o(D)$ for which some test outcome yields a posterior probability equal to $P_I(D)$ and at $P_o(D)$ equal to $P_I(D)$. Furthermore, for $P_o(D)$ less than $P_I(D)$, $\overline{\Delta r}$ is a monotonic increasing function of $P_o(D)$. For $P_o(D)$ greater than $P_I(D)$, it is a monotonic decreasing function of $P_o(D)$.
3. For a test with two possible outcomes, t^+ (positive) and t^- (negative), the maximum value of the expected reduction in clinical regret is equal to

$$\overline{\Delta r}_{max} = \frac{BC}{B+C}[P(t^+/D)+P(t^-/\sim D)-1]. \qquad (6)$$

where the conditional probabilities $P(t^+/D)$ and $P(t^-/\sim D)$ are known as the sensitivity and specificity of the test respectively.

By the EXPECTED REDUCTION IN CLINICAL REGRET criterion, if the cost of a test is greater than the expected reduction in clinical regret afforded by the test, the test is not indicated. Thus, if the projected generalized cost of a specific test is independent of the prior probability of disease and greater than $\overline{\Delta r}_{max}$ for that test, the test is not indicated regardless of what the prior probability of disease is. However, if the projected generalized cost of a specific test is constant and less than $\overline{\Delta r}_{max}$, there will be an interval of prior probability values, including the indifference probability, such that, if the prior probability of disease lies within this interval, the test is indicated. If the prior probability of disease lies outside this interval, the test is not indicated. If a test is of constant cost and is not indicated for $P_o(D)$ equal to $P_I(D)$, then it is not indicated at all. In general, the cost of a test may be a function of the prior probability of disease and a graphical comparison may be used to determine whether or not the test is indicated and, if it is, the interval of indication.

The above properties of the expected reduction in clinical regret and the concept of an interval of indication for a given test are illustrated in Figure 2. The width of the interval of indication is a function of B, the benefit of therapy, C, the cost of therapy in the absence of disease, the test cost, and the posterior probabilities given the test outcomes.

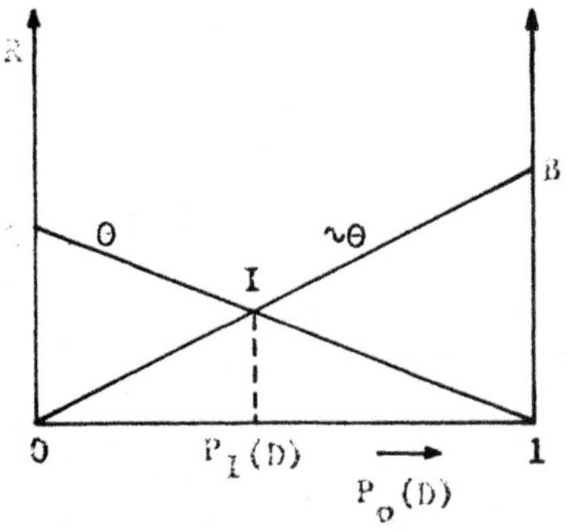

Figure 1

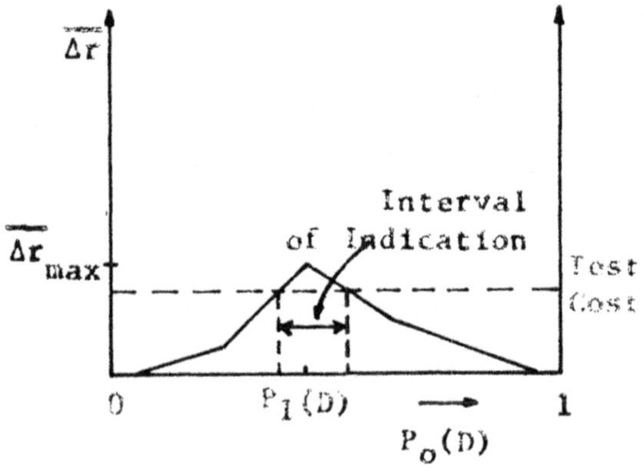

Figure 2

[1] J. A. Jacquez, Ed., Computer Diagnosis and Diagnostic Methods, Charles C. Thomas, Springfield, Illinois, 1972.

[2] W. B. Schwartz, G. A. Gorry, J. P. Kassirer, and A. Essig, Amer J of Med 55:459-472, 1973.

[3] S. G. Pauker and J. P. Kassirer, N Engl J Med 293:229-234, 1975.

[4] B. J. McNeil, E. Keeler, and S. J. Adelstein, N Engl J Med 293:211-215, 1975.

[5] J. P. Kassirer, Yale J of Biol and Med 49:149-164, 1976.

[6] E. C. Gregg, P. S. Rao, and H. L. Friedell, Invest Radiol 11:249-257, 1976.

[7] D. F. Ransohoff and A. R. Feinstein, Yale J of Biol and Med 49:165-168, 1976.

[8] B. W. Lindgren, Elements of Decision Theory, The Macmillan Company, New York, New York, 1971.

A TIME DOMAIN REPRESENTATION OF THE LEFT VENTRICLE*

Dennis J. Arena, William J. Ohley,
and Dov Jaron

Biomedical Engineering Program
Department of Electrical Engineering
University of Rhode Island
Kingston, Rhode Island 02881

Introduction

A quantitative analysis of the left ventricle is desirable in order to gain insight into left ventricular function and to determine the status of the left ventricle. The present work describes a new time domain representation of the left ventricle and presents a quantitative analysis of the simulated cardiac cycle. Thus far, the model has been used to study the function of a healthy left ventricle. With modifications in parameters, it is expected that the model will be capable of describing an infarcted left ventricle.

The Model

The left ventricle is modeled as a truncated ellipsoid approximated by a series of cylindrical shells. The assumptions used in the analysis are:

1) the ventricular chamber has a uniform pressure distribution at end-diastole,

2) the ventricle has uniform wall thickness at end-diastole,

3) the cylindrical shells contract sequentially, beginning at the ventricular apex,

4) the shells contract until a given stroke volume is produced; the radius of each cylindrical shell changes sinusoidally and in proportion to the end-diastolic radius of the shell,

5) if the radius of a cylindrical shell reaches its end-systolic radius before the other shells have completed contraction, it remains at that radius until all the shells reach their end-systolic radii,

6) stress is uniform throughout the wall of each cylindrical shell,

7) blood is an incompressible, Newtonian fluid,

----

*Supported in part by the Stride Memorial Foundation.

8. flow through each cylindrical shell is laminar.

The stress in the wall of each cylindrical shell is related to the pressure within the shell by the law of Laplace (1). Until contraction begins, the cylindrical shells exhibit passive compliant properties representing the compliance of the myocardium. The passive compliant properties are evaluated from the stress-strain relationship for myocardium in the passive state (2). The cylindrical shells contract sequentially beginning at the ventricular apex to simulate the mechanical action of the left ventricle. Before opening of the aortic valve, each contracting shell exhibits wall stress equal to the stress developed by active cardiac muscle in isometric contraction (3). The volume of blood expelled by the contracting shells is transferred to the shells which have not yet begun contracting. Distribution of the volume is in proportions that cause equal strain to the passive shells. Hence, during the simulated isometric contraction of the left ventricle, the end-diastolic volume is contained by the left ventricle but the geometry of the left ventricle changes. Whenever the radius of a shell changes, the wall thickness is adjusted to conserve the mass of the shell wall. Left ventricular flow begins when all the cylindrical shells are contracting. Viscous forces generated within the ventricular chamber are determined by applying the Hagen-Poiseuille equation to each cylinder (4). Inertial forces generated within the ventricular chamber are determined by calculating the change in momentum produced by each cylinder.

Results

The model was tested utilizing ten cylindrical segments. Parameters for the model were taken from the literature for a canine left ventricle. The model was terminated at the aortic valve using data for canine aortic input impedance (5). Figure 1 shows the shape of the ventricular chamber at end-diastole. It also shows the cylindrical shell approximation. The outline of the ventricular chamber at three instances of the cardiac cycle is displayed in Figure 2. Left ventricular flow and aortic pressure produced by the model are shown in Figure 3.

The pressure distribution throughout the ventricular chamber was determined for different instances in the cardiac cycle. As the myocardium in the vicinity of the ventricular apex begins contracting, pressures due to the active stress in the myocardium are developed in the apical area while pressures due to the compliance of the myocardium are generated in all the passive segments. As contraction progresses from the ventricular apex towards the aortic valve, additional segments develop active stress and fewer segments generate compliant stresses. The pressure distribution immediately before the opening of the aortic valve is displayed in Figure 4. During the ejection phase of the cardiac cycle, pressures are generated due to the viscous and inertial forces within the left ventricle (Figure 5).

Discussion

This representation of the left ventricle was developed by simulating the dynamics of the left ventricle through the cardiac cycle. The pressure and flow waveforms produced by the model are similar to those measured experimentally in dogs (Figure 3). The pressure distribution in the left ventricular chamber throughout the cardiac cycle is such that blood tends to move towards the aortic valve (Figures 4,5). The timing and sequence of cardiac contraction assumed in this model and the resulting geometric changes are responsible for establishing and maintaining the observed pressure distribution within the chamber.

The geometric configuration developed during the simulated isometric contraction phase of the cardiac cycle establishes a pressure gradient within the ventricular chamber clearly conducive to ejection (Figure 4). The potential energy distribution in the left ventricular chamber just prior to ejection has been used as an index of ejection efficiency. It is desirable that the direction of the gradient within the ventricle just prior to ejection be similar to Figure 4. Such a distribution has been shown to be absent in certain myocardial dysfunctions (6). The magnitudes of the pressure gradient obtained in our model for this instant in the cardiac cycle seem to be exagerated. Several factors may be responsible for this exageration. The value used for active isometric stress was 2.00 gr/mm^2, obtained from studies on papillary muscles (3). The applicability of results obtained from isolated papillary muscle preparations to the ventricular myocardium in-situ may be questioned. However, in studies on dog hearts in-vivo a value for total isovolumic tension in the left ventricle of 2.35 gr/mm^2 was obtained (7). Hence, we feel the value of stress used here was not unreasonable. Another characteristic of the representation which may be responsible for the large magnitudes is that equal stress is developed in all the myocardium just prior to ejection. There may be portions of the left ventricle which are not capable of developing full isometric tension just prior to ejection. Two assumptions used in the analysis should also be considered as possible causes of the exagerated pressure magnitudes: 1) the ventricular chamber has a uniform thickness at end-diastole; and 2) the radii of all the cylinders change by the same percentage over systole.

Our results also show that the pressure gradients during the ejection phase of the cardiac cycle are small. This indicates that the viscous and inertial forces generated within the left ventricle during ejection are small in comparison with the forces, producing static pressure, which are generated by the active myocardium.

Only the function of the healthy ventricle has been studied thus far using this model. However, the features of this model lend themselves to the investigation of various ventricular dysfunctions. Such studies are presently in progress.

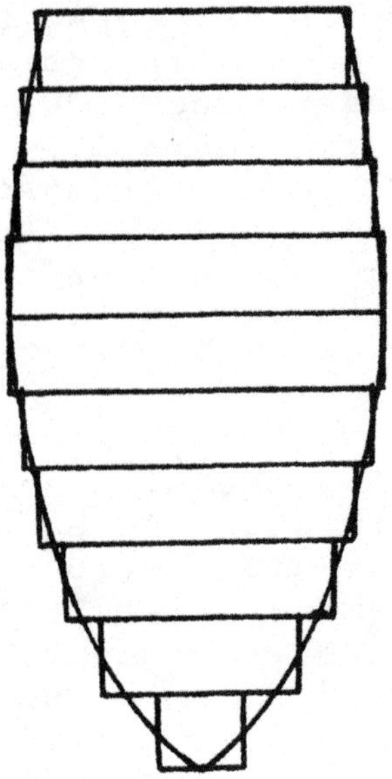

Figure 1. Outline of the ventricular chamber at end diastole and the cylindrical shell approximation.

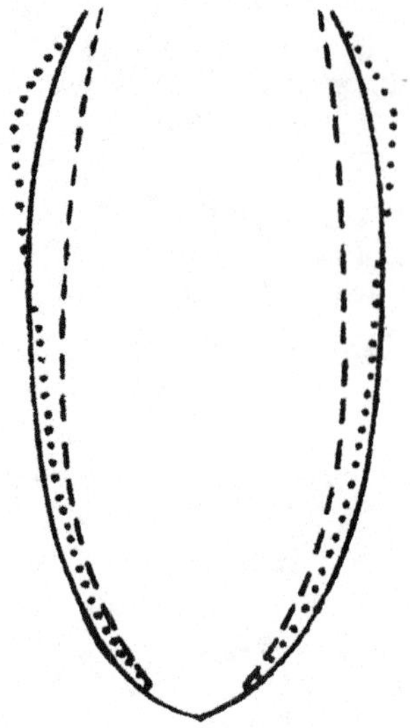

Figure 2. Outline of the ventricular chamber at three instances in the cardiac cycle.

——— end diastole
······ end isom. contr.
------ end systole

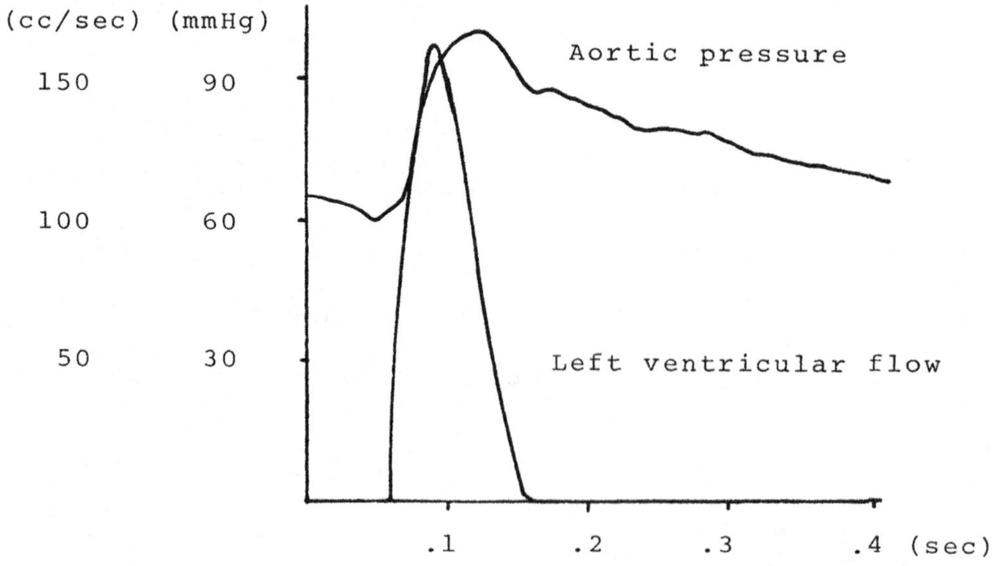

Figure 3. Sample flow and pressure waveforms generated by the model.

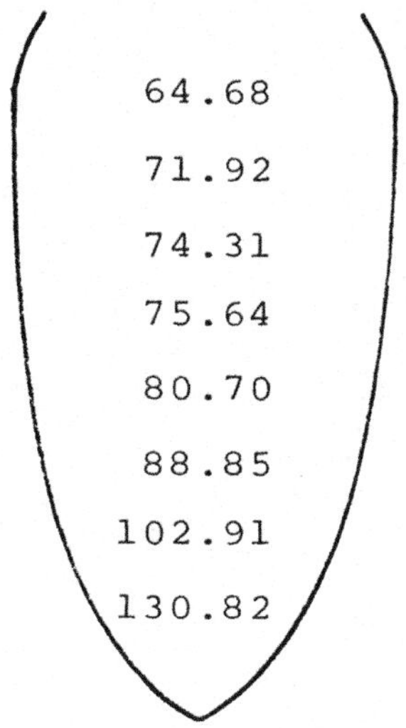

Figure 4. Left ventricular pressure distribution just prior to ejection in mmHg.

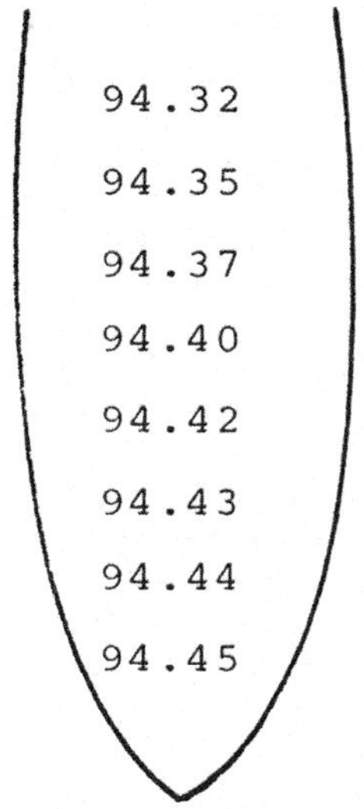

Figure 5. Left ventricular pressure distribution at an instant during ejection in mmHg.

References

1. Tallarida, R. J., Rusy, B. F., and Laughnane, M. H.: Left ventricular wall acceleration and the law of Laplace. Circ. Res. 4:217-223, 1970.

2. Weiss, M., and Forster, W.: A model for the assessment of left ventricular compliance: effect of hypertrophy and infarction. Cardiovasc. Res. 9:544-553, 1975.

3. Taylor, R. R.: Contractile properties of cardiac muscle in hyperthyroidism. Circ. Res. 27:539-549, 1970.

4. Pao, R. H. F.: Fluid Mechanics. John Wiley and Sons, New York: p. 186, 1966.

5. Patel, D. J., De Freitas, F. M., and Fry, D. L.: Hydraulic input impedance to the aorta and pulmonary artery in dogs. J. App. Physiol. 18:134-140, 1963.

6. Ray, G., Liu, Y. K., Chandran, K. B., Ghista, D. N., and Sandler, H.: Potential energy distribution prior to ejection as a basis for post surgical evaluation of left ventricular efficiency. Proc. 29th Ann. Conf. on Eng. in Med. and Biol. 18:126, 1976.

7. Taylor, R. R., Ross, Jr., J., Covell, J. W., and Sonnenblick, E. H.: A quantitative analysis of left ventricular myocardial function in the intact, sedated dog. Circ. Res. 21:99-115, 1967.

ADSORBER-ULTRAFILTER ARTIFICIAL KIDNEY CUM UREMIC PATIENT SYSTEM MODEL

by

U.R. Shettigar & D. Deepak &
Biomedical Engineering Division
Indian Institute of Technology
Madras-36, INDIA

Dhanjoo N. Ghista
Biomedical Research Division
NASA, Ames Research Center
Moffett Field, Calif.94035 U.S.A.

INTRODUCTION

The Adsorber-Ultrafilter Artificial kidney involves two stages of operation; namely, (i) blood ultrafiltration and (ii) adsorption of nitrogeneous metabolic waste products, excess cations, water and 'other uremic substances' from the filtrate, and re-infusion of the purifed fluid (see Fig. 1). This method has the following advantages over the conventional hemodialysis: (1) reduces cost and size by eliminating the dialysate, (2) eliminates the problems of trace element alterations associated with the usage of tap water as dialysate fluid, (3) permits removal of the 'middle molecules' from the uremic blood. Its advantage over Hemoperfusion is that it eliminates the necessity of coating the adsorbents, since the filtrate does not contain the formed elements and the plasma proteins.

MODELLING

This paper is chiefly concerned with the analytical simulation of activated carbon adsorber bed for the removal of urea, uric acid and creatinine, even though the developed model is applicable for any other uremic substance. The design analysis of the blood ultrafilter is not dealt with here, as it has been detailed in our earlier publication (Shettigar et al, 1977a). It is assumed that the ultrafilter is highly permeable to uremic substances, and its ultrafiltrate rate is between 15 to 25% of the inlet blood flow rate.

The adsorber is modelled as a two phase system: linear adsorption isotherms are assumed. The uremic patient is modelled as a three compartmental model. The integrated patient-artificial kidney system model is shown in figure 1. The governing equations comprise (i) mass balance equations for solute transfer in the 3 compartments of the uremic patient model, as per our earlier (more elaborate) analysis of solute transfer between the blood capillaries and tissues of the body compartments (Shettigar et al, 1976, 1977b), (ii) ultrafiltration rate equation, and (iii) flow continuity equations at the network junctions (fig. 1).

$$V_1 \frac{dC_1}{dt} = K_1(C_2-C_1)+K_2(C_2-C_1)+LC_3+Q_7C_7-Q_1C_1 \qquad (1)$$

$$V_2 \frac{dC_2}{dt} = K_1(C_1-C_2)+G_i \qquad (2)$$

$$V_3 \frac{dC_3}{dt} = K_2(C_1-C_3)-LC_3 \qquad (3)$$

$$Q_4 = 0.15\ Q_1;\ C_4 = C_1 = C_5 \qquad (4)$$

$$Q_6 = Q_5-Q_w;\ Q_7 = Q_4+Q_6 \qquad (5)$$

$$Q_6 = Q_5-Q_w;\ Q_7 = Q_4+Q_6 \qquad (6)$$

where (i) V,C and Q stand for volume, concentration and flow rate, repectively, (ii) K is the mass transfer coefficient, L is the convective flow of cerebro-spinal fluid and G_i is the generation rate of solute i.

The value of the solute concentration C_6, in the stream leaving the adsorber unit, is computed from the two phase model of the adsorber unit. For this purpose, the conservation equations are formulated separately for the activated carbon particle and the interstitial fluid in the adsorbed unit, thereby accounting for the fluid-particle and intraparticle diffusion effects. In the interstitial fluid phase, only the effects of axial dispersion of the solute are included.

The diffusion equations in the granule and interstitial fluid phase are as follows:

$$D_{ep} \left(\frac{\partial^2 C_p}{\partial r^2} + \frac{2}{r}\frac{\partial C_p}{\partial r}\right) - \rho_s \frac{100}{\alpha_p}\frac{\partial n}{\partial t} = \frac{\partial C_p}{\partial t} \qquad (7)$$

$$n = a_1 C_p + a_2 \qquad (8)$$

$$D_{eB} \frac{\partial^2 C_B}{\partial z^2} - U \frac{\partial C_B}{\partial z} - a_V k_m \left[C_B - C_p(r=R)\right] = \alpha_B \frac{\partial C_B}{\partial t} \qquad (9)$$

The boundary conditions for equations (7) and (9) are given by the following equations (10) and (11):

$$C_p(t=0,r) = \frac{\partial}{\partial r} C_p\ (t,r=0) = 0$$

$$D_{ep} \frac{\partial C_p}{\partial r} (r = R,t) = k_m\ [C_b - C_p\ (r=R,t)] \tag{10}$$

$$C_B\ (t=0,z) = 0, \quad \frac{\partial C_B}{\partial z}\ (t,z=Z) = 0$$

$$D_{eB} \frac{\partial}{\partial z} C_B\ (t,z=0) = U\ [C_B\ (t,z=0) - C_5] \tag{11}$$

where the subscripts p and B refer to the particle and bulk fluid phase in the adsorber respectively; a_1 and a_2 are the adsorption isothermal coefficients; D_e is the effective diffusivity; U is the axial plug velocity; a_v is the surface area of particle per unit volume of bed; k_m is the interface mass transfer coefficient; R is radius of particle; and n is the concentration of adsorbed solute on the pore wall, in equilibrium with C_p, the solute concentration in the particle pore.

Equations (7), (8) and (10) are solved analytically to give

$$C_p\ (s,T) = 2Sh^* \sum_{n=1}^{\infty} \frac{b_n^2\ \sin b_n}{b_n^2 - (1-Sh^*)\sin^2 b_n}$$

$$* \frac{\sin b_n s}{s} \int_0^T C_B(s)\ e^{-b_n^2\ (T-x)}\ dx \tag{12}$$

where b_n are the roots of the equation: $b \cot b-(1-Sh^*) = 0$, s is the reduced particle radius, Sh^* $(=k_m R/D_{ep})$ is the modified Sherwood number, $T = \alpha_p\ t\ D_{ep}/(100\rho_s\ a_1 + \alpha_p)\ R^2$ and ρ_s is the density of particle.

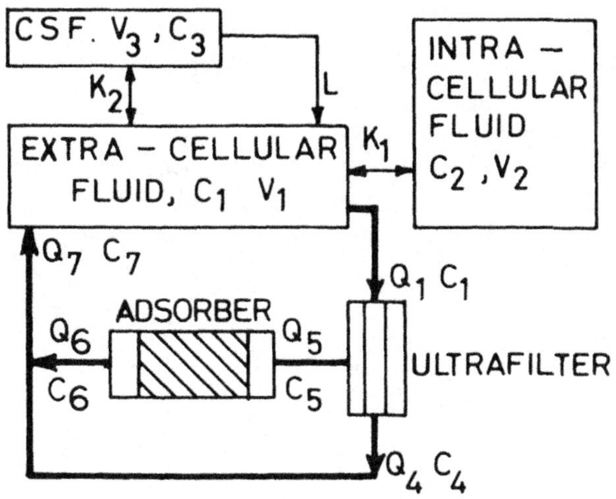

FIG.1. MODEL OF ADSORBER-ULTRAFILTER ARTIFICIAL KIDNEY TREATMENT OF UREMIC PATIENT.

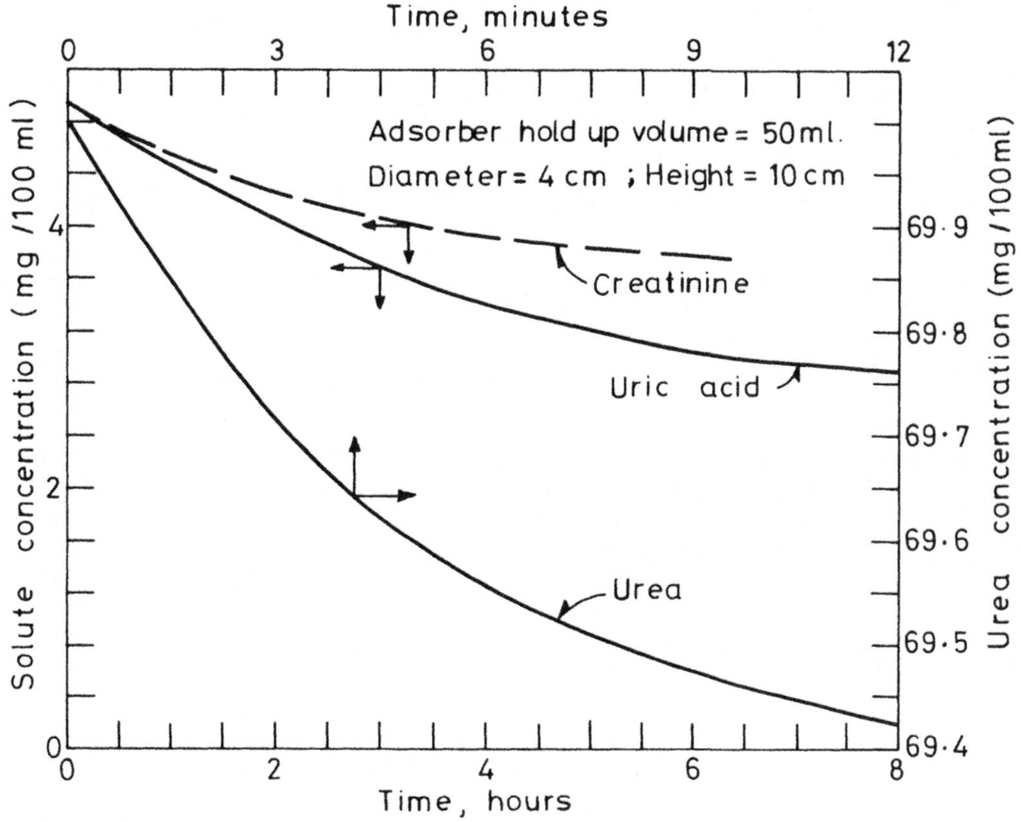

FIG.2. TRANSIENT CONCENTRATION PROFILES IN THE INTRACELLULAR COMPARTMENT.

Equations (1) to (6), (9), (11) and (12) are solved simultaneously by the finite difference method (using IBM 370 computer).

Fig. 2 shows the time variation of solute concentration in the extra-cellular fluid compartment of the uremic patient during the treatment. Concentrations of uric acid and creatinine decrease to their 'safe' values in about 6 to 7 hours of treatment, using activated carbon as the adsorbent. The decrease in urea concentration is small, however. Adsorbents, such as oxycellulose, may be used for effective removal of urea. The model developed in this study is useful to help arrive at the optimum design criteria for the adsorber.

References:

1. Shettigar, U.R., Prabhu, H.M., Ghista, D.N. (1977a), Design analysis of an ultrafilter, Medical and Biological Engineering (to be published).

2. Shettigar, U.R., Deepak, D., Ghista, D.N. (1976), Patient dialysis performance and treatment system model. Proc. of the Fourth New England Bioengineering Conference (New Haven). Pergamon Press.

3. Shettigar, U.R., Deepak, D., Ghista, D.N. (1977b), A model of the patient-dialysis treatment, to study the dynamics of metabolic waste concentrations and CSF pressure, and to hence provide guidelines for treatment. Medical and Biological Engineering (to be published).

LASER INTERFEROMETRY APPLIED TO MIDDLE EAR DIAGNOSIS

Allen Dean Drake
Otolaryngology Research Laboratory
Boston University Medical Center
720 Harrison Avenue
Boston, Massachusetts 02118

I. ABSTRACT

An instrument is described which measures submicroscopic displacements of the eardrum by laser interferometry. From these measurements the condition of the middle ear can be diagnosed.

II. INTRODUCTION

The middle ear is a mechanical linkage between the eardrum, where airborne sounds set up tissue vibrations, and the inner ear, where the vibrations are transduced into electrical nerve impulses. Any mechanical impediment to the passage of vibration through this linkage will therefore result in the impairment of hearing. Very often hearing losses can be attributed to diseases which have affected this linkage, and that can happen in a variety of ways. For example, an infection can produce an excess of fluid in the middle ear which itself can cause an increased effective mass of the small bones (ossicles) in the linkage and which can also block the eustachian tube, secondarily resulting in the build up of a high negative pressure. Both of these effects impair hearing. Or one of the ossicles may break (ossicular disruption). Or one of the ossicles may become stuck (otosclerosis). In each case there is a mechanical change associated with the particular disease and with the resulting hearing difficulty.

These changes in mechanical properties of the system have been exploited in tympanometry to diagnose middle ear pathologies. Tympanometry simply involves measuring the acoustic admittance of the system at two discrete frequencies while the external pressure on the eardrum is varied. Results, while far from perfect, have been helpful in providing the otologist with more diagnostic information, and most audiometry clinics now include a tympanometer.

Despite their usefulness, tympanometers are beset with several drawbacks, and the most notable of these is probably the necessity to procure and maintain a hermetic seal in the external ear canal. It is not only difficult for the technician to achieve, but it is also uncomfortable for the patient. Another drawback is the necessity to limit measurements to two frequencies due to the time involved (often two or three minutes) while a seal is maintained. Still another is the fact that the response is taken from the eardrum as a whole with no capability for examining discrete points on the drum.

Now the ear can detect sounds which cause the eardrum to move on the order of an angstrom, and so any technique which is to monitor a discrete point must be extremely sensitive. A narrow probe is necessary to go in a straight line down the curved ear canal, and it is preferable to have the instrument not contact the patient. A technique which meets all of these conditions and simultaneously overcomes some of the drawbacks of tympanometry is laser interferometry. Interferometry measures displacement instead of acoustic admittance, but in the presence of a standard acoustic stimulus, displacement is just as informative as acoustic admittance in determining the mechanical characteristics of the middle ear.

III. THE SYSTEM

The schematic of my system is shown in the figure, which displays a typical Michelson interferometer and associated electronics. Although other investigators [1] have placed small mirrors on the eardrum, it was found that the diffuse reflection from the drum itself is sufficient. This obviates the artifically increased mass of the drum due to the mirror, as well as the difficulty in attaching and aligning the mirror. The reference mirror, instead of being stationary, is mounted on two piezoelectric crystals which can be independently driven. The signal which drives one crystal is a 0.1 volt peak-to-peak 20kHz sine wave whose effect is detected from the photomultiplier tube's response and which is then used to provide a continuous calibration signal. This is necessary due to the slow drifts that are inherent in interferometers.

Across the second piezoelectric crystal behind the reference mirror is imposed a variable d.c. voltage for the purpose of optimizing the system's sensitivity. This is accomplished by optimizing the position of the light-fringe-dark-fringe border on the aperture in front of the photomultiplier tube. Some investigators [2] have found it expedient to perform this function in a feedback loop, but my drifts are slow enough to adjust the d.c. voltage by hand.

All the optical components are mounted on a heavy surface plate, which is in turn mounted an a system of four pneumatic inner tubes for vibration isolation. I use a 5mW helium-neon laser, and the purpose of the variable density filter is to equalize the light intensity in each arm of the interferometer. Displacement amplitudes as small as 0.5 angstrom have been detected.

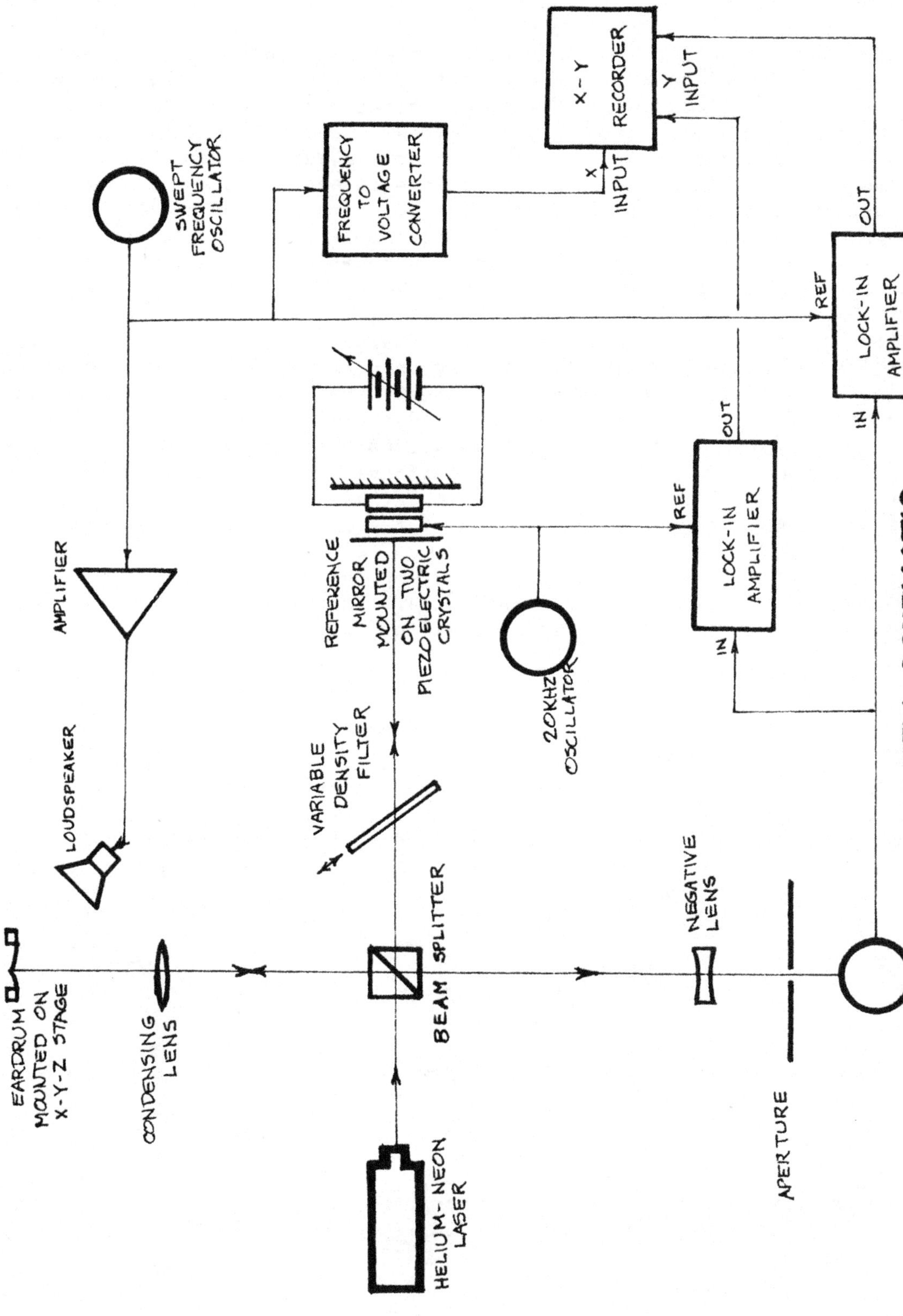

SYSTEM SCHEMATIC

IV. EXPERIMENTS

A human temporal bone including the intact eardrum is cemented in a brass holder which is mounted on an x-y-z translation stage so that different points on the surface of the drum can be examined. The muscle tendons connecting the temporal bone to the ossicles are cut to remove this unphysiological source of post mortem stiffness, and holes are provided in the bone holder for the purpose of varying the middle ear pressure and introducing fluids. During the experiments this pressure is set at several values, and the middle ear is filled to various levels with secretion simulating fluids while the loudspeaker is presenting a sweep of frequencies to the eardrum. Moreover, between some sweeps the long process of the incus is cut to simulate ossicular disruption, or the stapes is cemented to the frame of the oval window to simulate otosclerosis.

As the stimulus frequency is swept from 20Hz to 20kHz, one of the lock-in amplifiers follows and detects the response from the photomultiplier tube. This lock-in amplifier simultaneously detects in-phase and quadrature components of the signal, which is important in detecting differences in phase angle of the system's transfer function due to changes in the relative massive, resistive, or compliant components of the system. These changes are the most indicative feature of the particular middle ear disorder. The responses are then displayed as a function of frequency on an x-y recorder.

Experiments are presently in progress correlating various responses with particular simulated middle ear pathologies. In the future, responses will be correlated with middle ear pathologies in living patients.

V. ACKNOWLEDGMENTS

The author gratefully acknowledges support by a General Research Support Grant from University Hospital, by a grant from the Faculty Grants Committee of Boston University School of Medicine, and by generous contributions of time, ideas, and material by Geza J. Jako, M.D., Research Professor of Otolaryngology at Boston University.

VI. REFERENCES

[1] S.M. Khanna, J. Tonndorf, and W.W. Walcott, "Laser interferometer for the measurement of submicroscopic displacement amplitudes and their phases in small biological structures," J. Acous. Soc. Am., vol. 44, 1968, pp. 1555-1565.

[2] P.R. Dragsten, W.W. Webb, J.A. Paton, and R.R. Capranica, "Light-scattering heterodyne interferometer for vibration measurements in auditory organs," J. Acous. Soc. Am., vol. 60, 1976, pp. 665-671.

ULTRASONIC PREDICTION OF DELAYED OR NONUNION OF FRACTURES

Stanley A. Brown and Michael B. Mayor
Surgical Laboratories, Dartmouth-Hitchcock Medical Center, Hanover, N.H.03755

In managing patients with broken bones, the orthopaedist must rely on his previous experience in the interpretation of physical exams and radiographs. He tends to err on the side of conservatism and prolongs the disability in an attempt to avoid complications which might lead to delayed or nonunion. The availability of a device which could be used to objectively assess the response of the injured tissue and to predict the course of healing would be of significant value in making management decisions to hasten the recovery.

An ultrasonic system is described which has been used to study the healing of surgically produced fractures of the canine ulna. The transmission characteristics of a segment of bone are examined in vivo utilizing a pair of transducers. Two signal components have been identified and referred to as the "Bone Wave" and the "Callus Wave" (1,2). The bone wave is the first signal or pulse train which is detected by the receiver. This is the signal which has traversed the shortest path through the medium with the highest sonic velocity, i.e., cortical bone (3). The amplitude of the bone wave traversing a fracture defect is attenuated at the interfaces of abrupt mechanical discontinuity between bone and fracture. This attenuation is contrasted with that traversing a comparable segment in the contralateral intact ulna and used as an index of endosteal healing. A second signal component referred to as the "Callus Wave" propagates through the periosteal callus with a slower velocity and less attenuation than the signal traveling through bone (3). The signal indicates that the abrupt mechanical discontinuity between bone and fibrocallus is being obliterated by the formation of bony trabeculi at the extremities of the callus. Advanced healing and remodeling is characterized by increased callus wave velocity and attenuation, and ultimately by an increase in bone wave amplitude approaching that of control.

METHODS - A schematic representation of the ultrasonic system is shown in Figure 1. As we have described in detail elsewhere (2), the system utilizes a pair of 1/2 MHz ultrasonic transducers which are acoustically coupled to the subject's skin via focusing lenses, water filled plastic cones sealed with rubber tips and a hydrogel. The energy is conducted noninvasively to the bone through skin and soft tissue, with minimal signal attenuation (3). Adjustment of the incident angles of the cones permits examination of the propagation properties of longitudinal or compression waves at the first critical angle and transverse or shear waves at the second critical angle. These angles have experimentally been determined to be 23° and 38° respectively (2).

Short pulses of ultrasonic energy are generated by the transmitter transducer in response to electrical impulses generated by a Panametrics 5050p/r. The signal detected by the receiver transducer is amplified 40 db and displayed on a Tektronix 7313 oscilloscope. The sweep of the oscilloscope is triggered by the pulse generator delayed a length of time equal to the transit time through both probes and a portion of the osseous path length. Thus with a faster sweep rate an expanded signal could be displayed. The signals presented were recorded at 5 μsec per division with a signal delay of 96 μseconds. The dual storage capabilities of the oscilloscope were utilized in many instances. The control (contralateral or prefracture) signals were recorded at 23° on the upper trace and 38° on the lower trace. Many of the transfracture signals were recorded at two different oscilloscope gains on the same photograph to permit examination of weak and strong signal components on the same photograph. These were recorded at a high gain on the upper half and a lower gain on the lower half of the screen.

The system has been used to study the healing of surgically produced fractures of the canine ulna, which were produced with general anaesthesia and sterile technique. The midshafts were approached through lateral incisions and subperiosteal dissection. Cerclage notch defects were made in the bones and fractures produced by impacting the notch with a blunt wedge. The fractures were then reduced, the periosteum and skin closed with resorbable sutures and the fractures protected in padded plaster casts for three weeks. Followup radiographs and ultrasonic recordings were made with the animals anaesthetized.

RESULTS - Normal Healing Figures 2 and 3 are presented to demonstrate the radiographic and ultrasonic appearance of a normally healing fracture. In each figure the oscilloscopic tracings on the left (2a and 3a) are prefracture signals recorded prior to surgery. On the right hand side of the figures the oscilloscope recordings (b and c) made at the indicated period after surgery are shown. The tracings in the upper photographs (2b and 3b) were made at the first critical angle; the lower (2c and 3c) were made at the second critical angle. All of the postfracture oscilloscope photographs were made utilizing the dual recording technique, the upper tracing being amplified more than the lower.

The prefracture bone wave can be seen in Figure 2a as the first of four oscillations at the left of the tracings. The difference in starting times represents the difference in transit velocity of the compressional and shear mode of propagation. In comparing the bone waves in Figure 2a with those seen in the upper tracings of Figures 2b and 2c, one can see that the presence of the fracture, 20 days after surgery, resulted in a significant attenuation of the signal with a slight increase in delay. But perhaps more significant is the large amplitude signal that can be seen in the middle of the lower tracings in 2b and 2c. This large amplitude component, 900 mv and 2500 mv, in compression and shear respectively represents the callus formation. These recordings are the strongest callus waves recorded to date; the callus seen in the radiograph is the largest seen to date.

The radiographic and ultrasonic changes seen from 20 days to 7 weeks are typical for normal healing. The callus had bridged, increased in radiodensity and showed evidence of early remodeling. Ultrasonically, the changes in the bone wave suggested solidification of the endosteal callus, although the changes may be within experimental error. However, the changes in the callus wave were significant. The amplitude had decreased from 900 to 260 mv in compression and from 2500 to 440 mv in shear. The transit delay of each had also decreased. These changes were indicative of the radiographic changes noted.

By 16 weeks after surgery, the radiographs of the fracture and contralateral control were strikingly similar; the ultrasonic recordings were identical. As predicted by the 20 day recordings, the fracture had healed and remodeled during the 16 week study.

Delayed Healing - The recordings and radiographs in Figures 4-8 demonstrate the appearance of a delayed union. In each figure, the tracings on the left were made using the contralateral ulna as a control, while those on the right (b and c) were made across the fracture at the indicated number of weeks after surgery. In comparing the control recordings, one can see an increase in signal strength as indicated by the increase in millivolts per division noted below each figure. These changes were due to refinement in operator technique and some changes made in the system. Such changes were also reflected in the amplitude of the transfracture recordings, although they would not result in changes in wave form pattern.

At 3 weeks after surgery the fracture recordings (4b and 4c) demonstrated minimal delay of the bone wave, as would be predicted from the close apposition of the fragments, yet substantial attenuation of the signal. There is no evidence of a callus wave signal in compression (4b) while there is a hint of a pulse to the right of midscreen in 4c. The signal in 4c is too slow to be consistent with the callus wave theory and is probably a soft tissue artifact. Based on the 3 week recordings, one would predict minimal callus formation and delay of healing.

In comparing Figures 4c and 5c, one can detect a change in the bottom tracing indicative of a weak callus wave at four weeks. Although only 120 mv peak to peak, the signal has the appropriate delay to suggest slight callus formation. By 5 weeks, as seen in the lower tracings of Figures 6b and 6c, there is a suggestion of a callus wave in both modes of propagation. The signals are weak, but the delay is appropriate. Although the radiograph at 5 weeks showed a fracture line through the callus, and some rounding of the callus indicative of a nonunion, we predicted the fracture would heal based on the ultrasound.

The ultrasonic recordings at 7 weeks (7b and 7c) demonstrated a further increase in callus wave strength as compared with 5 weeks. The radiograph was still difficult to interpret, yet the ultrasound was quite convincing. By 16 weeks (Figure 8) the callus had bridged and begun to remodel. The ultrasonic recordings demonstrated a convergence of bone and callus waves yet the transfracture signal was significantly attenuated as compared with control. As predicted, the fracture was healing slowly, especially in comparison with that seen in Figures 2 and 3.

DISCUSSION - Callus wave signals similar to that shown in Figure 2 have been recorded as early as 10 days after surgery with 30 fractures subsequently showing signs of early callus formation and bridging. The absence of callus waves as in Figure 4 has been seen in four cases where subsequent examination revealed either delayed or nonunion. Had these cases not been part of the study proving the predictive value of ultrasound assessment of healing these fractures would have been protected in plaster for more than three weeks, to avoid excessive motion and the risk of nonunion. The availability of such predictive information should prove to be of value clinically.

ACKNOWLEDGEMENTS - This work was supported by USPHS Grant AM17749.

REFERENCES

1. Brown, S.A., Polymeric fixation and ultrasonic assessment of fractures and healing. Bull. N.Y. Acad. Med. In press.
2. Brown, S.A. and Mayor, M.B., Ultrasonic assessment of early fracture callus formation. Biomedical Engineering, 11:124-127,136, 1976.
3. Wells, P.N.T., Physical Principles of Ultrasonic Diagnosis, London, Academic Press, 1969.

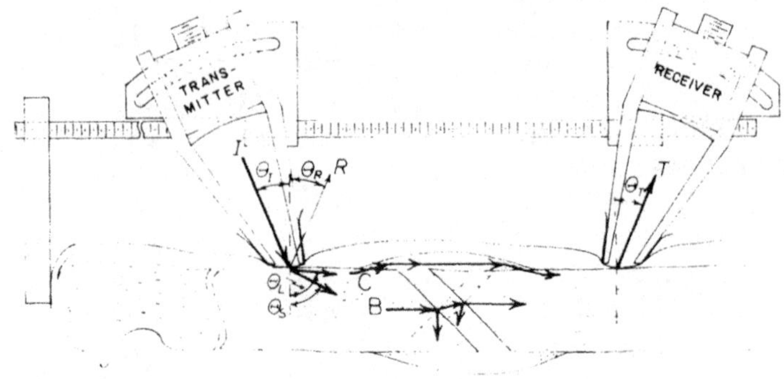

FIGURE 1
Ultrasonic system, depicting probe pair in position on ulnar fracture. Note reflections of "bone wave" at fracture site and path of "callus wave" through periosteal callus.

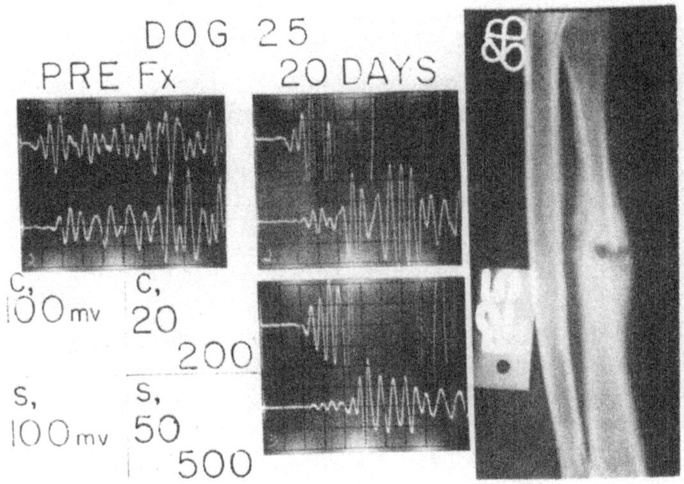

FIGURE 2
Radiograph and ultrasonic recordings, 20 days after surgery. Prefracture (2a) left; post fracture on right,(b,c). A large amplitude callus wave can be seen at 23° (upper) and 38° (lower) sets of recordings. Callus and callus waves are largest seen in series to date.

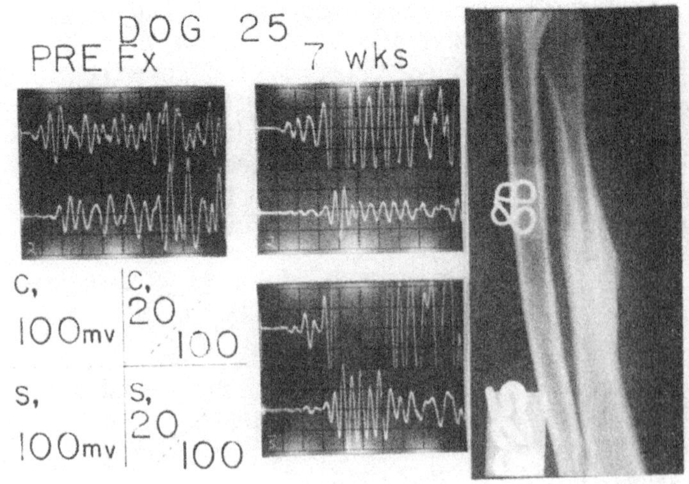

FIGURE 3
Radiograph and ultrasonic recordings at 7 weeks of same fracture as shown in figure 2. Prefracture recording (3a) shown on left, post fracture on right. A reduced callus wave signal, as compared with 20 days can be seen both at 23° and 38°, 3b and 3c, respectively

FIGURES 4-8: Sequential changes in ultrasonic recordings, and radiographs of a delayed union.

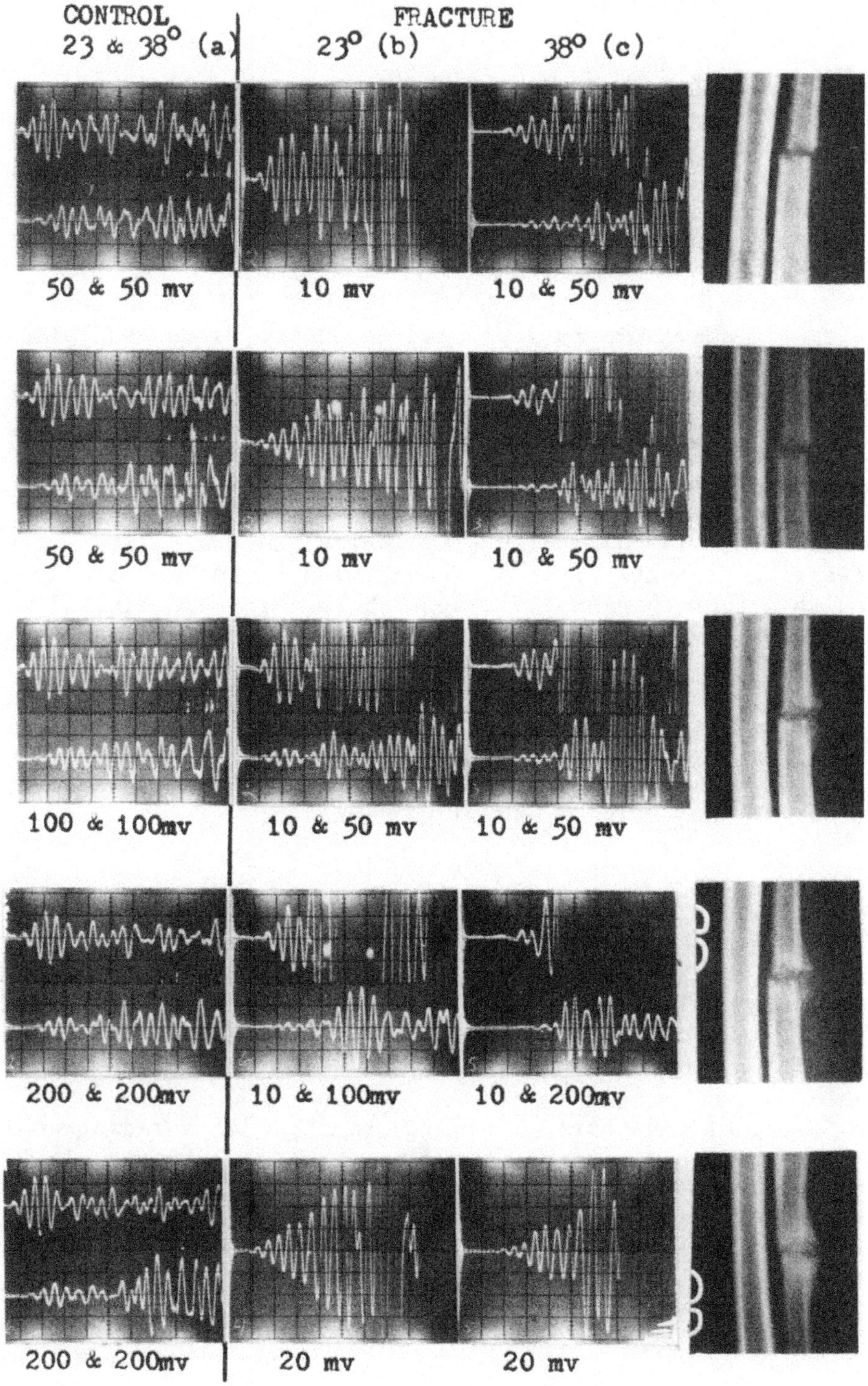

FIGURE 4
3 weeks

FIGURE 5
4 weeks

FIGURE 6
5 weeks

FIGURE 7
7 weeks

FIGURE 8
16 weeks

THE BONE DISTRIBUTION PROFILE AS A DIAGNOSTIC X-RAY TOOL

R. S. Bachtell, M.S., R. L. Sharp, C. Colbert, PhD

Radiological Research Laboratory
Wright State University
Dayton, Ohio 45431

F. J. Flynn, M.D., M. B. Ozonoff, M.D.

Newington Children's Hospital
Newington, Connecticut 06111

I. INTRODUCTION

The radiographic absorptiometry process developed by our Radiological Research Laboratory employs a scanning microdensitometer-computer combination to measure the mineral content of radiographed bones. The computer calculates the mineral (ash) weight of bone slices from scans which are orthogonal to the bone axis. To compute the bone mineral weight, we sum the weights of the individual slices.

An alternate method is to display graphically the bone slice mineral weight. Profiles (i.e., distributions) of bone mineral content provide the researcher with a tool to visually assess changes in bone mineral along the bone axis. They have already been employed by several investigators.[1,2]

This paper describes the development of a computerized method to display profiles of measured bone quantities such as mineral weight, bone area and mineral density. We will show examples of mineral density profiles of two normal subjects and one rickets patient.

II. EQUIPMENT DESCRIPTION

The computer-scanner combination consists of a modified Joyce-Loebl Mark III scanning microdensitometer, a Digital Equipment Corporation PDP-11/40 with two RK05 disk drives, GT-40 graphics terminal, LA30 DECwriter, card reader, paper tape reader/punch and a Control Data Corporation 9320 line printer.

The Joyce-Loebl relative optical density readings of a radiographic image are converted to 11-bit binary numbers by a Baldwin Shaft Angle Encoder before delivery to the computer. Two Slosyn digital stepping motors under PDP-11 control move the microdensitometer film table in orthogonal direction; the table X- and Y-position coordinates, generated continuously, are available to the PDP-11.

III. PROCESS DESCRIPTION

An operator positions the radiograph on the microdensitometer table, selects the scan region and sends the table position coordinates at diagonal corners of the scan region to the PDP-11 computer. These coordinates determine a scan rectangle within which the scans are located. The PDP-11 drives the stepping motors so that the bone is scanned line-by-line with a 0.15 mm × 1.0 mm aperture in a direction orthogonal to the bone axis. The scan line separation as well as the separation between optical density samples along the scan line may be selected by the operator.

The encoded optical density reading at each sample point is sent to the PDP-11 through the microdensitometer interface. After all of the optical density samples along a scan line have been received they are stored on an RK05 disk.

Each radiograph also contains the image of a small aluminum alloy wedge which is recorded along with the body part. A single scan is made in the up-ramp direction and the optical density samples are stored on disk in the same file as the bone samples.

To process the raw optical density data, the wedge optical density samples are written into core memory from the disk. A wedge algorithm computes the linear absorption coefficient of aluminum, the base optical density of the radiograph and a correction factor to compensate for difference in kilovoltage, exposure and other factors otherwise likely to perturb the radiological ash weight computed next.

The optical density samples of the bone scan lines are read into the core memory and each scan line is processed in turn. The edges of the bone are automatically detected and the radiological ash weight is computed from the optical density samples between the bone edges. Radiological ash density is computed by dividing the ash weight by the area of the bone image slice.[3,4] The important local quantities derived from each scan line--ash weight, bone scan area, and ash density--as well as a number of other quantities are stored as profiles in the memory. A series of program commands allows the display of any chosen profile on the GT-40 graphics terminal where numerical values of the selected quantity are plotted against scan line number. This profile lets us observe the change in the selected quantity along the bone. By using other options we can overlay several profiles, smooth the profiles with an n point moving average or store any profile display on the disk for later recall.

A further refinement allows us to call up and display the optical density data of any individual scan line. If a failure of the automatic edge detection algorithm should occur, we can determine the correct bone edges manually.

IV. APPLICATION OF BONE PROFILES

The ability to display bone ash profiles is a new and valuable tool which complements the calculation of the total ash content.

Figure 1 shows the distributions of radiological ash density in relative units along a 100-mm section of the radius of the left forearm plotted against scan line number for three females approximately 12 years old (two normals and one with hypophosphatemic rickets). Scan line 30 is near the distal end of the radius.

Figure 2 shows the corresponding area profiles of 1.0-mm wide slices. The width of the bone (proportional to the displayed slice area) decreases to a minimum and then increases sharply, depicting the flaring at the end of the radius. Like the other long bones of the body, the radius has a midshaft composed primarily of cortical or compact bone and flared ends of cancellous or spongy bone. The transition from compact to spongy bone is gradual and coincides with the increasing width.

Ash density profiles (Figures 1a and 1b) of the two normal female controls exhibit a monotonic decrease through the transition region. For these subjects, the ash density of the cortical bone is considerably higher than that of the spongy bone; the ratio of maximum to minimum ash densities is 2.1 and 1.7 for Subjects 1 and 2, respectively. However, the ash density profile of Subject 3, the hypophosphatemic rickets patient, behaves much differently. Although the bone width profile of the radius resembles those of the normal females, the ash density is relatively constant during the transition from cortical to spongy bone. Even though the ash density in the cortical region is less than that of the two normal subjects, the ash density of the spongy region is greater!

A very different situation exists in the ulna. Figure 3 shows the corresponding 100-mm profiles along the ulna from the compact bone to the spongy bone region. The ash density of Subject 2, the normal female, continuously declines, reaching its lowest value of about 110 relative units at the distal end. Subject 3, the rickets patient, exhibits a much lower ash density in the cortical region and a declining ash density in the transition region. Her ash density at the distal end is approximately 110, the same as the normal female.

The profiles shown in this paper constitute a very small sample. The significance of the ash density changes along the axis of different bones in both the normal and disease states will have to await the examination of a larger number of normal and pathological cases.

Supported in part by grants: AM-16475 from the National Institute of Arthritis, Metabolism and Digestive Diseases; ES-00831 from the National Institute of Environmental Health Sciences; DI-44068 from the National Science Foundation; and by the Osteogenesis Imperfecta Foundation.

REFERENCES

1. T.J. Hahn, V.C. Boisseau and L.V. Avioli, "Effect of Chronic Corticosteroid Administration on Diaphyseal and Metaphyseal Bone Mass," JCE&M, Vol. 39 (1974), pp. 274-282. These investigators make measurements at two locations on the bone rather than a continuous profile.

2. B.A. Sigmon, "Distribution of Bone Mineral in the Femur." *In* Proceedings of Bone Measurement Conference, U.S. Atomic Energy Commission CONF-700515, Chicago, Ill. (May 1970), pp. 428-435.

3. C. Colbert, "Radiographic Absorptiometry," 2nd Ed. (Wright State University, 1976). PhD Dissertation, 1974.

4. C. Colbert and R. S. Bachtell, "Precision of Measuring Bone Mineral from Radiographs." Submitted for publication.

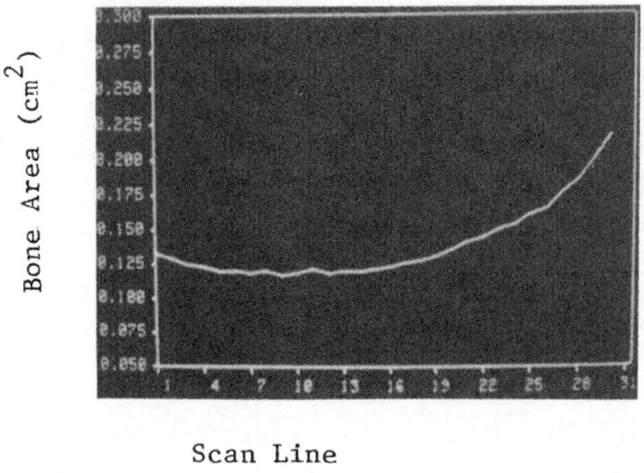

a. Subject Number 1. Normal female,age 12.7 years.

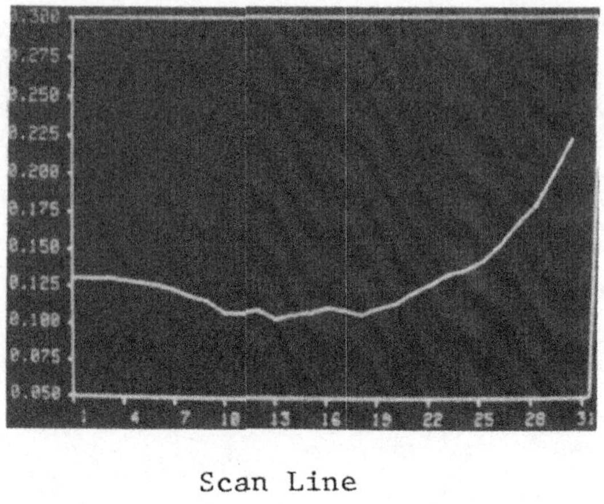

b. Subject Number 2. Normal female, age 11.6 years.

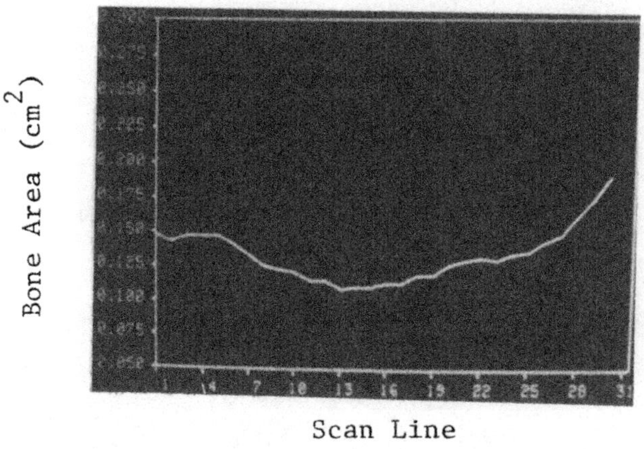

c. Subject Number 3. Female, age 12.1 years. Diagnosis: hypophosphatemic rickets.

Figure 2. Profiles of the area of 1.0-mm wide slices along a 100-mm length of the left radius. The scan lines are separated by 3.3 mm; the last scan line is near the distal tip.

NOTE: The bone area on these graphs is proportional to the local bone width.

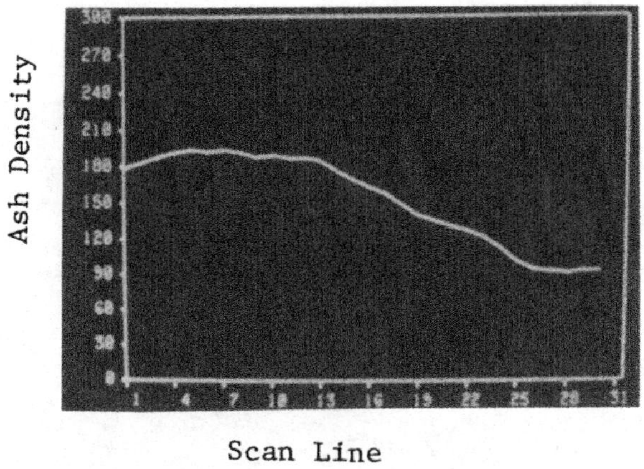

a. Subject Number 1. Normal female, age 12.7 years.

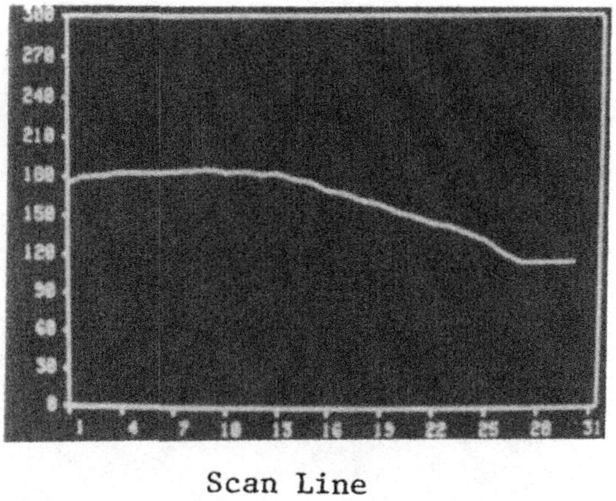

b. Subject Number 2. Normal female, age 11.6 years.

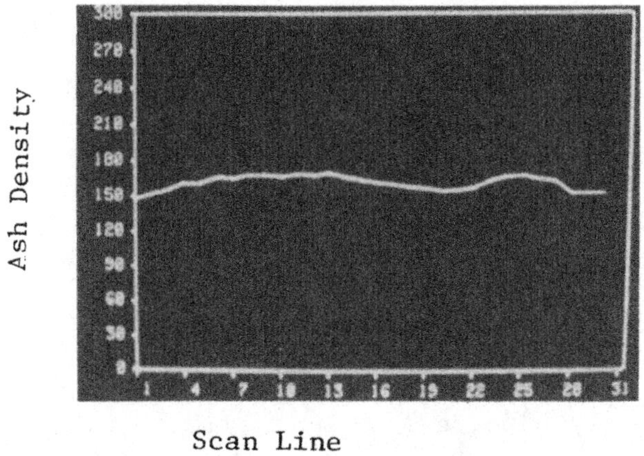

c. Subject Number 3. Female, age 12.1 years. Diagnosis: hypophosphatemic rickets.

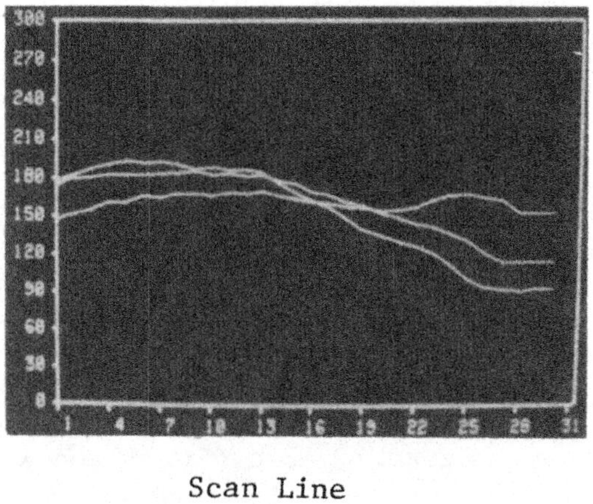

d. Composite display of the three profiles.

Figure 1. Ash density profiles along a 100-mm length of the radius of the left forearm. The scan lines are separated by 3.3 mm; the last scan line is near the distal tip. Each profile has been smoothed by a three point moving average.

NOTE: Radiological ash density on these graphs is defined as the local radiological ash weight divided by the local bone slice area.

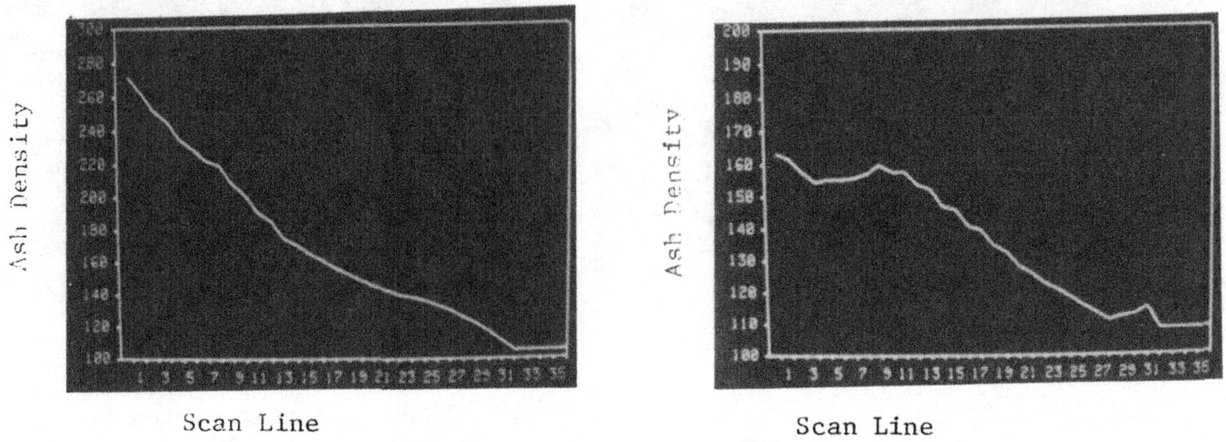

a. Subject Number 2. Normal female, age 12.7 years.

b. Subject Number 3. Female, age 12.1 years. Diagnosis: hypophosphatemic rickets.

Figure 3. Profile of the ash density along a 100-mm length of the ulna. The scan lines are separated by 2.9 mm. The last scan line is near the distal end. Each profile has been smoothed by a five point moving average.

NOTE: Radiological ash density on these graphs is defined as the local radiological ash weight divided by the local bone slice area.

AN ULTRASONIC DEVICE FOR USE IN THE HEMATOLOGY LABORATORY

J. A. Rooney and D. J. Paduan
Physics Department, University of Maine
Orono, Maine 04473

Abstract - An ultrasonic device has been developed whose output is indicative of the viscoelastic properties of the fluid in which the transducer system is immersed. Applications of the device for hematocrit determination and coagulation monitoring are discussed.

INTRODUCTION

At present ultrasound is being used extensively in medicine both in diagnostic and therapeutic applications. Most of these ultrasonic devices operate in the megahertz frequency range and currently they are not being widely used for clinical laboratory procedures. In this paper we wish to describe an ultrasonic system operating at 20 kHz which is designed for use in hematological laboratories. To date we have demonstrated its usefulness for determination of hematocrit values and also the monitoring of a coagulation process.

APPARATUS AND PROCEDURES

In ultrasonic work, one objection to using oscillator-amplifiers with pre-set tuning is that the transducer is often required to work into a wide range of acoustic loads which may be reactive. Also, particularly in some of our applications, the load and/or transducer characteristics may change in time. It is then desirable to design a generator-transducer which will keep itself in mechanical resonance independent of any such changes. This can be done by driving the transducer by a signal derived from the transducer motion alone, unaffected by its electrical characteristics. To accomplish this end, we use an electrical circuit wired to the transducer and obtain a signal proportional to the motion of the transducer. Such a system has the advantages that it is convenient and eliminates problems of mechanical bonding and electrical screening of the transducer.

The transducer is made part of a capacitance bridge[1] shown in Figure 1 since the transducer has a large bulk capacitance due to its construction. The electrical model of the transducer is that of a large bulk capacitance, as in Figure 1, in parallel with a series RCL circuit. Use of this bridge thus permits the nulling of the bulk capacitance of the transducer. If the transducer is then driven at resonant frequency, the equivalent circuit is just a resistive component whose magnitude is related to the loading of the transducer.

The capacitance bridge as seen in Figure 1 is driven by the triangular waveform output of a 565 phase locked-loop, PLL. The bridge voltage, after differential amplification and wave shaping with a comparator, is applied to the phase locked-loop input. The transducer is driven by the phase locked-loop and the PLL tracks the resonance frequency of the transducer as that frequency shifts with the loading of the transducer. This insures that the equivalent circuit for the transducer remains resistive. The output of the differential amplifier is also applied to a full wave rectifier. The rectified signal is then applied to a low pass filter, the output of which is a dc level whose magnitude is proportional to the amplitude of the bridge voltage, which in turn is related to the magnitude of the resistive component of the transducers impedance. This signal has its largest value when the horn tip is unloaded. This rectified, filtered bridge voltage is summed with an offset voltage so that the net output of the circuit is zero when the horn tip is unloaded, and increases in magnitude with increasing loading of the horn tip. If care is taken to immerse the horn tip to the same depth in fluid samples, the output voltage is proportional to the viscosity-density product of the fluid and this voltage may be calibrated.

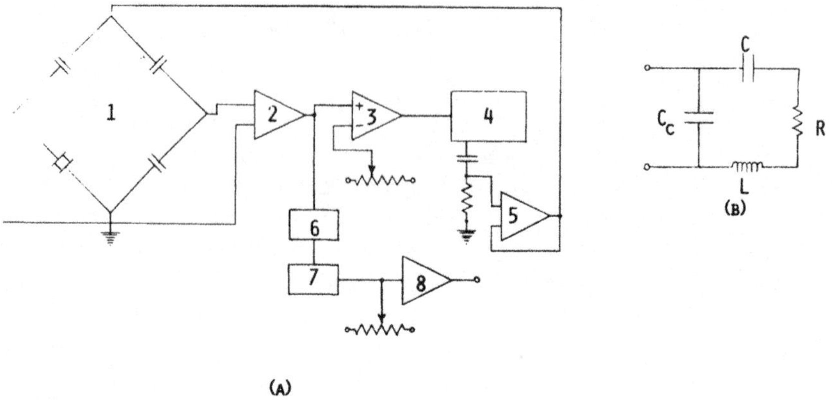

FIGURE 1: a) Block diagram of device. The circuit consists of a capacitance bridge with the transducer in one branch (1), differential amplifier (2), comparitor (3), phase locked-loop (4), filter and bridge driver (5), full wave rectifier (6), low pass filter (7) and a summing amplifier (8). b) Equivalent circuit for transducer.

In our experimental arrangement we use a stainless steel exponential horn driven by a barium titanate transducer resonant near 20 kHz. When the horn oscillates in air, its amplitude of the oscillations is restricted only by the mechanical resistance of the system and by the small damping of the surrounding air. When the tip of the horn is immersed in liquid, the amplitude is further restricted by the retarding force which the liquid exerts on the vibrating horn. This retarding force consists of a resistive component which damps the motion and a reactive component which for a liquid adds to the effective mass of the transducer. Each component will be proportional to $(f, \eta, \rho)^{\frac{1}{2}}$ where f, η and ρ are respectively the frequency of oscillation, solution viscosity and solution density. Thus, if the damping force on the horn can be measured, the value of $\eta\ \rho$ can be determined. In our arrangement, as indicated, this measurement is accomplished by always operating the system at resonance so that the bridge output is related to the resistive component of the force.

In order to gain some insight into the theory of the system, let us calculate the mechanical impedance presented by a liquid to a cylinder immersed in the liquid and executing linear oscillations. The theory is rather complex and requires numerical calculations but essentials of the theory can be seen by considering the case of a flat plate of infinite extent immersed in a viscous fluid[2]. Let the plate be the y-z plane and let it execute sinusoidal oscillations in the y direction given by $\dot{y}_0 = \dot{y}_1 \cos\omega t$ where $\dot{y}_0$ is the instantaneous velocity and $\dot{y}_1$ is the peak velocity of vibration. Then the transverse motion in the liquid at a distance x from the plate is

$$\dot{y} = \dot{y}_1 \cos(\omega t - \beta^{-1}x)\exp(-\beta^{-1}x) \qquad (1)$$

here $\omega = 2\pi f$ and β is the boundary layer thickness which equals $(2\,\eta/\omega\rho)^{\frac{1}{2}}$. The coefficient of viscosity is defined by Eq. (2).

$$F = -\eta\,(\partial\dot{y}/\partial x)_{x=0} \qquad (2)$$

Here F is the tangential force per unit area on one side of the plate. Taking the derivative of Eq. (1) and substituting into Eq. (2) gives for sinusiodal vibrations

$$F = \dot{y}_0\,(\tfrac{1}{2}\,\omega\,\eta\,\rho)^{\frac{1}{2}}\,(1 + i) \qquad (3)$$

where $i = \sqrt{-1}$. Thus the impedance $F/\dot{y}_0$ comprises two equal parts of magnitude $(\frac{1}{2}\,\omega\,\eta\,\rho)^{\frac{1}{2}}$, one real, the other imaginary and both dependent upon frequency. The real part is a dissipative impedance corresponding to mechanical resistance and is measured from the bridge output. The imaginary part is an equivalent addition of mass per unit area to the system.

EXPERIMENTAL RESULTS

The device as described has been constructed, tested, and calibrated using a series of glycerine-water solutions of differing viscosity density products and good agreement was found between experimental results and the derived theory. Based on these results the device has been applied to problems of the hematology laboratory, specifically determination of hematocrit values and prothrombin times.

In the past several different methods have been used to determine hematocrit. The oldest and most commonly employed technique is the separation of the erythrocytes from the plasma by centrifugal force. The volume of the red cell sediment is then compared to the total volume of blood and expressed as a percentage. While the method is satisfactory for processing a moderate number of samples it does not lend itself to large scale use because of the total time involved in the measurement. Other techniques have been developed to overcome the problem of time required for a centrifugal hematocrit determination. One of these methods measures the electroconductivity of the blood, but the accuracy of this method is affected by the protein and ion concentration in the blood as well as the type of anticoagulant used[3]. Techniques using light transmission and radioactive tracers have not widely accepted.

Using the ultrasonic device described earlier, we have been able to accurately ultrasonically determine the hematocrit. The procedure followed is to observe the bridge output for a variety of centrifugally determined hematocrit values with the results as seen in Figure 2. As seen, the output from the device can be used to accurately determine hematocrit values and overcomes many of the problems associated with other techniques.

The second application of interest was determination of prothrombin times and monitoring a coagulation process. The prothrombin time is not only indicative of prothrombin activity but also measures the effectiveness of other clotting factors including factors V, VII, and X. While the prothrombin time is not used extensively for diagnostic purposes, it has proved useful for assistance in controlling dosages of anticoagulants in various clinical states. At present two different devices can be used for determination of prothrombin times in addition to the standard loop technique. The first device measures the conductivity of the fibrin clot while the second measures changes in optical absorption. Both devices have disadvantages in that neither continuously monitors the coagulation process and in the optical technique the plasma must be separated from the blood.

To perform the test with a system similar to that described we have used a modified Quick test. In this procedure 0.1 ml of 0.02 M CaCl and 0.1 ml of prepared thromboplastin are mixed in a small vessel maintained at 37°C. To this mixture is added 0.1 ml of plasma and the time from mixing to clot formation is recorded. The most popular method at present for determining the onset of clot formation is to pull a wire loop through the mixture until the clot is pulled from the liquid. Using our system, the tip of our exponential horn is immersed in the thromboplastin-CaCl mixture, the plasma is added and the bridge voltage is amplified, rectified and recorded on a strip chart recorder. The ultrasonic device thus continuously monitors the coagulation process as seen in Figure 3. As we see there is essentially no change in the bridge voltage until the onset of coagulation, the coagulation proceeds rapidly and then the bridge voltage reaches a plateau.

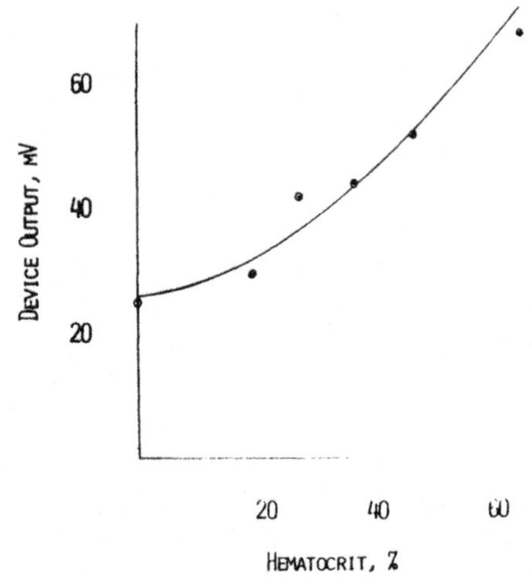

Figure 2: Voltage output of ultrasonic unit as a function of hematocrit.

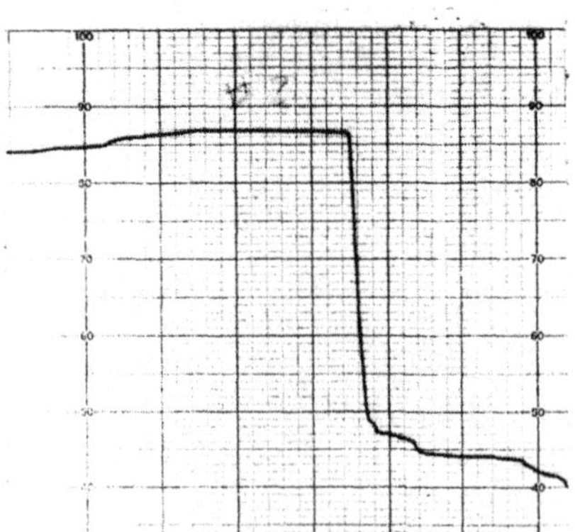

FIGURE 3

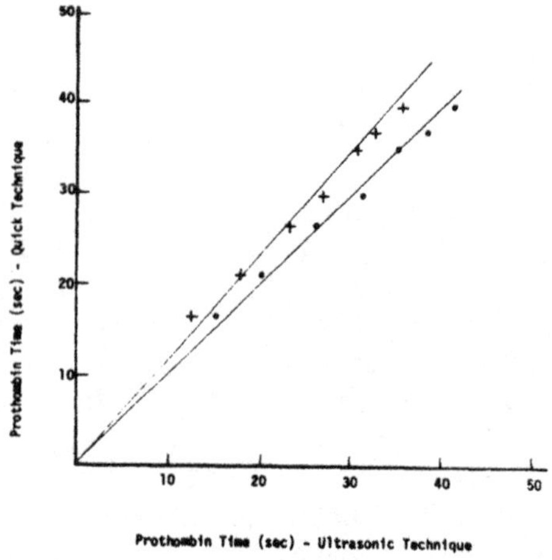

FIGURE 4

Figure 3: Voltage output as a function of time during a coagulation process. Vertical scale, 25 mV full scale; horizontal scale, 1 sec per division.

Figure 4: Prothrombin time as determined by loop and ultrasonic techniques.

In order to validate the results obtained using our system we have compared prothrombin times determined by the ultrasonic technique with those obtained using the loop procedure. To obtain various prothombin times we diluted the control plasma with physiological saline. The normal range of values for prothrombin time is from 12 to 17 seconds. Using dilution of plasma techniques we were able to measure times to about 40 seconds. Results of the comparison are shown in Figure 4. The crosses correspond to prothrombin times as measured with the ultrasonic device using the onset of clotting as the end point. The circles represent data using the time for reaching the plateau as an end point. For perfect correlation between the techniques we would expect to see a straight line at 45°. As we note in Figure 4 the criterion of clot onset gives slightly low prothrombin times as the result of the sensitivity of our technique. As we might expect there is a much better correlation between completion of clot formation as measured by our device and the loop technique.

We believe we have demonstrated the effectiveness of our device for use in hematology. It appears there is wide application for such a device and that the ultrasonic unit has the capability of simplifying existing tests as well as creating new and unique procedures.

ACKNOWLEDGEMENT

The authors wish to thank the American Red Cross for use of blood samples. This work was partially supported by National Institutes of Health Grant GM-0514.

REFERENCES

1. E. E. Fill, "A Simple Bridge for Use With Piezoelectric Transducers", I.E.E.E. Trans. Biomed. Eng., BME-16, 165-166 (1969).

2. H. Lamb, Hydrodynamics, New York, Dover (1945).

3. R. E. Davis, R. Bresland and R. Green, "Evaluation of Electronically Measured Hematocrit Values", Lab Practice 15, 1376-1378 (1966).

DESIGN OF A NO-TOUCH INFANT APNEA MONITOR

Robert B. Northrop and Bruce M. Decker

Biomedical Engineering Laboratories, U-157, The University of Connecticut, Storrs, CT 06268

Abstract

This paper describes the rationale and design for an ultrasonic Phase-lock apnea monitor (UPLAM) for infants which requires no physical contact with the infant in an incubator. Ultrasound is used to measure precisely the distance from the transducers to the baby's chest and abdomen. Small changes in this distance, which are proportional to respiratory effort, are converted to a voltage z which is processed to select a "normal" range of amplitudes. If the cyclical changes in z drop below a lower threshold for a predetermined time, the apnea alarm is given. The alarm may also be activated if z exceeds the upper threshold for a predetermined time (convulsive seizure detection).

Introduction

Premature and low birth-weight infants, and infants with low Apgar scores frequently have problems with their respiration in the first few days after birth. Spontaneous breathing can stop (apnea) causing anoxia and subsequent permanent CNS damage if respiration does not resume in time. It is important to monitor reliably and detect apnea in high-risk babies in order to prevent damaging anoxia and/or death.

The new type of apnea monitor described in this paper should prove reliable and have a number of advantages over apnea monitors now in use.[1] The ultrasonic phase-lock apnea monitor (UPLAM) system has three component subsystems: 1) An ultrasonic phase-lock system which precisely measures the distance from the fixed transducers to the infant's chest and abdomen, detecting respiratory movements. Its output z is a voltage which is the input to the second subsystem. 2) A window circuit which selects only those voltage changes due to normal breathing. Large amplitude artifacts are not counted. A multiplexer is used to display the window's upper and lower threshold voltages, its output, and z on an oscilloscope to facilitate window adjustment. The window output continuously resets a countdown alarm system. If apnea occurs, the counters are allowed to count to a preset value which actuates the alarms. 3) The third subsystem is an automatic lock loss circuit which monitors the output of the integrator in the phase-lock system. If a large change occurs in the airpath length (as when changing diapers), the integrator may saturate, causing the loop to crash. If the integrator output approaches ± saturation, the lock loss circuit opens the phase-lock loop, sets the system VCO to the nominal center frequency, and puts zero initial conditions on the integrator for 200 ms, then closes the loop again.

The phase-lock system which detects respiratory movements is a closed-loop regulator which acts to keep the number of wavelengths of ultrasound in the airpath constant in spite of small changes in airpath length Δx due to respiration. The transmitted and received phases are compared; the phase detector output is integrated and drives the VCO generating the output frequency. It can be shown[2] that the integrator output is proportional to Δx. The integrator output is bandpass-filtered to select respiratory movement frequencies. The filter output is the input to the window circuit.

The center frequency of the UPLAM is 39 kHz. The transducers are fixed inside the top of the incubator and directed at the baby's abdomen and chest. The sound intensity used in innocuous, measured at < 73 dB SPL, 21 cm from the transducers.

Description of the Ultrasonic Phase-Lock System (PLS)

A block diagram of the UPLAM PLS is shown in Fig. 1. The transmitting and receiving transducers (TRX and RCX) are inexpensive Calectro J4-815, 40 kHz crystal units. The output of RCX is conditioned by a low-noise preamplifier (PRA) with a gain of 15,000 followed by a bandpass filter with low and high corner frequencies of 29 and 43 kHz, respectively. The PRA output is infinitely clipped by a μA 760 comparator (COM-1) whose TTL output triggers a one-shot multivibrator (OSH-1) with a 1 μs dwell time. OSH-1's output pulses serve as one input to a simple, digital phase detector (PHD) which also receives a reference input from the 1 μs pulses from OSH-2. OSH-2 is driven from the TTL output square wave from the system's voltage-controlled oscillator (VCO).

The PHD consists of a NAND gate RS flip-flop whose output states are made symmetrical around 0 V. Under open-loop conditions the PHD output is a square wave of the transmitted (VCO) output) frequency whose duty cycle, hence average value, depends on the fractional wavelength phase difference between the VCO output and the received signal. The PHD does not sense integer wavelength phase differences between the reference and returned signals.

Under closed-loop (CL) conditions, when the PLS is locked, the steady-state PHD output is a symmetrical, 50% duty cycle square wave with zero average value. During a transient change in airpath length (Δx) caused by the infant's breathing there is a transient change in the duty cycle of the PHD output (V_e), hence the average value of V_e is momentarily nonzero. This average value is integrated by a resettable op-amp integrator (INT). The integrator output (z) controls the VCO output frequency. It has been shown[2] that under steady-state conditions, small changes in z are directly proportional to Δx, due to the closed-loop, phase tracking behavior of the system. The integrator output is conditioned by a bandpass filter (BPF) to form an output $\hat{z}(t)$ proportional to $\Delta x(t)$.

The summing op-amp (SUA) adds a dc level (z_o) to the integrator output, z. Thus when the loop is opened using the MOS switches MSW-1 and MSW-2, z_o sets the VCO frequency. POA is a ML 318 op-amp which drives the output transducer.

Window and Alarm Circuits

The conditioned output of the PLS, $\hat{z}(t)$, is processed by a pulse-height window shown in Fig. 2. The amplitude of $\hat{z}$ is proportional to the respiratory efforts of the infant, being determined by the time-varying changes in airpath length between the transducers and the infant's abdomen and chest. A window output pulse (WOT) occurs whenever $\hat{z}(t)$ exceeds the lower threshold ($\emptyset_1$), does not exceed the upper threshold ($\emptyset_2$), and recrosses $\emptyset_1$ with negative slope. The window output pulse resets the alarm counters (CTR) which are continuously counting clock pulses up to a preset value which will trigger OSH-5 and the alarm flip-flop (AFF).

If The PLS output $\hat{z}(t)$ does not cross $\emptyset_1$, which is the case in apnea, the counters count up from the last reset pulse (WOT=RST) to the preset value which triggers the apnea alarms.

On the other hand, if $\hat{z}(t)$ exceeds $\emptyset_2$, which could occur if the infant makes convulsive movements, no RST pulses are produced by the window. The alarm will sound if superthreshold $\hat{z}(t)$ continues for the alarm count-up period. Thus UPLAM can detect convulsive behavior as well as apnea in infants. If it is desired to detect apnea alone, $\emptyset_2$ can be raised to a maximum level $\hat{z}(t)$ cannot cross.

Lock-Loss Control (LLC)

The LLC, shown in Fig. 3, senses when the integrator in the PLS approaches ± saturation. The outputs of comparators COM-5 and COM-6 are ORed such that when $|z|>12V$, OSH-7 changes the state of the 7410 control flip-flop (CFF) to HI, causing MSW-1 and MSW-2 to close, putting the PLS in the open-loop mode. Under OL conditions, the VCO output is a constant 39 kHz, determined by z_o. 200 ms after the OL mode is initiated, OSH-8 and OSH-9 reset CFF to allow the PLS to relock. Becuase the transient distrubance in airpath length which caused INT saturation is generally short-lived, after several tries the LLC establishes normal, closed-loop operation of the PLS.

UPLAM Performance

Northrop and Nilakhe (1977) have shown that system sensitivity is given by the relation

$$S = \frac{\Delta z}{\Delta x} = \omega_{TL}/K_3\bar{x}$$

where ω_{TL} is the steady-state UCO frequency under CL locked conditions given an average transducer-to-baby distance $\bar{x}$. K_3 is the VCO constant in r/s volt. The measured value of S, using a 0.5 Hz sinusoidal $\Delta x(t)$, was 0.92 V/mm at $\bar{x}$ = 21 cm.

Fig. 4A shows $\hat{z}(t)$ along with $\emptyset_1$, $\emptyset_2$, and WOT pulses, the subject being a 4.5 kg New Zealand rabbit anesthetized with ketamine HCl, 45 mg/kg. The rabbits's abdomen was covered with a sheet of polyethelene to reflect the ultrasound; it was found that the thick rabbit fur was an almost perfect ultrasound absorber at the frequencies used. Fig. 4B illustrates respiration of the same rabbit as it came out of the anesthesia. Breathing effort is more pronounced, hence certain peaks exceed $\emptyset_2$ and these cycles do not reset the alarm counters. Time base in Fig. 4; 0.2 sec/cm. Vertical scale; 0.5 V/cm.

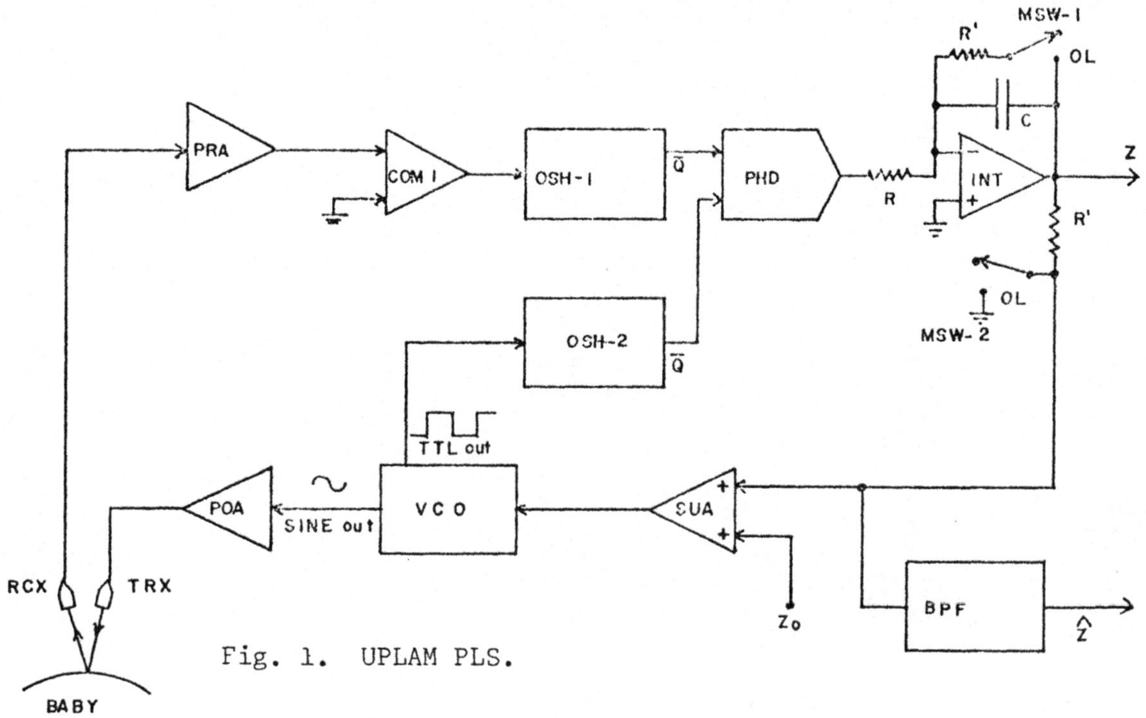

Fig. 1. UPLAM PLS.

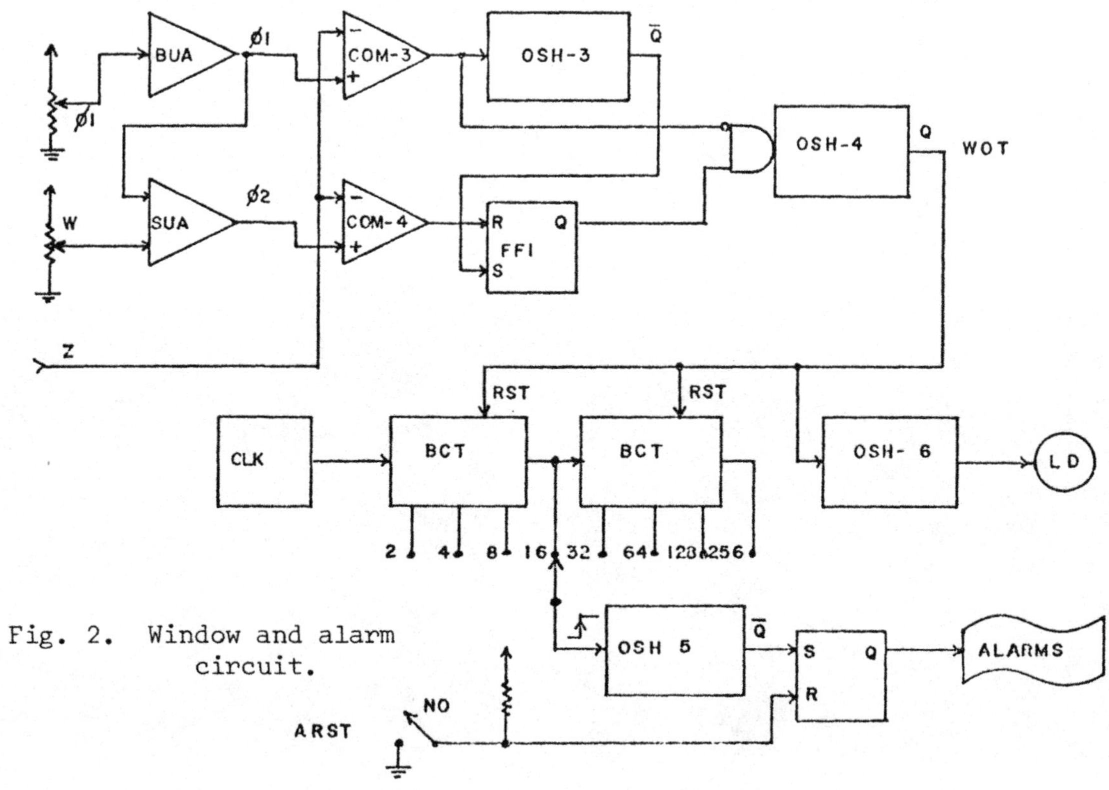

Fig. 2. Window and alarm circuit.

References

1. "Infant Apnea Monitors". Health Devices, Nov. 1974. pp. 3-23.

2. Northrop, R. B. & S- S. Nilakhe. A no-touch ocular pulse measurement system for diagnosis of carotid occlusion. IEEE Trans. on Biomed. Engineering. 24: March, 1977, in press.

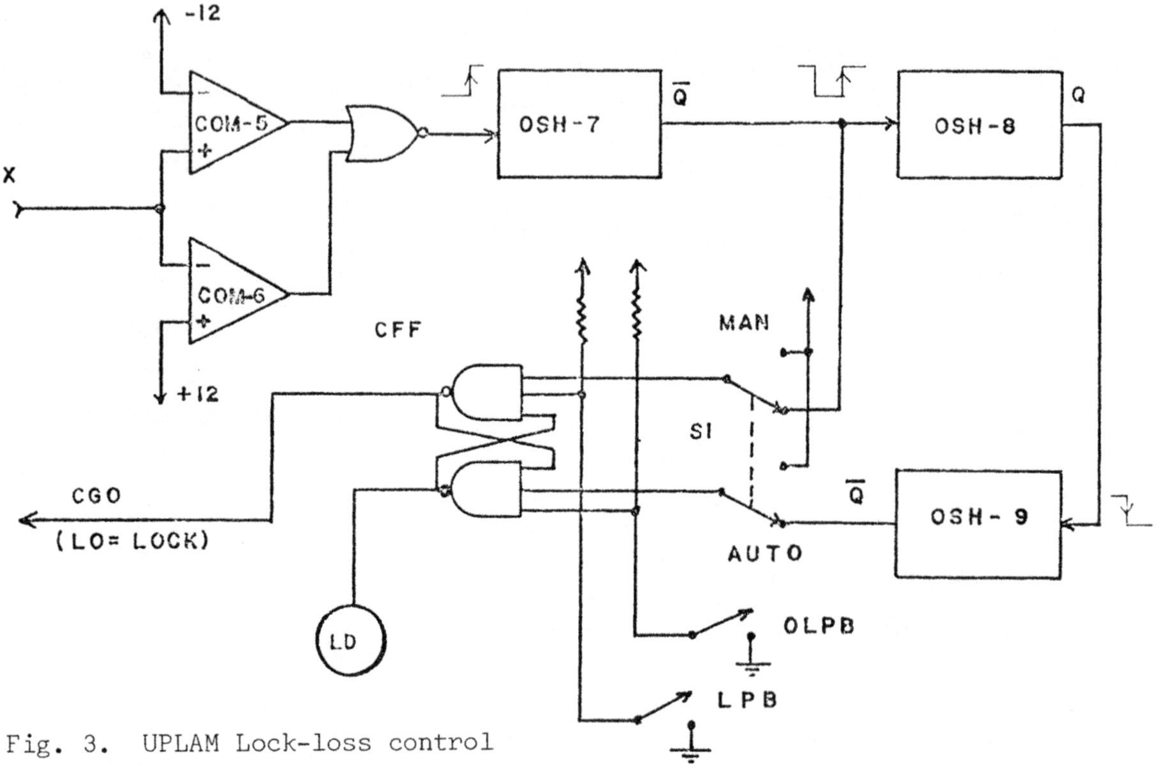

Fig. 3. UPLAM Lock-loss control

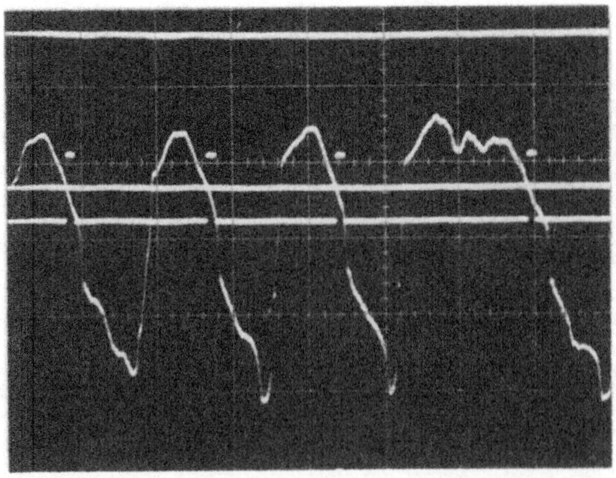

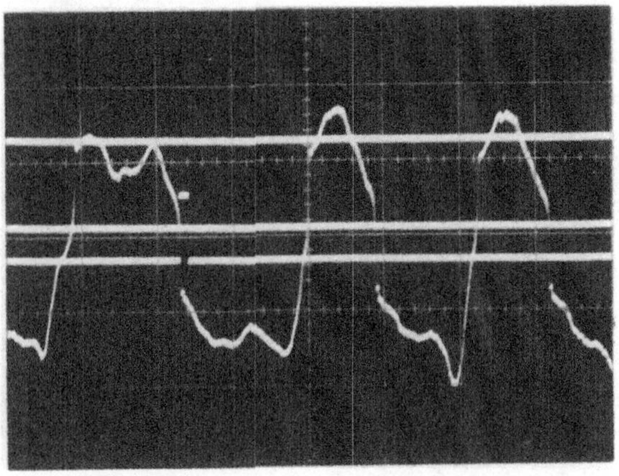

Fig. 4A. See text for description. Fig. 4B

A CARDIAC PACEMAKER CAPTURE-FAILURE MONITORING SYSTEM

James V. O'Connor
S.U.N.Y. Downstate Medical School
450 Clarkson Avenue, Brooklyn, New York 11203

I. INTRODUCTION

This paper presents a cardiac pacemaker monitoring system which uses any ECG lead configuration as an input. It detects and informs the user of any pacemaker complication, with the exception of pacemaker induced arrythmias.

Before considering complications which can follow pacemaker implantation some general information about pacemakers should be mentioned. The most commonly used pacemaker is the demand (QRS inhibited) type. If the pacemaker senses the patient's heart rate is less than the pacemaker's own pre-set rate, then it fires. The usual rate of the pacemaker is 70 ± 2 impulses per minute, therefore, for a normally functioning pacemaker two conclusions can be drawn:

(1) The patient's heart rate should never be less than 68 beats per minute.
(2) The pacemaker rate should never exceed 72 impulses per minute.

The most common complications of cardiac implantation and their physiological consequences are:

(1) Broken Pacemaker Lead: There will be no pacemaker spike and the patient's heart rate may drop to a ventricular rate. The monitoring system should give an alarm signal if the patient's rate is less than 68 beats per minute.
(2) Dislodged Pacemaker Lead: In this case a pacemaker spike may be recorded but due to lead positioning no ventricular depolarization will follow.
(3) Increased Pacing Threshold: The pacemaker is outputting a spike but the amplitude is insufficient to cause ventricular contraction.
(4) The Pacemaker Fails To Sense The QRS Complex: The QRS and pacemaker spike will not be synchronized.
(5) Runaway Pacemaker: The pacemaker rate exceeds 72 impulses per minute.
(6) Cardiac Arrythmias: These may be caused by the pacemaker and will not be detected by the monitoring system.

Cases (2) and (3) are the true capture-failure mode where the pacemaker is outputting and the ventricles are not contracting in response to it.

All the above are potentially life-threatening and all, except cardiac arrythmias, will be detected by the monitoring system.

There are, therefore, four types of failure states which this system will detect:

(1) Patient's heart rate less than 68 beats per minute.
(2) Pacemaker impulse rate greater than 72 impulses per minute.
(3) Pacemaker output fails to cause ventricular contraction.
(4) Pacemaker spike does not cause ventricular contraction but the ventricles are contracting on their own.

II. SYSTEM DESIGN

The monitoring system can take its input from any ECG monitoring system. Any lead configuration can be used, but the most often-used configuration in a cardiac care or intensive care unit is a pre-cordial lead that gives a configuration similar to V5. The first stage of the system (see Fig. 1) is a simple voltage follower. This is simply to buffer the input of the monitoring system from output of the ECG monitor. The output of the voltage follower splits and is used in two separate channels, one for the pacemaker spike and one for the QRS complex.

The first task of the system is to separate the pacemaker spike and the QRS complex from the ECG. The shortest pacemaker spike duration is .8 msec, the longest is 1.1 msec. This represents a frequency range of 910 to 1250 HZ. A band-pass filter can be designed with the center frequency at 1100 HZ, Q=2, and a gain of 2.

For the QRS complex a band-pass filter with center frequency of 20HZ, Q=2 and gain =2 is used.

Once the filter passes the appropriate signal component, it must be detected. A window comparator is used, which allows for the detection of a positive or negative going signal. The window comparator will give a 5 volt output if the input signal is between the upper and lower thresholds. What is in fact needed, is a 5 volt output when the input is greater than either threshold value. This means using an inverter at the output of the window comparator.

The output of the inverter will be high as long as a QRS or pacemaker spike is detected. In order to get pulse widths that are more reasonable, a monostable multivibrator (MSMV) is used. If A input to the MSMV is low and a negative to positive edge is seen at the B input, the output will stay high for a selected amount of time. In this case 700 nses. These pulses are the digital inputs to the monitoring system. Recall that the system must detect four failure modes. They are:

(1) Patient's heart rate less than 68 beats per minute:
Each QRS will cause the MSMV to time out for .88 seconds. The output of the MSMV will go low when the rate drops below 68 beats per minute. The alarm condition exists when $\overline{Q}$ goes high (Labled ALARM I on Fig. 1).

(2) Pacemaker rate greater than 72 impulses per minute:
A MSMV is timed out for .83 seconds. If another pacemaker spike occurs within that time it will clock the flip-flop and its output will go high (Labled ALARM II on Fig. 1).

(3) Pacemaker output fails to cause ventricular contraction:
This means that two pacemaker impulses occur without an intervening QRS complex. Each pacemaker impulse clocks Q1 high and each QRS sets Q1 low. If a second impulse comes before the QRS, then Q2 will go high and the alarm sounds. (Labled ALARM III on Fig. 1).

(4) The QRS and pacemaker impulse are not synchronized:
Normally the delay between the pacemaker impulse and the QRS is 20 msec. To determine if the pacemaker and QRS are disassociated a timer is used. The time out is set for 30 msec, the pacemaker channel is the trigger and the QRS is the reset. The output will remain low as long as QRS follows the impulse within 30 msec. Therefore, when the output goes high the alarm sounds. (Labled ALARM IV on Fig. I).

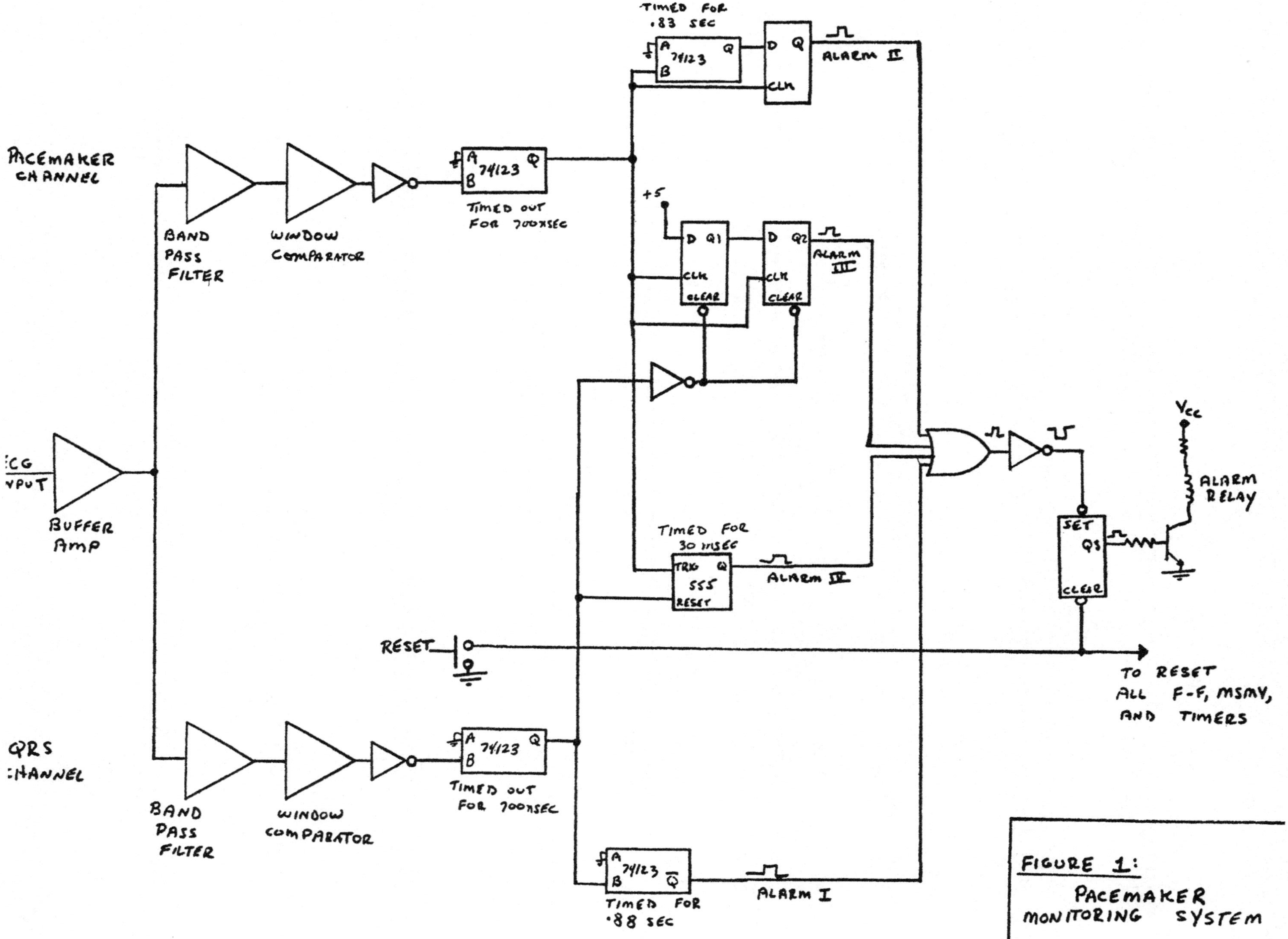

FIGURE 1: PACEMAKER MONITORING SYSTEM

The alarm will sound when any of the above conditions have occurred. The four alarm conditions are inputs to an OR gate, the output of which sets Q3 high. This, in turn, activates the alarm relay through a transistor driver. The alarm will continue to sound until the system is reset by the operator. The particular type of failure can be displayed by turning on an appropriately labled LED. This can be accomplished by taking each of the four alarm conditions as one input to four AND gates, the other input of which, is Q3. Thus, when an alarm occurs, its appropriate AND gate will go high and can drive the LED.

III. CONCLUSIONS

A monitoring system has been presented that can detect any complication of cardiac pacemaker except arrythmia induced by the pacemaker itself, which is best detected by ECG. The system uses any ECG as an input, will sound an alarm that must be reset by an operator and identifies the type of failure.

References

1. C.J.Tegtmeyer, "Complications of Cardiac Pacemakers." Am. Fam. Physician 14 (1):66-75, July 76
2. G.R.Held and D. Fisk, "Complications of Cardiac Pacemakers." S.D. J. Med. 29 (5): 31-33, May 76
3. Z. Vera, et al, "Cardiac Pacemakers: Indications and Complications" Heart Lung 4 (3):444-451, May-June 75
4 Z. Vera, et al, "Complications of Cardiac Pacemakers: Diagnosis and Management" Geriatrics 30 (1SZ):38-44, Jan 75
5 D.J.Escher, "Types of Pacemakers and Their Complications" Circulation 47 (5): 1119-1131, May 73

INTRAOPERATIVE ASSESSMENT OF CEREBRAL ACTIVITY DURING OPEN HEART SURGERY

William B. Shaffer, C.C.P., Dorothy F. Farrell, C.C.P., Paul Barash, M.D.
Judith M. Gatehouse, C.C.P., Susan W. Iverson, B.A., C.P.
Departments of Surgery and Anesthesiology, Yale University School of Medicine
333 Cedar Street, New Haven, Connecticut 06510

During cardiopulmonary bypass (CPB), disturbances in neurological status may be reflected by changes in cerebral activity. The etiology of this dysfunction is thought to be micro-emboli (air or particulate matter) decreases in perfusion pressure, blood flow, and hypoxia.[1,2] Other factors to be considered in relation to the incidence of neurological damage include: patient's age, the duration of cardiopulmonary bypass, the pre-operative physiologic and neurologic status, nature of the surgery, history of previous cardiac surgery, type of oxygenator and the priming solution used.[3] Previously, electroencephalograms (EEG) were used to monitor cerebral activity and deemed a highly accurate method of diagnosing intra-operative neurological deficits. However, EEG has not been widely utilized during anesthesia and surgery.[4] Since the voltages recorded from the scalp are minute, they are easily lost in the noisy electrical atmosphere of the operating room (e.g. electrocautery). The recordings are voluminous and do not readily lend themselves to intrepretation for trend recording. In many cases a special technician is required to operate equipment and collect EEG data.

Recently, we have utilized a Cerebral Function Monitor(R) (CFM)* to assess cerebral activity during open heart surgery. In this system, an amplified EEG signal obtained from two electrodes placed over both parietal regions (plus a neutral ground electrode) is passed through a wide band filter that rejects frequencies below 2 Hz and those above 15 Hz, thus, minimizing artifacts and electrical interference.[5,6,7] Following logarithmic compression and peak to peak rectification, the output (micro-volts) is (essentially a straight line drawn through the peaks of the compressed signal) then displayed on a write-out recorder (6 or 30cm/hr). The lower edge of the trace (height above the baseline) indicates the actual level of activity. The variability in EEG amplitude is seen in the width of the trace. The cerebral activity is recorded as part of a four channel write-out recording (Figure 1). Key to the system for long term trend recording is the ability to measure electrode impedance and thus insure adequate conductivity and eliminate artifact. A study was undertaken to assess the value of such monitoring during open heart surgery.

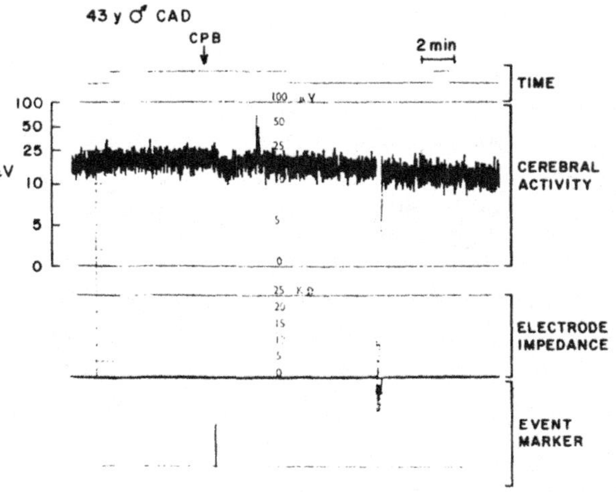

Figure 1. A typical 4-channel recording showing a time marker, cerebral activity channel, electrical impedance channel and an event marker. This trace of a 43 year old male with coronary artery disease (CAD) undergoing coronary artery bypass graft surgery demonstrates no change in cerebral activity at the onset of CPB.

*Devices Limited, New Brunswick, New Jersey 08903

METHODS AND MATERIALS

Fifty patients with congenital and acquired heart disease (ages 2 months to 64 years) undergoing cardiac surgery were evaluated (Table I).

TABLE I

PROCEDURE	NO. PATIENTS
Coronary Artery Surgery	31
Coronary Artery Surgery and Valve Replacement	7
Valve Replacement	5
Correction of Congenital Heart Defect	5
Left Ventricular Aneurysmectomy	2
	50

Intramuscular premedication consisted of secobarbital (4mg/kg) and morphine (0.1mg/kg)(children) or diazepam (0.15mg/kg) and morphine (0.15mg/kg)(adults) one hour prior to the induction of general anesthesia. In children, anesthesia was maintained with nitrous oxide and halothane. In adults, anesthesia was maintained with nitrous oxide, halothane and pancuronium. The premedication and anesthesia were obviously adjusted to the patient's physiologic status.

Following induction of general anesthesia, and prior to cardiopulmonary bypass, the CFM electrodes were placed and cerebral activity recorded (30cm/hr) until the conclusion of surgery. After CPB is initiated, mean arterial blood pressure is maintained by regulation of flow from the oxygenator and peripheral vascular resistance of the patient. Peripheral constriction is controlled by a vasopressor (phenylephrine) or anesthetic vasodilator (halothane) which also served as the maintenance anesthetic while on cardiopulmonary bypass. Four types of oxygenators were employed: Temptrol(R) (Bentley), Veriflo(R) (Travenol), H-1000 Bubble Oxygenator(R) (Harvey) and Kolobow(R) (Sci-Med). For this study the ideal pump flow was calculated as $2.2L/min/m^2$ and the mean blood pressure was 60-100mmHg. In all cases, a 40μ micro-emboli filter was used in the arterial circuit. Hemodilution (hematocrit 22-28%) was employed in all but 2 cases. Temperature during cardiopulmonary bypass was decreased to 28°C in 36 patients (1 patient to 26°C). Two children underwent surface induced profound hypothermia with circulatory arrest.[8] Arterial oxygen tension was maintained above 150mmHg and $PaCO_2 \cong$ 30-35mmHg. Monitoring, in addition to the CFM, consisted of electrocardiogram, arterial, pulmonary artery, and/or left atrial pressures, cardiac output (thermodilution), esophageal and rectal temperatures, blood gases (arterial and venous), serum potassium, and urinary output.

RESULTS

Prior to CPB, no changes were noted in cerebral activity. However, immediately following onset of CPB, 3 distinct patterns emerged. An increase in cerebral activity (>5 micro-volts) was observed in 10% of patients (5/50)(Figure 2). A decrease in electrical activity (>5 micro-volts) was noted in 28% (14/50). No initial changes were observed in 62%

(31/50). However in the latter category, a secondary decrease was noted in 75% (24/32) which was related to a decrease in temperature (Figure 3) (see discussion). At the conclusion of surgery, cerebral activity returned towards normal. The units of change in cerebral activity are measured as micro-volt seconds (decrease or increase [μV x seconds]). Other than specific techniques of temperature reduction and surface induced profound hypothermia decreases range from 100μV seconds to 1850μV seconds. In this series, two patients were noted to have clinical cerebral impairment post-operatively. One sustained a mild psychosis which resolved several days post-operatively. The cerebral function record showed no change in activity. The other patient suffered multiple strokes during the peri-operative period and expired 17 days following surgery. This patient's cerebral function trace revealed a prolonged 6 V decrease with "burst suppression patterns" (indicating cerebral depression). At the conclusion of surgery, he still had a marked depression of cerebral activity (7-8μV) (Figure 4).

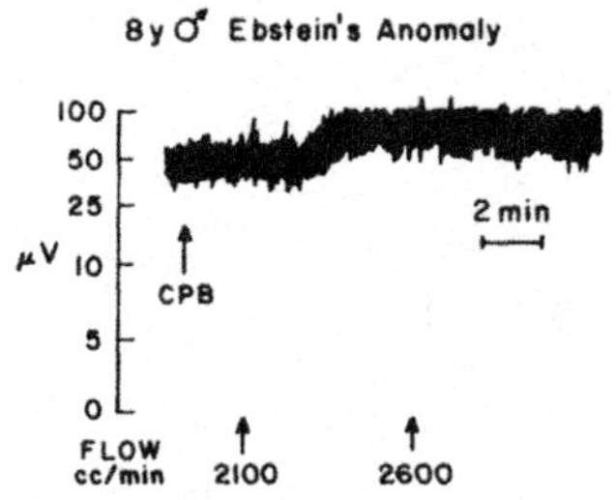

Figure 2. An 8 year old male with Ebstein's anomaly showing 12.5 micro-volt with increased perfusion flows during CPB.

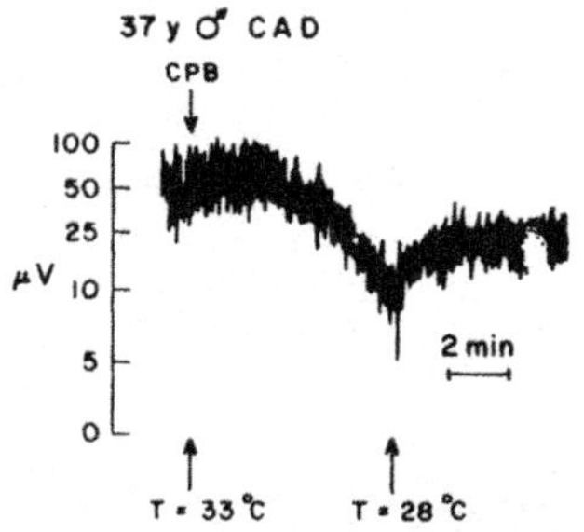

Figure 3. A 37 year old male undergoing coronary artery surgery showing a typical transient decline in cerebral activity associated with a decrease in temperature to 28°C.

DISCUSSION

The reported incidence of neurologic sequalae following cardiac surgery is quite variable. While Javid estimated the incidence as 53%, Branthwaite observed that 10/140 patients suffered neurologic dysfunction following open heart surgery.[1,9] However, 62.9% of all his patients had a significant alteration in cerebral activity during CPB. Stockard noted an excellent correlation between changes and hypotension (<50mmHg) during CPB.[2] When measures are taken to diagnose these problems, the incidence can be reduced.[1,2,3] Early in this series, we observed that perfusion pressure at the onset of CPB significantly affected the level of cerebral activity. Even transient perfusion pressures below 50mmHg were associated with marked decreases in cerebral activity. However, when cerebral function monitoring was added to our protocol, perfusion techniques were altered to assure that arterial pressure did not fall below 50mmHg. Thus, most records were similar to Figure 1.

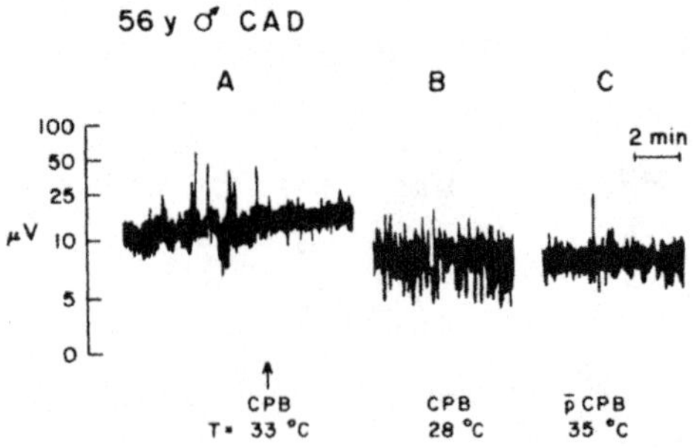

Figure 4. A 56 year old male undergoing coronary artery surgery who had neurologic dysfunction postoperatively. Normal activity prior to CPB (A). During CPB, at 28°C, the patient developed a burst suppression pattern with a very low level of activity (4μV)(B). Following CPB the patient's cerebral activity has not been restored to normal (C).

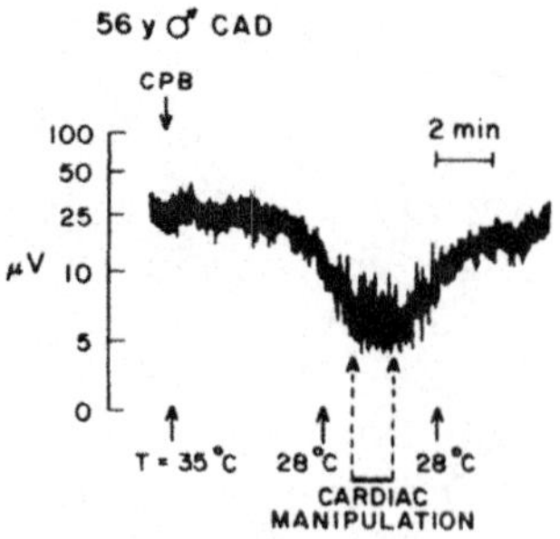

Figure 5. A 56 year old male undergoing coronary artery surgery showing the expected decrease in electrical activity at 28°C. However, with cardiac manipulation a further decrease was observed which returned to baseline levels following cessation of this maneuver.

In 5 cases, an increase in activity was noted (Figure 2). We hypothesize a cardiac output may have been reduced either secondary to the intrinsic heart disease or the cardiac depressant effects of anesthesia. With the initiation of cardiopulmonary bypass, flow was increased above this minimal level. Alternatively, at the onset of CPB, there are marked changes in the concentration of anesthetic agents which may accounted for the patient being at a very "light" level of general anesthesia. Occasionally, this CFM trace was the only sign of poor cerebral perfusion. In one case, when the heart was manipulated during CPB, an unexpected drop in cerebral activity was observed which was promptly restored by having the surgeon cease manipulation (Figure 5).

In our series, a decrease in cerebral activity was deemed an appropriate physiologic response to a decrease in temperature (<28°C)(Figure 3) or elective circulatory arrest (cessation of perfusion (Figure 6).

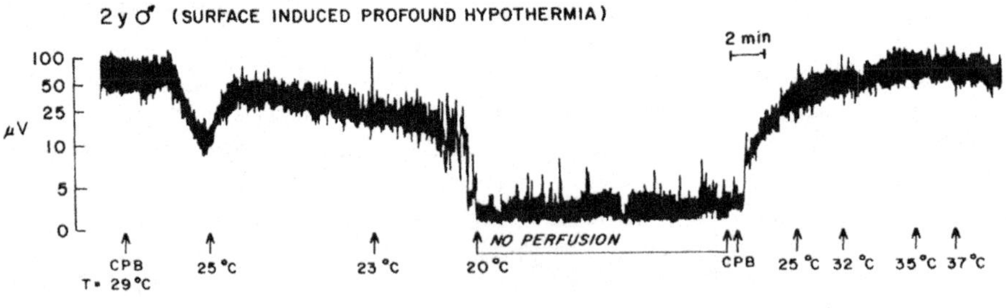

Figure 6. A 2 year old male for repair of the ventricular septal defect (VSD) using surface induced profound hypothermia. Note the marked decrease in activity at 25°C. During this protocol, there is a period of 15 minutes of circulatory arrest (no perfusion) at 20°C during which the VSD is closed. Note virtually, no cerebral activity at this point. When cardiopulmonary bypass is reinstituted there is a marked increase in cerebral activity which is progressive as temperature is restored at 37°C.

There are disadvantages associated with the use of the CFM. The monitor is not designed to detect focal cerebral disturbances. Since the sensitivity of the instrument is limited, the nature, extent and location of injury cannot be easily determined. Most anesthetic drugs do not appear to alter the CFM trace.[10] Levels of cerebral activity which are pathologic may occur beyond the filter range of the CFM. Although the CFM trace may indicate considerable cerebral dysfunction, clinical neurologic sequelae do not inevitably result.[1,5]

In a daily clinical use, the Cerebral Function Monitor can serve as a first line diagnostic device. When cerebral impairment is suspected, further evaluation should be undertaken both intra- and post-operatively.

REFERENCES

1. Branthwaite MA: Detection of neurological damage during open heart surgery. Thorax 28:464-472, 1973.

2. Stockard JJ, Bickford RG, Schauble JF: Pressure dependent cerebral ischemia during cardiopulmonary bypass. Neurology 23:521-529, 1973.

3. Branthwaite MA: Prevention of neurologic damage during open heart surgery. Thorax 28:464-472, 1975.

4. McDowall DG: Monitoring the brain. Anesthesiology 45:117-134, 1976.

5. Schwartz MS, Colvin MP, Prior PF, et al.: The Cerebral Function Monitor. 28:611-618, 1973.

6. Maynard DE, Prior PF, Scott DF: Device for continuous monitoring of cerebral activity in resuscitated patients. Br Med J 4:545-546, 1969.

7. Prior PF, Maynard DE, Sheaff PC, et al.: Monitoring cerebral function: Clinical experience with new devices for continuous recording of electrical activity of the brain. Br Med J 2:736-738, 1971.

8. Barash PG, Berman MA, Stansel HC, et al.: Markedly improved pulmonary function after open heart surgery in infancy utilizing surface cooling, profound hypothermia, and circulatory arrest. Amer J Surg 131:499-503, 1976.

9. Javid H, Tufo HM, Najafi H, et al.: Neurological abnormalities following open heart surgery. J Thorac Cardiovasc Surg 58: 502-509, 1969.

10. Branthwaite MA: Factors affecting cerebral activity during open heart surgery. Anaesthesia 28:619-625, 1973.

CHANGES IN MOTONEURON EXCITABILITY FOLLOWING SPINAL CORD SECTION IN THE CAT

Jonathan Delatizky, J. Thomas Mortimer, and John C. Baker*

Applied Neural Control Laboratory
Department of Biomedical Engineering
Case Western Reserve University
Cleveland, Ohio 44106

*Department of Orthopaedic Surgery
Cleveland Metropolitan General Hospital
3395 Scranton Road
Cleveland, Ohio 44109

Functional electrical stimulation is a common functional aid for patients with varied neuromuscular pathologies. There is some evidence that it may be of therapeutic value in neuromuscular rehabilitation. Several investigators have reported changes in performance persisting after stimulation is discontinued (Liberson, et.al., 1961; Vodovnik and Rebersek, 1973). Most evaluations of these long-term changes have been based on performance in a particular task, e.g. gait or production of voluntary force (Vodovnik and Rebersek, 1973, for example). These methods are difficult, if not impossible, to relate quantitatively to the pathological neurophysiology of the patient. Since long-term changes are reflected at the level of the lower motoneurons, a technique for measuring their excitability is required. Application of such tests must not place undue psychological stress on the patient.

In humans, the H-reflex has been used frequently to measure motoneuronal excitability (Braddom and Johnson, 1974, for a review). It is clear from these studies that lower motoneuron excitability is altered in spasticity and other pathologies. Existing techniques using H-reflexes have many limitations (Braddom and Johnson, 1974); in particular variability between patients makes comparison difficult except in large samples. As a simple animal model of hyperactive lower motoneurons, chronic spinal cats have been used. Before applying electrical stimulation to one side of an antagonistic muscle group (to simulate functional electrical stimulation), it has been necessary to characterize spinal motoneuron excitability during the development of hyperactivity (spinal spasticity). In this way, a baseline for evaluating changes on subsequent application of electrical stimulation has been established. It has proved necessary to look at composite activity in the excitatory pathway from primary muscle spindle afferents to spinal motoneurons, rather than just the classic monosynaptic response.

Methods

Four adult cats (2.0 - 2.5 Kg) were used. In each case, an H-reflex recovery curve was recorded under light barbiturate anesthesia before spinalization as control data. Two weeks later, aseptic spinal cord section at T12 level by blunt dissection was performed under Halothane anesthesia. There was no evidence of damage to major spinal cord blood supply using this procedure. Following recovery from surgery, tests were performed at various intervals for a total period of four to six weeks.

Conscious spinal cats were restrained and tested with one hindlimb immobilized in a special frame allowing control of joint angles. Test conditions were maintained as constant as possible in each test. The animal's bladder was voided prior to each session. H-reflexes were elicited and recorded in the small foot flexor muscles (physiologically extensors) following the technique of McLeod and van der Meulen (1967). The tibial nerve was stimulated near the medial malleolus through uninsulated needle electrodes and evoked myoelectric signals were recorded using similar needles inserted in the small foot flexors. Following amplification, evoked myoelectric signals were displayed on a storage oscilloscope. Rectified myoelectric signals were integrated during the occurrence of the reflex for a time period set at the beginning of each session and the resultant signal was also displayed on the oscilloscope. Peak to peak and integrated H-reflex amplitudes were recorded.

Recovery curves were obtained from responses to equal stimuli separated by an adjustable delay. Seven delays - 20, 30, 40, 70, 100, 250 and 400 msec were chosen to characterize the response. Stimulus amplitude was chosen to elicit peak-to-peak reflex amplitudes 5% of the direct motor response to a supramaximal stimulus. For each delay, the conditioning and test reflexes were averaged, and their ratio determined. At least six determinations were made for each point.

In one of the four cats, additional recordings were made from the severed S1 ventral root following decerebration to eliminate the need for anesthesia. Stimulation in this case was applied through a cuff electrode around the tibial nerve. The ventral root was exposed by a lumbar laminectomy and gross recordings were made through a hook electrode. A pool of warm mineral oil was maintained around the exposed spinal cord and roots.

Results

Families of recovery curves are shown for two of the cats in Figure 1. Curves from the other two animals were similar to these. These data have not been presented as they were not as complete. Curves drawn from integrated reflexes fall very close to these and have not been included. In both cases, the curves rose from control levels soon after surgery and peaked after about ten days. Subsequently, a decrease occurred that was largely complete after four to six weeks. Control curves were generally similar in shape to those in the literature (McLeod and van der Meulen, 1967; Mayer and Mawdsley, 1965). It is clear that after some time has passed, the curves are at least qualitatively similar. The relation of amplitude to the stimulus reflex threshold was recorded for each experiment and the results for one animal are presented in Table 1. The stimuli were approximately uniformly suprathreshold. The variability of the reflex itself made this 'calibration' difficult to achieve. For those determinations where these parameters were recorded accurately, the mean value consisted of 4.23% (S.D. 0.92%) of the motor pool. Reflexes were accompanied by small direct motor responses (10% of the reflex or less) indicating that the stimulus was only just above threshold for motor fibers.

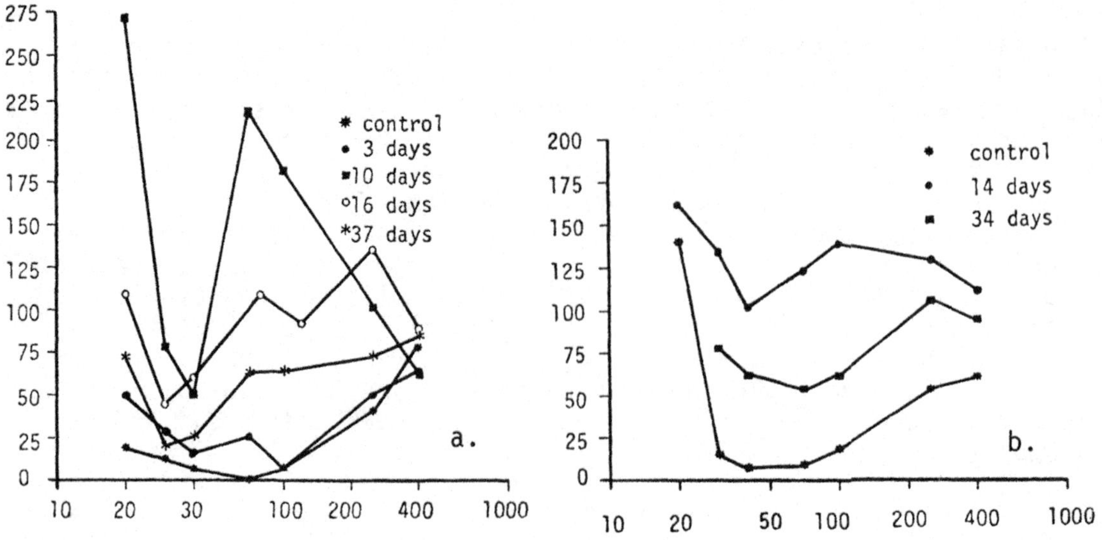

Figure 1. Recovery curves recorded from two animals at the indicated period after spinal cord section. Abscissa: delay between stimuli in milliseconds. Ordinate: ratio of test to conditioning reflex expressed as percentage.

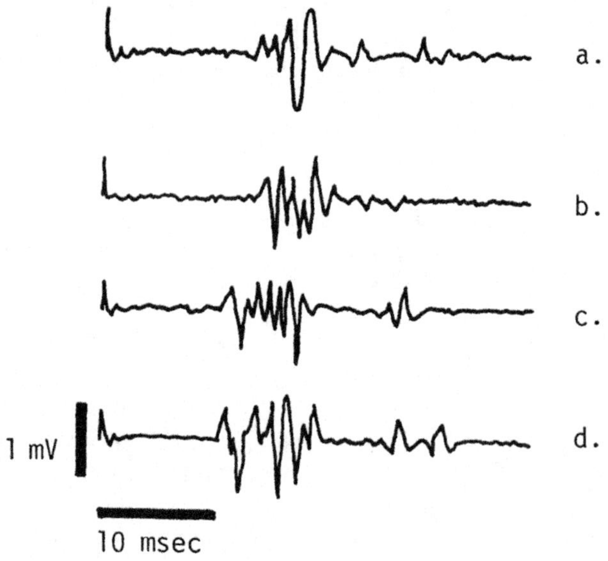

Figure 2. Response of same animal as in Figure 1a to different stimulus amplitudes. a, b: 1.05 threshold; c, d: 1.18 threshold. Note lower threshold of later components. Time and amplitude calibration same for all four records. (51 days)

Days after Spinalization	Stimulus/Threshold Ratio
Control	1.2
3	1.31
4	1.25
14	1.33
16	1.47
37	1.18
44	1.26

Table 1. Relationship of stimulus to threshold for a sequence of tests (cat number 256).

The H-reflex is notoriously variable and most methodological papers in the literature indicate the need for multiple determinations and cautions in technique (e.g. Hugon, 1973). All points shown in the figures are means of at least six determinations, and as many as fifteen in some cases. Standard deviations are not shown; coefficients of variation are on the order of 0.5 for most points except where the test reflex was very small, in which case the coefficient increased, often being greater than unity. Together with the rise in recovery curves, the recorded responses became more polyphasic with the passage of time after spinalization. Figure 2 shows responses to increasing stimulus amplitude in one experiment. Responses were observed for a longer time period after the stimulus and the lowest threshold responses occurred later (typically 14-16 msec) as opposed to 9-10 msec in normals. At higher stimulus amplitudes, the earlier response reappeared. The same pattern was observed in the ventral root recordings from cat 257 (last curve in Figure 1b). Thus, the response is not due to slow non-neural conduction within a mixed group of muscles, but represents a central phenomenon.

Discussion

The results presented above indicate that motoneuronal excitability, following spinal cord section, rises rapidly to a very high level within ten days. A more gradual settling to a fairly stable excitability level occurs over the next 25 to 30 days. Development of clinically manifested spasticity is slower, more occurring during the later 'settling period' than at the time of the peak. The reasons for this are unknown, but may reflect an early depression of the fusimotor system.

The polyphasic responses observed are consistent with observations in the literature that polysynaptic excitatory transmission from primary muscle spindle afferents occurs in the cat (Tsukahara and Ohye, 1964; Pacheco and Guzman-Flores, 1968). The occurrence of a bimodal latency distribution in human H-reflexes (Trontelj, 1973) may indicate similar phenomena in man. The responses suggest that the polysynaptic pathways become facilitated to a greater extent than the direct monosynaptic connections. Since the recovery curves were based on the maximum amplitude of the early reflex burst, they actually represent the excitability of the composite excitatory pathway.

The results of this study suggest that long-term quantitative evaluation of motoneuronal excitability may best be found in two functional areas: a) by following recovery curves in the same animal during the stimulation period; b) by noting relative changes in the efficacy of the monosynaptic and polysynaptic pathways (i.e. early and late responses). Since reasonable stabilization occurs by six weeks after spinalization, this appears to be a good time to begin objective assessment of functional electrical stimulation.

Acknowledgments

This study was supported in part by Orthopaedic Research and Educational Fund, Rainbow Babies and Childrens Hospital RG330 and in part by NIH-NINCDS Contract Number NO1-NS-2-2314. Thanks to Dr. Patrick Crago for his helpful comments.

References

1. Braddom, R.L. and Johnson, E.W. (1974). H-reflex: Review and classification with suggested clinical uses. Arch. Phys. Med. Rehabil. 55: 412-417.
2. Hugon, M. (1973). Methodology of the Hoffmann reflex in Man. New Developments in Electromyography and Clinical Neurophysiology, (ed. J. E. Desmedt) 3:277-293, Karger, Basel.
3. Liberson, W.T., Holmquest, H.J., Scott, D., Dow, M. (1961). Functional Electrotherapy: Stimulation of the peroneal nerve synchronized with the swing phase of the gait of hemiplegic patients. Arch. Phys. Med. Rehab. 42:101.
4. McLeod, J.G. and van der Meulen, J.P. (1967). Effect of Cerebellar ablation on the H-reflex in the cat. Arch. Neurol. 15:421-432.
5. Mayer, R.F. and Mawdsley, C. (1965). Studies in man and cat of the significance of the H wave. J. Neurol. Neurosurg. Psychiat. 28: 201-211.
6. Pacheco, P., and Guzman-Flores, C. (1961). Intracellular recording of extensor motoneurons of spastic cats. Exp. Neurol. 25:472-481.
7. Trontelj, J.V. (1973). A study of the H-reflex by single fibre EMG. J. Neurol. Neurosurg. Psychiatr. 36:951-959.
8. Tsukahara, N., and Ohye, C. (1964). Polysynaptic activation of extensor motoneurons from group Ia fibres in the cat spinal cord. Experientia, 20:628-629.
9. Vodovnik, L. and Rebersek, S., (1973). Improvements in Voluntary control of paretic muscles due to electrical stimulation. Neural Organization and its Relevance to Prosthetics (ed. W.S. Fields and L.A. Leavitt) 101-116.

IN VITRO AND *IN VIVO* LINEAR A C POLARIZATION IMPEDANCE STUDIES ON POROUS STIMULATING ELECTRODES

B.N. Hung, R.B. Beard, M. Brownstein,
H. Carim*, S.E. Dubin, Biomedical
Engineering and Science Institute
Drexel University, Philadelphia, Pa.
A.S. Miller
Department of Pathology, School of
Dentistry, Temple University,
Philadelphia, Pa.

I. INTRODUCTION - The transvenous technique has become widely accepted with preference for the transveneous electrode being almost unanimous.(1-4) One of the complications associated with the use of transvenous electrodes is the lack of retention of the electrode.(3,4) Electrode polarization which increases the energy needed at the electrode interface increases with the electrode current density. Thus although the threshold decreases the smaller the electrode size, the polarization voltage increases.(5,6) In order to increase the efficiency of stimulating electrodes the unexcitable scar tissue must be minimized and the electrode overpotential decreased to permit stimulation at lower voltages and currents. A materials approach to this problem is the use of porous stimulating electrodes which greatly decreases the unexcitable tissue and electrode polarization,(9-12) with tissue ingrowth into the pores to enhance retention of the electrode.

An insight into the controlling process at the electrode interface can be obtained from impedance data.(13) The phase angle; ϕ of the Faradaic impedance can be used to determine the type of controlling process. The Faradaic impedance is obtained by the following procedure. First, the solution impedance is subtracted from the measured impedance, and then the electrode polarization impedance is converted to its equivalent admittance. The remaining admittance is the Faradaic admittance which is then converted back to its equivalent impedance. The phase angle is equal to

$$\phi = \arctan \frac{X_{sfar}}{R_{sfar}} = \arctan \frac{\delta\omega^{-\frac{1}{2}}}{\theta+\delta\omega^{-\frac{1}{2}}} \qquad (1)$$

where ω is the angular frequency; R_{fars}, X_{fars} are the Faradaic resistance and reactance for series representation; θ is the charge transfer resistance and δ the Warburg coefficient. If the reaction is charge controlled, $\theta > \delta\omega^{-\frac{1}{2}}$, ϕ will go to zero as $\theta \to \infty$. If the reaction is diffusion controlled $\theta < \delta\omega^{-\frac{1}{2}}$, ϕ approaches $\pi/4$ as $\theta \to 0$. A phase greater than 45° is accounted for by an adsorption process.

DeLevie(16) has treated the pore of a porous electrode as a transmission line, where the series component of a section of the distributed system is R the ohmic resistance of the electrolyte solution inside the pore measured per unit pore length, and z the parallel component of a section is the impedance of the electrolyte-electrode interface taken per unit pore length. The pore exhibits the following impedance, Z_o, at the orifice of the pore.

$$Z_o = (RZ)^{\frac{1}{2}} \text{cotanh } \rho\ell \qquad (2)$$

*3M Medical Products Div., St. Paul, Minn.

where ℓ is the pore length and $\rho = (R/Z)^{\frac{1}{2}}$. For a semiinfinite pore the cotanh term can be dropped.

When there is a double-layer charging and a slow electrode reaction taking place as independent, parallel processes. The impedance of the electrolyte-electrode interface is

$$Z = \frac{(1-j\omega\theta c)}{1+(\omega\theta c)^2} \qquad (3)$$

where c is the double layer capacitance.

The plot of the reactance vs. the resistance as the frequency is varied is a circle for Z.

II. EXPERIMENTAL PROCEDURES

1. Transient Galvanostatic - In the transient galvanostatic technique the electrode potential variation with time is recorded for a current pulse. The electrode under test and a counter electrode are placed in a medium approximating the ionic concentration of subcutaneous tissue fluid. The potential of the test electrodes referred to a reference electrode is measured as shown in Figure 1. *In-vitro* and *in-vivo* measurements have been taken with reference to a Ag/AgCl 16mm Beckman biopotential electrode.

2. A C Impedance Measurements - The A.C. bridges include the Wayne Kerr B201 for 0.1-5MHz and a B221 for 0.25 to 50KHz in conjunction with a Wayne Kerr SR268 source detector. Reagent grade chemicals and doubly distilled water are used for the chloride solutions. The bovine serum albumin (BSA) solutions are made up from Fraction V BSA Nutritional Biochemicals Corp.

The total measured impedance between the electrodes include the solution and/or tissue impedance. The test cell for *in-vitro* measurements consists of a porous palladium stimulating electrode with a geometric surface area of approximately $0.28cm^2$ and a large flat palladium counter electrode whose impedance is significantly lower than the test electrode.(7) Two stimulating electrodes spaced 2mm apart are used for *in-vivo* measurements.

The impedance of the test electrode is obtained by subtracting the solution, and a small counter electrode impedance from the total measured impedance at each frequency. The background electrolyte is corrected for by determining the impedance at a sufficiently high frequency, e.g., 1MHz where the electrode impedance can be neglected. This technique is described in greater detail elsewhere.(13)

3. Surface Area and Porosity - The Brunauer, Enmett, and Teller (B.E.T.) gas absorption method has been used to measure the surface area of porous compacts and powder. The pore size and porosity is determined by the capillary pressure technique of Dodds and Lloyd.(12,14)

4. Implantations - All implanted electrodes are sterilized in a Bard ethylene oxide gas sterilizer. Electrical connections are made by teflon coated multistrand stainless steel wire No. 31 A.W.G. which are percutaneously brought out of the rabbit and dogs. Stimulating electrodes porous and platinum-iriduim are placed in triceps and thigh muscle of rabbits. All wires are gathered behind the back neck to prevent interference by the rabbit. Dog implants are subcutaneous.

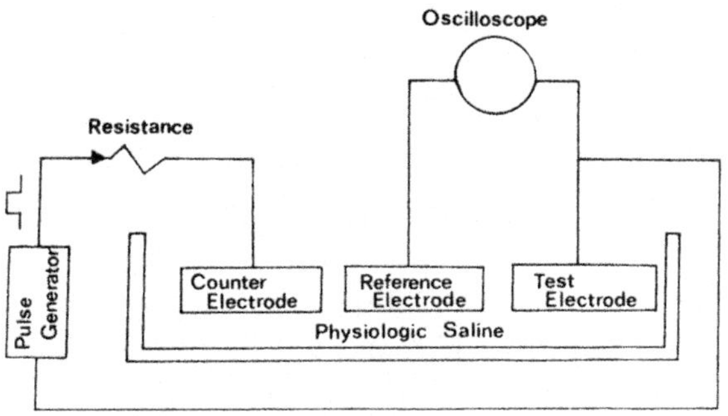

Figure 1

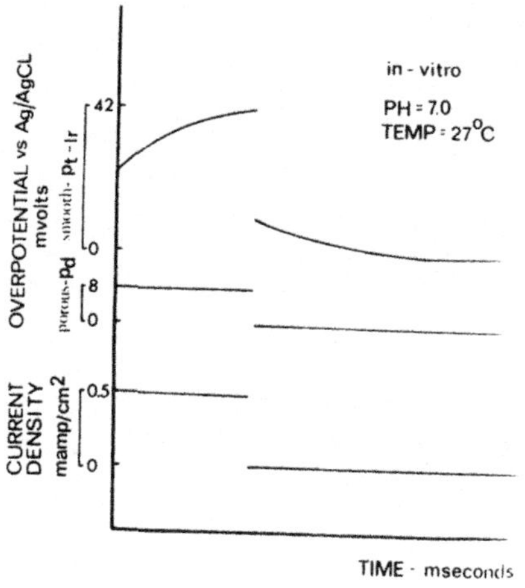

Figure 2

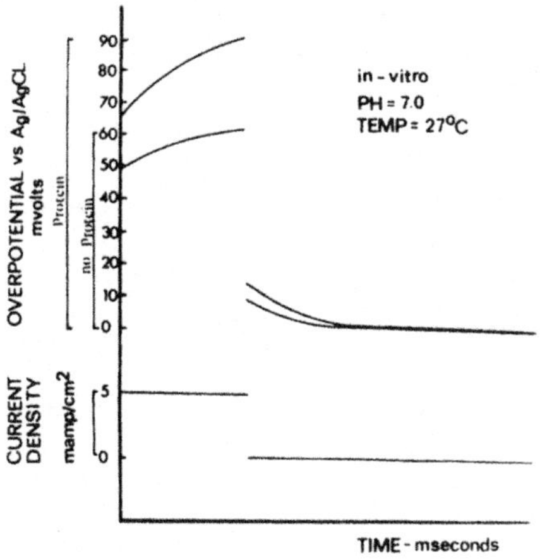

Figure 3

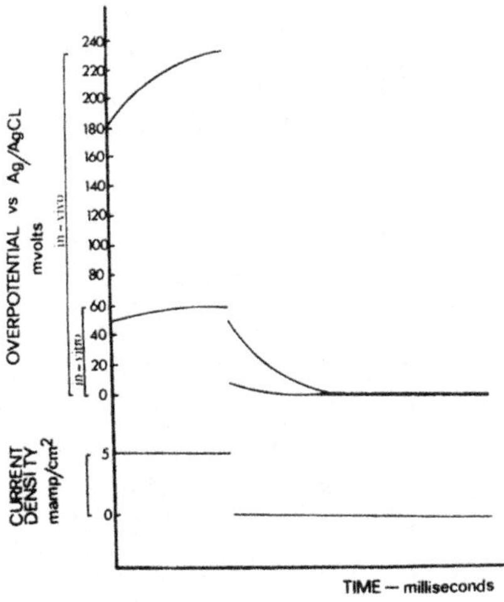

Figure 4

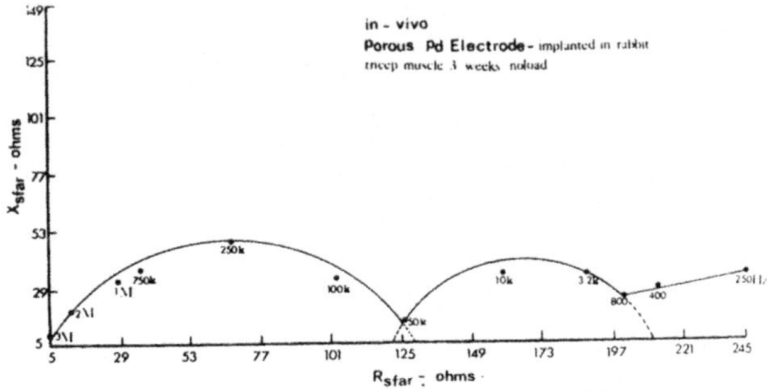

Figure 5

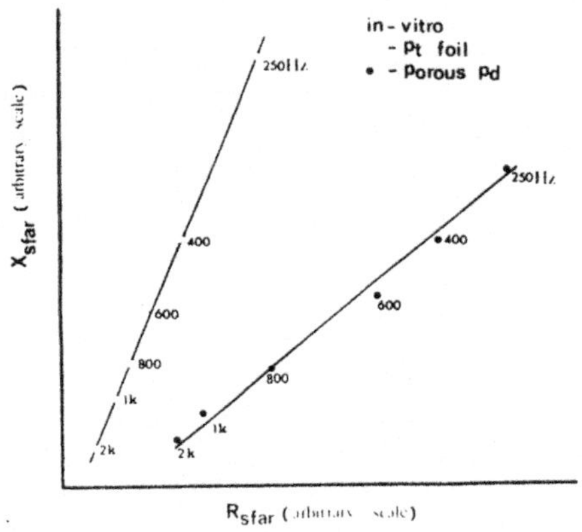

Figure 6

5. Porous platinum and palladium compacts have been fabricated isostatically and axial using powder metallurgy techniques,(12) with pore formers such as polymethyl methocrylate and amonium carbonate.

Tissue ingrowth studies for porous stimulating electrodes have been undertaken by removing the electrode "en block" and fixing in formulin. Procedures developed at Clemson University by Hulbert, Klawitter, et al. (15) were used.

III. RESULTS AND DISCUSSION - The overpotential measured by the transient galvanostatic technique for smooth platinum-iridium electrodes and porous palladium electrodes are plotted in Figures 2 through 4. Similar results have been obtained for porous platinum as for porous palladium.(11) Figure 2 for a 0.5ma/cm^2 pulse in a physiological saline solution for porous palladium and smooth Pt- electrodes demonstrate the improvement in overpotential of the porous over the smooth interface. Figure 3 illustrates the influence of protein in increasing the overpotential by 30%. It is also noted that there is a marked increase in polarization with a ten fold increase in current density. The overpotential is seen to be greater for porous electrodes *in-vivo* than in *in-vitro* in Figure 4.

Figure 5 the complex impedance plane plot of the series Faradaic polarization reactance versus resistance for platinum foil in 0.1MKCl produced a phase angle of 78° and for the porous palladium at 27°C in 0.9% NaCl, ϕ=37°. This is to be expected since the controlling step on polished and palladium at low current density and a pH around 7,0 is adsorption controlled.(16) The angle from equation 2 for the porous electrode should be approximately one half of 78° which it is. This checks also with the results obtained for platinum by DeLevie.(16) *In-vivo* implantations give the complex impedance results shown in Figure 6 for 3 weeks after implantation. It is seen that a circle if obtained indicating a double layer charging and slow electrode reaction. The size of the circles increase as well as a shift in the characteristic frequency towards higher frequencies with implantation time.

Additional *in-vitro* impedance measurements are being taken in various solutions of components found in body fluids such as BSA and mucopolysaccharides in order to better understand the *in-vivo* results.

IV. ACKNOWLEDGEMENTS - This study was supported in part by the Biomaterials Program of NSF Grant No. GH 3378 and a NHLI Training Grant HL 05417.

V. REFERENCES

1. Furman, S., and Escher, D.J.W.: "Principles and Techniques of Cardiac Pacers", Harper and Row, 1970, N.Y.

2. Yanoz, M.D., Attai, L.A., and Furman, S.: J. of Thoracic and Cardiovascular Surgery, Vol. 68, pp. 43-46, July 1974.

3. Buffle, P.J.: J. of Thoracic and Cardiovascular Surgery, Vol. 72, pp. 226-31, Aug. 1976.

4. Fiandio, O., et al.: Am. Heart Journal, Vol. 91, No. 4, pp. 468-474, April, 1974.

5. Myers, G.H., and Parsonnet, V.: "Engr. in the Heart and Blood Vessels", Wiley, 1969.

6. Klafter, R., et al. IEEE Trans., B.M.E., Vol. 23, July 1976.

7. Beard, R.B., et al. IEEE Trans., B.M.E., Vol. 19, May 1972.

8. Parker, B., Burman, S., et al.: Proc. 27th ACEMB, Oct. 1974.

9. Hung, B.N., et al.: Proc. 28th ACEMB, p. 36, New Orleans, Sept. 1975.

10. Hung, B.N., et al.: 2nd Annual Meeting Soc. Biomaterials, Phila., April 1976.

11. Sturm, L.: Master's Thesis, Drexel University, 1974.

12. Hung, B.N.: Master's Thesis, Drexel University, 1975.

13. Carim, H.M.: Ph.D. Dissertation, Drexel University, 1976.

14. Dodds, J.A., and Lloyd, P.J.: Powder Technology, Vol. 5, 1971/72.

15. Hulbert, S.F., Morrison, S.J., and Klawitter, J.J.: J. Biomed., Water Res., Vol. 6, pp. 347, 1972.

16. DeLevie, R.: "Electrochemical Response of Porous and Rough Electrodes", Advances in Electrochemistry and Electrochemical Engr., Vol. 6, Delahayectitor Interscience.

17. DeRosa, J.F., and Beard, R.B.: IEEE Trans. B.M.E., accepted for publication in May 1977.

18. Hench, L.L., and Ethridge, E.C.: Biomaterials - The Interfacial Problem, Advanced in Biomedical Engr., Vol. 5, pp. 36-139, 1975.

19. Johnson, P.F., and Hench, L.L.: J. Biomed. Mater. Res., Vol. 10, pp. 907, 1976.

THE FEASIBILITY OF USING A TEMPERATURE DEPENDENT BIPOLAR TRANSISTOR ARRAY FOR FLUID VELOCITY MEASUREMENTS*

Walter S. Kuklinski, Ganesan Sadasiv and Dov Jaron

Biomedical Engineering Program
Department of Electrical Engineering
University of Rhode Island
Kingston, Rhode Island 02881

Introduction

There are instances in both clinical and research situations where it is desirable to measure blood pressure and velocity simultaneously[1-3]. The aim of this investigation was to determine the feasibility of fabricating a velocity sensor using techniques compatible with those used in the production of catheter tip pressure sensors[4]. This velocity sensor could then be combined with a catheter tip pressure sensor on a single integrated chip.

The prototype sensor that was used in this investigation employed the thermal method of velocity measurement. This method is based on the principle that the rate of heat transfer and hence temperature of a heat source immersed in a fluid is a function of the velocity of that fluid. The performance of the prototype velocity sensor was evaluated by recording its response to steady-state flows and then comparing these results with those predicted by a theoretical analysis of the sensor. The theoretical analysis yielded a transfer function relating the output voltage of the sensor to the velocity of the fluid flowing past the sensor.

Sensor Fabrication

The npn bipolar transistor used as the velocity sensor was fabricated from a 10 ohm-cm resistivity, n-type 111 orientated, silicon wafer. The metalization used in this sensor was gold over nickel rather than the more usual gold over aluminum. This was done to facilitate the incorporation of this velocity sensor in a probe that would also contain a piezo-resistive pressure sensor[4]. After the wafer was processed, metalized and diced, it was mounted on a ceramic substrate using a conductive epoxy (EPO TEK 410-E). The transistor was placed at the edge of the substrate so that the thermal and hydrodynamic boundary layers would both start at the same point, hence reducing the complexity of a theoretical analysis of the sensor. One mil (.001 in) gold wires were used to make connections between the sensor and contact pads. The pads had been vacuum deposited on the ceramic substrate. The wires were attached to both the sensor and substrate contact pads using a thermo-compression bonding process.

*Supported in part by NSF Grant Eng 74-21085, a grant from the Stride Memorial Foundation and a grant-in-aid from the University of Rhode Island.

Theoretical Derivation of Transfer Function

The following simplifying assumptions were made in deriving the transfer function:

1. Forced convection is the primary heat transfer mechanism.
2. Fluid flow over the sensor is laminar.
3. The heat flux is uniform over the surface of the sensor.
4. The thermal and fluid dynamic properties of the fluid are not functions of temperature.
5. The sensor surface can be modeled as a semi-infinite plane.
6. The average sensor surface temperature equals the temperature of the junction region of the transistor.

Using these assumptions the average value of the temperature of the sensor surface ($\bar{T}$) as a function of velocity can be written in the form:

$$\bar{T} = T_a + C_1 \dot{q}_o'' v_\infty^{-\frac{1}{2}} \tag{1}$$

where T_a = ambient temperature,

$\dot{q}_o''$ = heat flux (watt/cm^2),

v_∞ = velocity in the free stream, and

C_1 = constant related to the thermal and fluid dynamic of the test fluid.

The equation relating the transistor gain (β) and temperature of the junction region (T_s) is given as[5]:

$$\beta = C_2 T_s^{\,n} \tag{2}$$

where C_2 and n are constant for a specific device.

The transistor sensor was used in a circuit whose output voltage (V_{out}) was a linear function of gain:

$$V_{out} = C_3 \beta \tag{3}$$

By combining equations 1 through 3 with the assumption that $T_s \simeq \bar{T}$ the overall transfer function becomes:

$$V_{out} = C_2 C_3 \left(T_a + C_1 \dot{q}_o'' v_\infty^{-.5}\right)^n \tag{4}$$

By using the appropriate values for the constant C_1 the response of the sensor in any fluid, such as saline or blood, could readily be obtained.

Methods and Results

A number of calibration experiments were performed to determine the constants C_1 through C_3 and n in equation (4). A second series of experiments designed to determine the relationship between fluid velocity and sensor output voltage (V_{out}) was then performed. The sensor assembly was placed in the center of a rectangular flow channel and its response to oil flows of various magnitudes recorded. Oil was used as a test fluid for the prototype sensor because of its electrical insulation property. These results for a number of operating points of the sensor are shown in Figure 1. Note that because of the assumption that forced convection is the primary heat transfer mechanism the theoretical curves exhibit behavior near zero velocity not seen in the experimental data.

The time constant of the sensor was measured by introducing a step change in velocity and then recording the time required for the sensor output voltage to reach 63% of its final value. This data is shown in Figures 2 and 3. Note that the sensor responds more quickly to increase in velocity, "cool down" time constant (Figure 2), than to a corresponding decrease in velocity, "heat up" time constant (Figure 3).

Discussion

Three criteria were considered in evaluating the performance of the velocity sensor: (1) Reproducibility of the output voltage vs. velocity data; (2) Agreement between measured and theoretically predicted output voltages and (3) Applicability of the sensor's time response to blood flow measurements.

A number of trials were performed when recording the output voltage vs. velocity data. For all velocities and operating points the greatest percent error between trials were on the order of 3%.

While the general trend of decreasing output voltage for increasing velocity was found in both theoretical and experimental data, the theoretical transfer function predicted higher voltages at low velocities and lower voltages at high velocities that were recorded experimentally, as seen in Figure 1. To improve the low velocity behavior of the transfer function both conductive and convective heat transfer mechanisms would have to be considered.

The disagreement between theory and experiment of the sensor's output voltage at high velocities was probably related to the assumption that the thermodynamic and hydrodynamic boundary layers start at the same point. The current flow in the sensor was concentrated under the emitter region, producing a small unheated starting length. By neglecting this unheated starting length the temperature of the sensor surface and output voltage would be underestimated.

To utilize this sensor for measurement of physiological signals the time response will have to be improved. The time constants of the prototype sensor were both on the order of 15 seconds. To decrease these time constants the effective thermal mass of the sensor will have to be reduced. Another method of decreasing the time constants is by incorporating the sensor in a feedback loop that would attempt to maintain a constant sensor temperature. A constant temperature mode of operation would significantly reduce the time constants of the sensor. Our calculations indicate that by using a constant temperature mode of operation in addition to reducing the effective thermal mass of the sensor the time constants could be reduced by approximately two orders of magnitude.

The results of this study indicate that by incorporating the changes mentioned above a second generation sensor capable of measuring physiological flow signals can be constructed.

References

1. Schenk, W.G., et al: A biophysical approach to hemodynamic problems in some surgical disease states. Surgery 51:26, 1962.

2. Jacobs, R.R., et al: On line computation of instantaneous vascular impedance. Surg. Forum 22:182, 1971.

3. Jacobs, R.R., et al: Human application of a catheter with tip-mounted pressure and flow transducers. Rev. of Surg., March-April, 1972.

4. Samaun, et al: An IC pressure sensor for biomedical instrumentation. IEEE Transaction on Bio-Med. Engr. BME-20, 101, March, 1973.

5. Blaser, E.M., et al: A miniature digital pressure transducer. 24th ACEMB, 211 October, 1971.

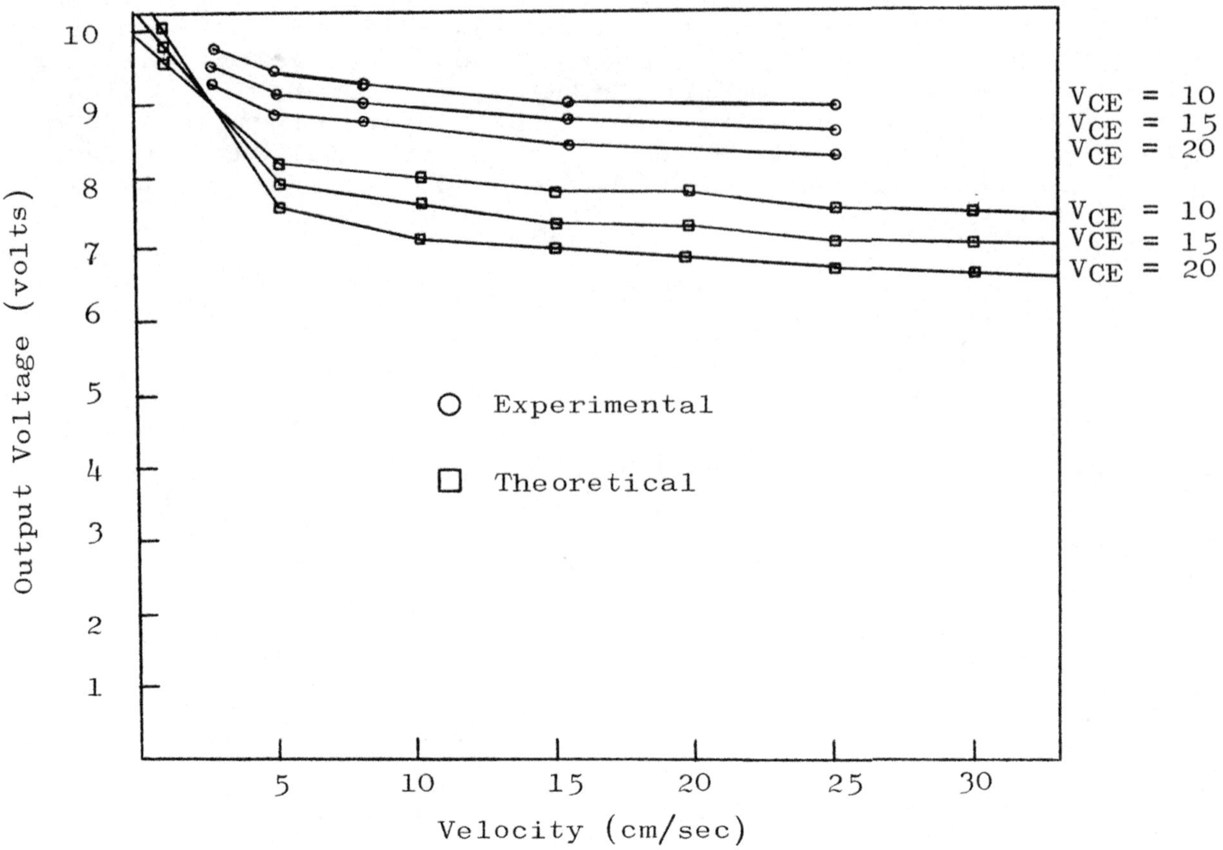

Figure 1. Output Voltage vs. Velocity I_C = 15 ma

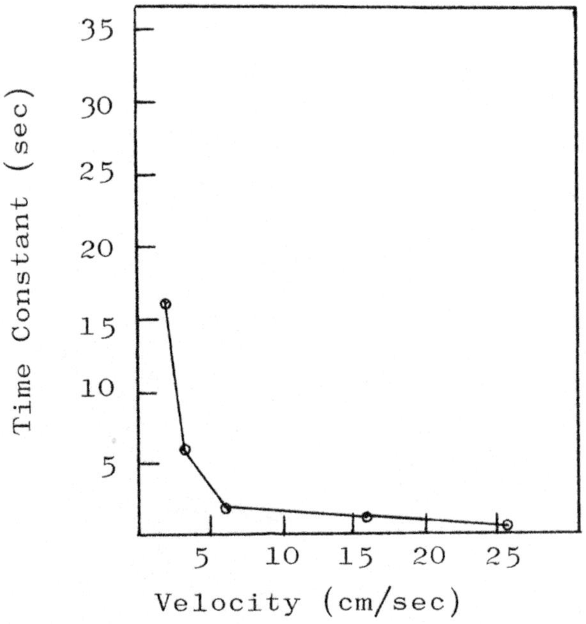

Time constant vs. velocity

Figure 2. "Cool down"

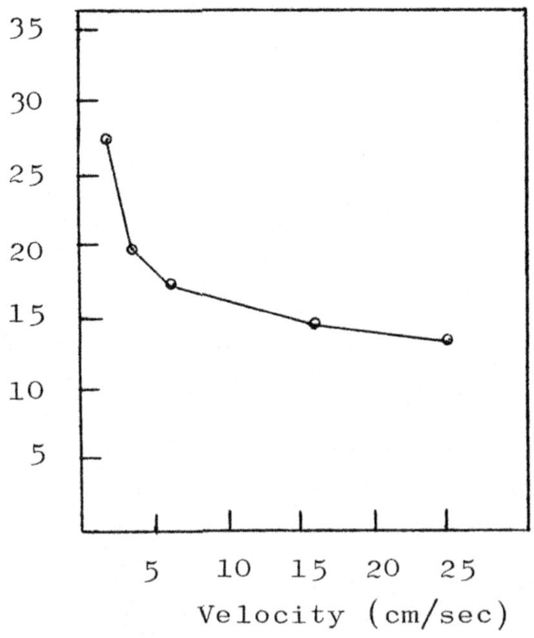

Time constant vs. velocity

Figure 3. "Heat up"

EVALUATION OF NEEDLE-MANOMETER AND NEEDLE-DIFFERENTIAL MANOMETER SYSTEMS IN THE MEASUREMENT OF PRESSURE DIFFERENCES

John K-J. Li and Abraham Noordergraaf, Department of Bioengineering, D2, University of Pennsylvania, Philadelphia, Pa. 19104.

1. Introduction.

Of fluid-filled blood pressure measurement systems, the needle-mano -meter combination possesses superior frequency response over the catheter -manometer combination, mostly because of the difference in their complian -ce (8,9). However, the best approach to the measurement of small pressure differences has not been clear. Two avenues are open: subtraction of the outputs of two needle-manometer systems, or the use of a differential mano -meter. This communication will compare the two approaches.

2. Theory.

A differential manometer containing a single diaphragm may be describ -ed statically, assuming linearity and negligible hysteresis, by

$$y_+ = m_+ x + c_+ \quad \text{and} \quad y_- = m_- x + c_- \qquad \text{(a)}$$

for the positive (+) and negative (-) sides, respectively. y= output volt -age, x= input pressure, m= slope and c= intercept or deviation from zero. The static difference between the two sides is then

$$\Delta y = (y_+ - y_-) = (m_+ - m_-)\, x + (c_+ - c_-) \qquad \text{(b)}$$

where $c_+ - c_-$ is known as the zero imbalance and $(m_+ - m_-)\, x$ as the static imbalance. For a perfect differential manometer it is required that

$$m_+ - m_- = 0 \quad \text{and} \quad c_+ - c_- = 0 \qquad \text{(c)}$$

in addition to linearity and negligible hysteresis. For a more precise description, taking into account hysteresis and nonlinearity, second or higher order equations are necessary to substitute for equations (a). Equations (a), (b) and (c) also apply to two needle-manometer systems in the measurement of pressure differences. The requirements (c), thus describe two perfectly matched manometer systems.

The dynamic response can be described using available theory (9,11), applied to each side.

3. Calibration of Manometer Systems.

Static and dynamic tests were carried out using a water column(250cm) and 'pop' tests respectively, for three manometers (two P23Gbs, one MS-10) and a differential manometer (MP-45). A sinusoidal pressure generator was also used for the testing of dynamic imbalance of the manometer systems (P23Gb-P23Gb combination, P23Gb-MS10, and MP-45). The manometers were con -nected to 22 Ga. needles, 2cm in length. Results are shown in Table 1. All manometer systems were found to satisfy the above theory, respond lin- early and show negligible hysteresis.

The table shows that for the instruments investigated, two carefully matched systems easily outperform the differential manometer with respect to frequency response and imbalance.

4. Conclusions.

The understanding of many important hemodynamic events require the accurate determination of small pressure differences (1-6,10,13). However, a differential manometer capable of high sensitivity, low imbalance and high frequency response, has not been available commercially. Thus many have turned to the matching of two single manometer systems (2,3,7,10), or made their own differential manometer (6,12,13). It is reported here that the matched manometer systems (P23Gb-P23Gb) have low imbalance and high frequency response. The sensitivity was lower however, compared to the differential manometer (MP-45). The unmatched system (P23Gb-MS10)was found to be inferior to the two systems mentioned above.

Table 1. Response of Manometer Systems.

Manometer System(s)	Sensitivity (uV/V/cmHg)*	Resonance Frequency (Hz)**	Static Imbalance (%)	Dynamic Imbalance (Hz, ±5 %)
P23Gb	~ 20	> 200		
MS10	~ 50	> 100		
MP-45	~ 500	~ 60	~ .4	~ 15
P23Gb-P23Gb			~ .1	> 30
P23Gb-MS10			~ .8	> 15

* output voltage per volt of excitation per cmHg of applied input pressure.
** all transducers were connected to 22 Ga. needles 2 cm in length.

References.

1. Bergel D.H., D.A. McDonald and M.G. Taylor. Method for measuring arterial impedance using a differential manometer. J. Physiol. 141:17, 1958.
2. Cox R.H. Determination of the true phase velocity of arterial pressure waves in vivo. Circ. Res. 29:407, 1971.
3. Fry D.L., D.M. Griggs & J.C. Greenfield. In vivo studies of pulsatile blood flow: the relationship of the pressure gradient to the blood velo-city. In: Pulsatile Blood Flow (ed. E.O. Attinger), pp101-114, McGraw-Hill, New York, 1964.
4. Gabe I.T., J. Karnell, I.G. Porjé and B. Rudewald. The measurement of input impedance and apparent phase velocity in the human aorta. Acta. Physiol. Scand. 61:73, 1964.
5. Greenfield J.C. and D. Fry. Measurement errors in estimating aortic

blood velocity by pressure gradient. J. Appl. Physiol. 17:1013, 1962.
6. Gregg D.E. and H.D. Green. Phasic blood flow in coronary arteries obtained by a new differential manometric method. Proc. Soc. Exp. Biol. 41:597, 1939.
7. Lapointe A.C. and F.A. Roberge. Mechanical damping of the manometric system used in the pressure gradient technique. IEEE Trans. BME 21(1): 76, 1974.
8. Li J.K-J., A.G.W. van Brummelen and A. Noordergraaf. Fluid-filled blood -pressure measurement systems. J. Appl. Physiol. 40:839, 1976.
9. Li J.K-J. Fluid-filled blood pressure measurement systems. M.S. thesis University of Pennsylvania, Philadelphia, 1974.
10. Ling S.C., H.B. Atabek, W.G. Letzing and D.J. Patel. Nonlinear analysis of aortic flow in living dogs. Circ. Res. 33:198, 1973.
11. McDonald D.A. Blood Flow in Arteries. Arnold Ltd., London, 1974.
12. Müller O. and J.P. Schllingford. A manometer for differential and single pressure measurements. J. Physiol. 127:2p, 1954.
13. Porjé I.G. and B. Rudewald. Hemodyamic studies with differential pressure technique. Acta. Physiol. Scand. 51:116, 1961.

Acknowledgement.

This study was supported in part by N.I.H. Grant HL 10,330.

HUMAN PERCEPTION OF TRANSIENT ELECTRIC SHOCKS*

Brian C. Hodgkin, Ph.D. Biomedical Research Institute
University of Maine, Portland, Me. 04102

Introduction:

The level of perception of steady state current at 60 hertz is well established to be 0.7-1.0 mA rms for gripping contacts and about 0.35 mA for tapping contacts. Perception threshold of transient shocks, such as those received when a person touches an object charged to a different potential from himself, has not previously been established.

The potential induced on an object in an electric field depends on the source voltage, the capacitance between the source and the object, and the capacitance between the object and ground. Perception of transient shocks depends on the duration as well as magnitude of the shock. Duration is a function of circuit capacitance and resistance.

The objectives of the current study were: a) to determine the appropriate electrical parameters for specifying a perception threshold, b) to establish a perception threshold for transient shocks, and c) to develop a standard measurement procedure.

Methods:

Assuming induction effects are negligible, the waveshape of transient shocks from charged objects is an exponential decay with an initial voltage equal to the induced potential. This waveshape is also present across a series resistor through which a capacitor is charged. Experimentally, a constant voltage source in the form of a bank of dry cells was used to charge a capacitor through the hand of the experimental subject. The heel of the hand was placed on one metal plate and the other plate was tapped with the forefinger.

Experimental capacitances, C_i, ranged from 0.001 to 0.04 μF. Time constant, or duration, of shock depended on C_i and subject hand resistance. In order that hand resistance be low and as nearly constant as possible, the forefinger and heel of the hand were moistened with saline. In selected subjects, the finger was not moistened and was wiped dry between tests in order to compare results to subsequent moist finger results. Current was calculated from the potential across a 100 ohm series resistor measured using a storage oscilloscope. Charge and peak instantaneous energy were calculated as CV and $CV^2/_2$ respectively.

The general procedure was as follows. A capacitance was selected. A forefinger was dipped in saline and wiped across the plate to which the heel of the hand was then applied. The second plate was tapped once. Between taps, C_i discharged through a parallel 10 megohm resistance. Voltage, current, and whether or not a shock was perceived were recorded. By successive voltage changes, the interval between perceived and non-perceived voltages was diminished until a threshold could be specified.

Nine subjects between 21 and 84 years of age were tested of which two were females in the mid-twenties.

*Experiments conducted in the Department of Surgery, Johns Hopkins University and supported by the Edison Electric Institute.

Results:

In each subject, peak voltage of perceived shocks was established repeatedly to within a few volts. Perception current was not so reliably determined. Both voltage and current thresholds decreased with increasing capacitance, C_i. In Figure 1 perception threshold is shown as a function of C_i both for moist and dry fingers. Although the trend is the same, considerably greater voltages were required if fingers were dry. Figure 2 shows results for current, charge and energy.

For moist fingers the most nearly constant parameter was charge (0.475 μC) for C_i of 0.005 μF or less, was energy (22.4 μJ) for C_i between 0.005 and 0.01 μF, and was voltage (54.7V) for C_i greater than 0.01 μF. For dry fingers, based on limited measurements, the knee of the curve is broader so charge was nearly constant (1.03 μC) for C_i of 0.003 μF or less, energy (242 μJ) for C_i between 0.005 and 0.01 μF, and voltage (178V) for C_i of 0.02 μF and greater.

Discussion:

In establishing a threshold, it is desirable to be able to specify a parameter which is constant over a wide range of conditions. It is also desirable that measurements be simple and reproducible. It is usually difficult to estimate current flow since some resistance must be assumed. Shock duration also is a function of subject resistance. Transient currents are relatively difficult to measure.

It was shown that three parameters--charge, energy, and voltage--can be used to specify perception threshold. Voltage is easily measured; charge and energy are easily calculated from voltage and capacitance. Capacitance might or might not be difficult to determine depending on available measuring equipment and whether or not the source is alternating or direct. If it is alternating and of known voltage and frequency, only two simple measurements are necessary--voltage from the object to ground and short circuit current from the object to ground.

Let C_1 be the capacitance from the source to the object and C_2 the capacitance from the object to ground. Assume the object is resistively isolated from ground. Let V_s be the source voltage and V_2 the open circuit voltage from object to ground. The short circuit steady state current from the object to ground is $I_{sc} = j\omega C_1 V_s$ from which C_1 can be calculated. C_2 can then be calculated by measuring V_2 and using the relationship $C_1 (V_s - V_2) = C_2 V_2$. Charge equals C_2V_2 and energy equals $CV_2^2/2$.

The experimental conditions are valid for alternating sources if the time constant is small relative to the source period. For example, the time constant for a resistance of 10K ohms and a capacitance of 0.04 μF is 0.4 ms while the period of a 60 Hz source is about 16.7 ms. The results are therefore valid for objects in power frequency fields as well as for electrostatically charged objects and capacitor discharges.

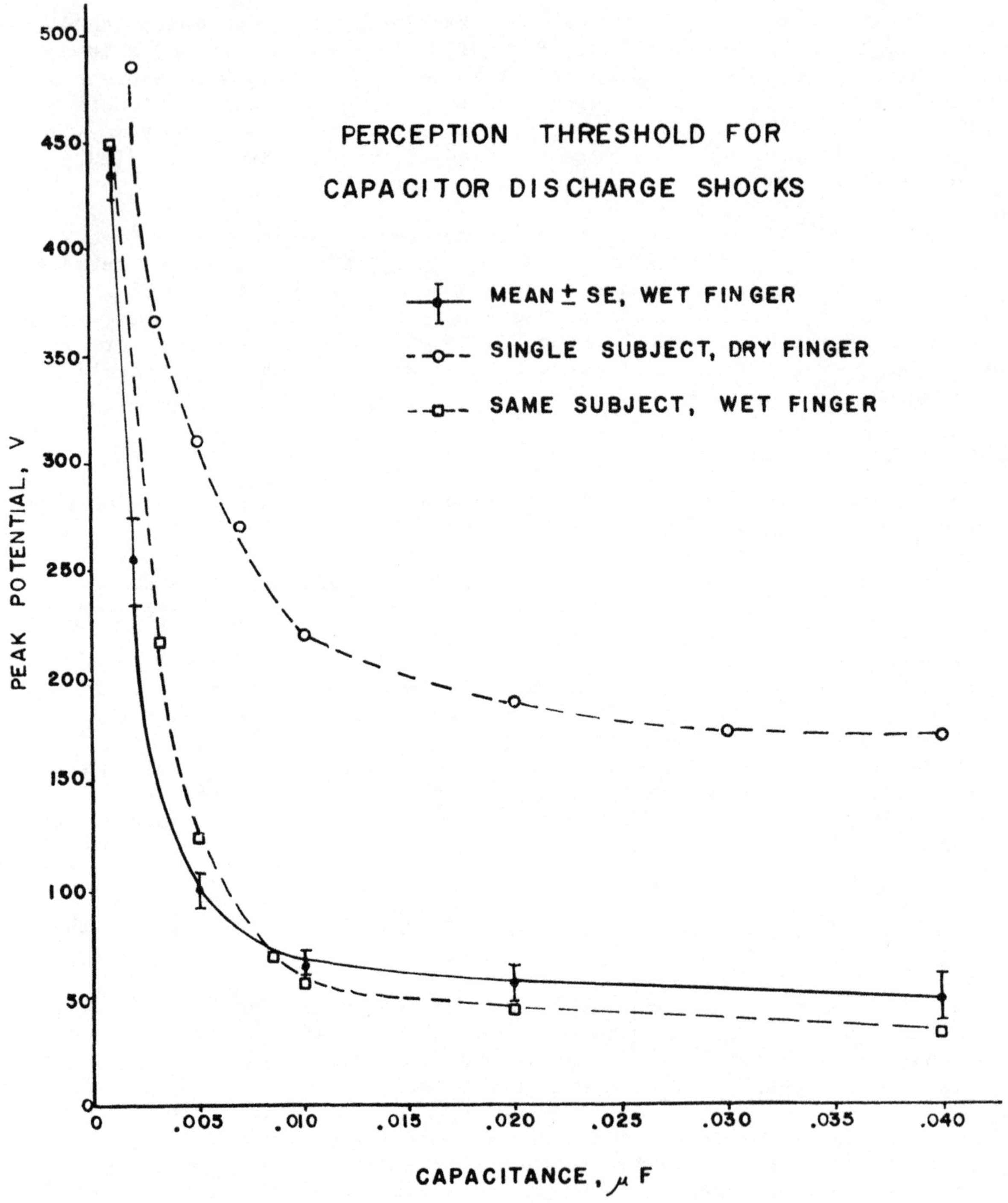

Figure 1 - Voltage perception threshold. Single subject wet finger results are also included in average.

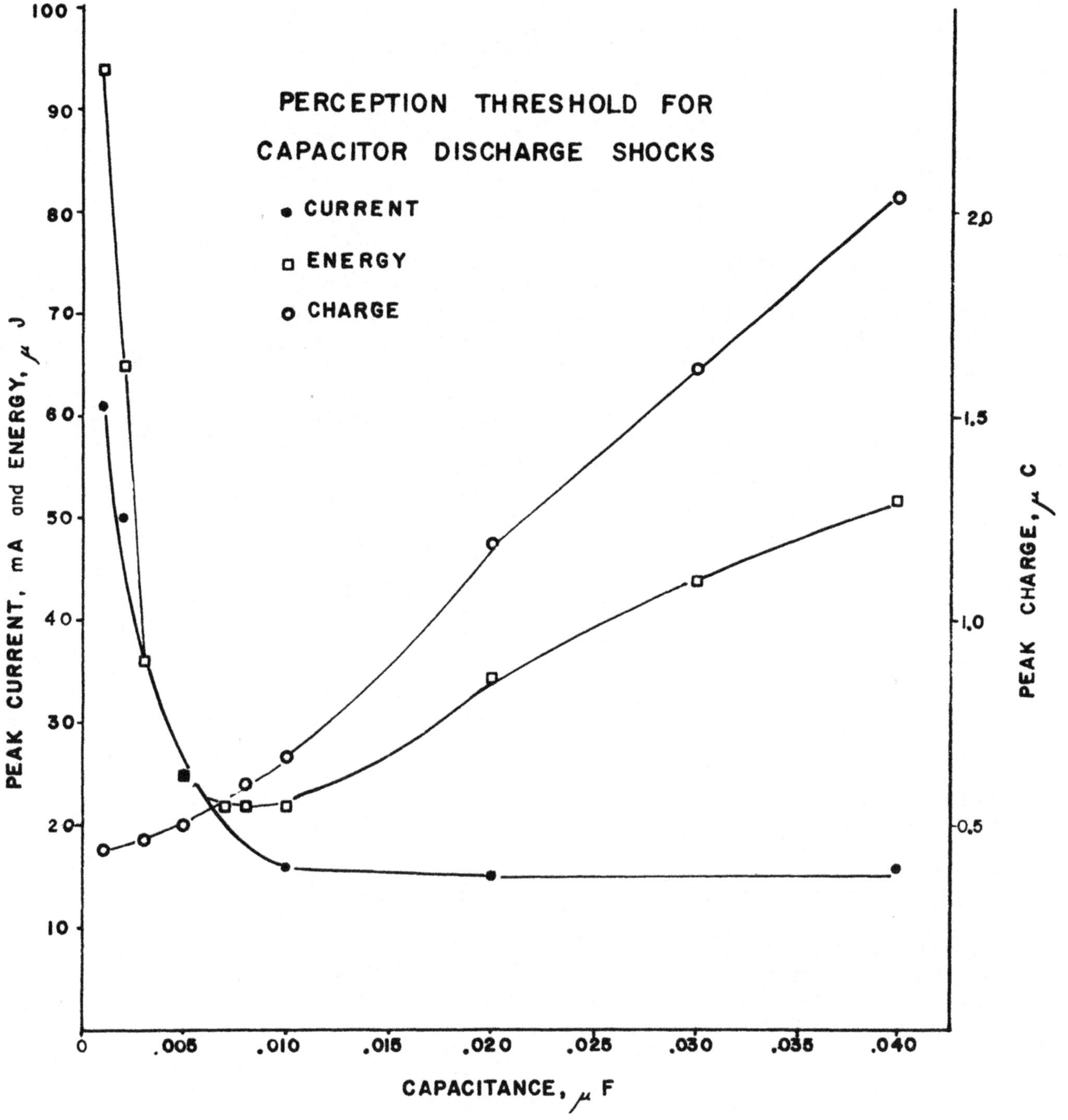

Figure 2 - Current, energy and charge perception threshold. Averages for wet fingers.

AUTHOR INDEX

*Indicates primary author

Printed in Dunstable, United Kingdom

91342996R00172